FOUNDATIONS AND FUNDAMENTAL CONCEPTS OF MATHEMATICS

THIRD EDITION

Howard Eves

University of Maine
University of Central Florida

DOVER PUBLICATIONS, INC.
Mineola, New York

Bibliographical Note

This Dover edition, first published in 1997, is an unabridged and unaltered
republication of the third edition originally published by PWS-Kent Publishing
Company, Boston, in 1990.

Library of Congress Cataloging-in-Publication Data

Eves, Howard Whitley, 1911–
    Foundations and fundamental concepts of mathematics / Howard Eves. —
3rd ed.
        p.   cm.
    Originally published: 3rd ed. Boston : PWS-Kent, c1990.
    Includes bibliographical references and index.
    ISBN 0-486-69609-X (pbk.)
    1. Mathematics—Philosophy. 2. Mathematics—History. I. Title.
[QA9.E89   1997]
510'.1—dc21   97-8057
                                                               CIP

Manufactured in the United States of America
Dover Publications, Inc., 31 East 2nd Street, Mineola, N.Y. 11501

To the Memory of Pride and Kabar
Two Great Friends and Companions

FOREWORD

Mathematical procedures are widely recognized as increasingly necessary and powerful tools in scientific explorations, industrial developments, and many of our personal activities in this age of scientific and technological innovations. This third edition of *Foundations and Fundamental Concepts of Mathematics* provides a welcome opportunity for college undergraduates to obtain an overview of the historical roots and the evolution of several areas of mathematics.

The selection of topics conveys not only their role in this historical development of mathematics but also their value as bases for understanding the changing nature of mathematics. The continuing rapid growth of mathematics makes it impossible to cover all aspects of the subject in a single course or year of courses. Indeed, Henri Poincaré (1854–1912) is frequently cited as the last universal mathematician, the last person who understood all of the aspects of mathematics that were known in his or her time. Our present situation is further complicated by the fact that the scope of the mathematical sciences has at least doubled during the last seventy-five years.

The topics included in *Foundations and Fundamental Concepts of Mathematics,* Third Edition, have special significance for mathematics majors and prospective teachers of mathematics. In particular, the emphasis on axiomatic procedures provides an important background for studying and applying more advanced topics. The inclusion of the historical roots of both algebra and geometry provides an essential background for prospective teachers of school mathematics.

The readable style and sets of challenging exercises from the popular earlier editions have been continued and extended in this third edition, making it a very welcome and useful version of a classic treatment of the foundations of mathematics. It is always a happy occasion to welcome the reappearance in revised form of a truly satisfying book.

Bruce E. Meserve
Professor Emeritus, *University of Vermont*

PREFACE

The story of the development of mathematics is made up of two intertwined strands. One strand narrates the growing content of mathematics and the other the changing nature of mathematics. Almost everyone realizes that mathematics must have arisen from some very slender beginnings far back in time and then gradually grown into its present enormous structure, but not so many realize that the nature and very meaning of mathematics have also changed and evolved over the ages. The story of the growing content of mathematics constitutes the subject matter of books and courses on the history of mathematics, and the story of the changing nature of mathematics constitutes the subject matter of books and courses devoted to the foundations, philosophy, and fundamental concepts of mathematics. For a properly rounded picture of the development of mathematics, both stories (which are highly interlinked) should be studied, and each should constitute a "must" course for all serious students of mathematics and all prospective teachers of mathematics.

Accordingly, two companion volumes were conceived, one to be devoted to the history of mathematics and one to the foundations and fundamental concepts of mathematics. The task was begun under the enthusiastic encouragement of Carroll V. Newsom, then (among many other things) the very able mathematics editor of Rinehart Book Company. The unavoidable bits of overlap of some of the topics of the two books was not considered as constituting any great sin of duplication, for these overlaps would be viewed somewhat differently in the two treatments. In each case, the presentation was to be on a level understandable by the able undergraduate college mathematics student. The first of the two books to be written was the *History* book, followed a few years later by the *Foundations* book. Thus started a fine association with Rinehart Book Company, in those wonderful days when the author–publisher relationship was a true partnership.

Unfortunately, the intimate union of the two books as companion volumes was fated not to continue. Rinehart was swallowed by Holt (to become Holt, Rinehart and Winston), and the two books somehow lost their twinship; and when Holt decided to give up its mathematics offerings, the *History* book was given up for adoption by Saunders College Publishing and the *Foundations* book was allowed to expire. Thus the intended union of the two books came to an end and the expired volume lay dead for a number of years, in spite of a wide wish of many of the mathematical fraternity that it be brought back to life. The resurrection was finally instigated by Steve Quigley, senior mathematics editor of PWS-KENT Publishing Company.

It is perhaps pertinent to quote here, with some slight amendment, a few paragraphs from the preface to the original edition of the *Foundations* book.

There is little doubt that man's rapid progress in recent decades in the control and understanding of nature, in providing himself abstract tools of creation that border upon the miraculous, and in his actual comprehension of the powers and limitations of the human mind is a direct consequence of mathematical triumphs of the last few centuries, especially the nineteenth and twentieth. Thus scholarly endeavor in virtually all areas of human activity requires to an increasing extent a knowledge of mathematics and the ability to use it; except for the study of language, mathematics may well be the most basic component of a so-called general education.

Yet even students of mathematics are usually denied until their most advanced years of study an understanding of the meaning and nature of mathematics. They labor under false definitions and impressions, which, unfortunately, continue to be promulgated in most elementary courses. Until they have reached the graduate level, students generally have heard little of mathematical structure; they have virtually no acquaintance with the common collections of axioms, with the postulational method, and with the nature and use of mathematical systems and models; it is probable that they have little genuine knowledge of the real number system and have had little more than a superficial experience in working with such a fundamental concept as *set*.

This book has been written, therefore, in an attempt to make available to able undergraduate students an introductory treatment of the foundations of mathematics. A course of study utilizing this work as a text would, it is hoped, rectify a great deal of the curricular deficiency just described. The contents were selected with considerable care in order that the exposition might have value not only to mathematical scholars and scientists but also to philosophers, historians, and others; it is especially hoped that potential teachers of mathematics will have an opportunity to study under competent instruction such material as this book contains.

The treatment is strongly historical, for the study is concerned with fundamental mathematical ideas; a genuine understanding of ideas is not possible without an analysis of origins. Even the order of topics, as revealed by the table of contents, provides in a rough way a chronological development of the basic concepts that have made mathematics what it is today. Obviously, the treatment at many points is far from exhaustive, for the exposition has been designed to be compatible with the level of maturity and understanding of the able undergraduate student; the bibliography contains many suggestions for those who desire to learn more about a particular topic.

The book will have accomplished its purpose for many readers if it does no more for them than explain the nature of geometric and algebraic systems. It is hoped, however, that much more will be accomplished.

Perhaps some students of mathematics will for the first time see the forest without becoming confused by the trees; and a few students may obtain from the work a glimpse of the meaning and opportunities of mathematical creation, the finest testimony to humankind's inherent genius.

Finally, it is a pleasure to repeat former thanks to The Cambridge University Press, The Macmillan Company, McGraw-Hill Book Company Inc., Mrs. R. E. Moritz and Mrs. C. J. Keyser for graciously granting permission to quote from certain works. Much of the task of writing the book was made easier by the understanding and cooperation of Professor Spofford H. Kimball, at the time chairman of the Department of Mathematics at the University of Maine. A large part of the manuscript was read and critically discussed from the philosophical point of view by Dr. Charles G. Werner, then of the University of Miami. An undoubted debt is also owed to Professor Raymond L. Wilder, with whom Carroll Newsom first studied such matters as are contained in the book, and to Professor Clayton W. Dodge of the University of Maine for his invaluable counsel in the preparation of the second edition of the work. And, of course, special thanks go to Steve Quigley for his energetic efforts in bringing the book back to life.

Howard Eves
Fox Hollow
Lubec, Maine

CONTENTS

1 Mathematics Before Euclid 1

1.1 The Empirical Nature of pre-Hellenic Mathematics 1
1.2 Induction Versus Deduction 5
1.3 Early Greek Mathematics and the Introduction of Deductive Procedures 9
1.4 Material Axiomatics 13
1.5 The Origin of the Axiomatic Method 15
Problems 17

2 Euclid's *Elements* 26

2.1 The Importance and Formal Nature of Euclid's *Elements* 26
2.2 Aristotle and Proclus on the Axiomatic Method 29
2.3 Euclid's Definitions, Axioms, and Postulates 32
2.4 Some Logical Shortcomings of Euclid's *Elements* 37
2.5 The End of the Greek Period and the Transition to Modern Times 41
Problems 45

3 Non-Euclidean Geometry 51

3.1 Euclid's Fifth Postulate 51
3.2 Saccheri and the *Reductio ad Absurdum* Method 54
3.3 The Work of Lambert and Legendre 58
3.4 The Discovery of Non-Euclidean Geometry 60
3.5 The Consistency and the Significance of Non-Euclidean Geometry 65
Problems 70

4 Hilbert's *Grundlagen* 79

4.1   The Work of Pasch, Peano, and Pieri   79
4.2   Hilbert's *Grundlagen der Geometrie*   82
4.3   Poincaré's Model and the Consistency of Lobachevskian Geometry   88
4.4   Analytic Geometry   92
4.5   Projective Geometry and the Principle of Duality   98
      Problems   104

5 Algebraic Structure 113

5.1   Emergence of Algebraic Structure   113
5.2   The Liberation of Algebra   118
5.3   Groups   124
5.4   The Significance of Groups in Algebra and Geometry   128
5.5   Relations   132
      Problems   136

6 Formal Axiomatics 147

6.1   Statement of the Modern Axiomatic Method   147
6.2   A Simple Example of a Branch of Pure Mathematics   150
6.3   Properties of Postulate Sets—Equivalence and Consistency   154
6.4   Properties of Postulate Sets—Independence, Completeness, and Categoricalness   158
6.5   Miscellaneous Comments   162
      Problems   166

7 The Real Number System 173

7.1   Significance of the Real Number System for the Foundations of Analysis   173
7.2   The Postulational Approach to the Real Number System   179
7.3   The Natural Numbers and the Principle of Mathematical Induction   183
7.4   The Integers and the Rational Numbers   191
7.5   The Real Numbers and the Complex Numbers   196
      Problems   202

8 Sets 212

8.1   Sets and Their Basic Relations and Operations   212
8.2   Boolean Algebra   216
8.3   Sets and the Foundations of Mathematics   221
8.4   Infinite Sets and Transfinite Numbers   224
8.5   Sets and the Fundamental Concepts of Mathematics   229
      Problems   236

9 Logic and Philosophy 243

9.1 Symbolic Logic 243
9.2 The Calculus of Propositions 250
9.3 Other Logics 257
9.4 Crises in the Foundations of Mathematics 262
9.5 Philosophies of Mathematics 266
      Problems 271

Appendix 275

A.1 *The First Twenty-Eight Propositions of Euclid* 275
A.2 *Euclidean Constructions* 276
A.3 *Removal of Some Redundancies* 284
A.4 *Membership Tables* 289
A.5 *A Constructive Proof of the Existence of Transcendental Numbers* 290
A.6 *The Eudoxian Resolution of the First Crisis in the Foundations of Mathematics* 292
A.7 *Nonstandard Analysis* 294
A.8 *The Axiom of Choice* 296
A.9 *A Note on Gödel's Incompleteness Theorem* 299

Bibliography 303

Solution Suggestions for Selected Problems 316

Index 329

MATHEMATICS BEFORE EUCLID

1.1 The Empirical Nature of Pre-Hellenic Mathematics

The thesis can be advanced that mathematics arose from necessity. The annual inundation of the Nile Valley, for example, forced the Egyptians to develop some system for redetermining land markings; in fact, the word *geometry* means "measurement of the earth." The need for mensuration formulas was especially imperative if, as Herodotus remarked, taxes in Egypt were paid on the basis of land area. The Babylonians likewise encountered an urgent need for mathematics in the construction of the great engineering structures for which they were famous. Marsh drainage, irrigation, and flood control made it possible to convert the land along the Tigris and Euphrates rivers into a rich agricultural region. Similar undertakings undoubtedly occurred in early times in south-central Asia along the Indus and Ganges rivers, and in eastern Asia along the Hwang Ho and the Yangtze. The engineering, financing, and administration of such projects required the development of considerable technical knowledge and its attendant mathematics. A useable calendar had to be computed to serve agricultural needs, and this required some basic astronomy with its concomitant mathematics. Again, the demand for some system of uniformity in barter was present in even the earliest civilizations; this fact also furnished a pronounced stimulus to mathematical development. Finally, early religious ritual found need for some basic mathematics.[1]

Thus there is a basis for saying that mathematics, beyond that implied by primitive counting, originated during the period of the fifth, fourth, and third millennia B.C. in certain areas of the ancient orient as a practical science to assist in engineering, agricultural, and business pursuits and in religious ritual. Although the initial emphasis was on mensuration and practical arithmetic, it

---

[1] See A. Seidenberg [1] and [2]. (References by author's name only are to the Bibliography at the end of the book.)

was natural that a special craft should come into being for the application, instruction, and development of the science and that, in turn, tendencies toward abstraction should then assert themselves and the subject be studied, to some extent, for its own sake. In this way a basis for the beginnings of theoretical geometry grew out of mensuration, and the first traces of elementary algebra evolved from practical arithmetic.[2]

In our study of early mathematics we are restricted essentially to that of Egypt and Babylonia. The ancient Egyptians recorded their work on stone and papyrus, the latter fortunately enduring because of Egypt's unusually dry climate; the Babylonians used imperishable baked clay tablets. In contrast to the use of these media, the early Indians and Chinese used very perishable writing materials like bark and bamboo. Thus it has come to pass that we have a fair quantity of definite information, obtained from primary sources, about the science and the mathematics of ancient Egypt and Babylonia, while we know very little indeed, with any degree of certainty, about these fields of study in ancient India and China.

It is the nature, rather than the content, of this pre-Hellenic mathematics that concerns us here, and in this regard it is important to note that, outside of very simple considerations, the mathematical relations employed by the Egyptians and by the Babylonians resulted essentially from "trial and error" methods. In other words, to a great extent the earliest mathematics was little more than a practically workable empiricism—a collection of rule-of-thumb procedures that gave results of sufficient acceptability for the simple needs of those early civilizations. Thus the Egyptian and Babylonian formulas for volumes of granaries and areas of land were arrived at by trial and error, with the result that many of these formulas are definitely faulty. For example, an Egyptian formula for finding the area of a circle was to take the square of eight ninths of the circle's diameter. This is not correct, as it is equivalent to taking $\pi = (4/3)^4 = 3.1604\cdots$. The even less accurate value of $\pi = 3$ is implied by some Babylonian formulas.[3] Another incorrect formula found in ancient Babylonian mathematics is one that says that the volume of a frustum of a cone or of a square pyramid is given by the product of the altitude and half the sum of the bases. It seems that the Babylonians also used, for the area of a quadrilateral having a, b, c, d for its consecutive sides, the incorrect formula $K = (a + c)(b + d)/4$. This formula gives the correct result only if the quadrilateral is a rectangle; in every other instance the formula gives too large an answer. It is curious that this same incorrect formula was reproduced 2000 years later in an Egyptian inscription found in the tomb of Ptolemy XI, who died in 51 B.C.

In general, simple empirical reasoning may be described as the formulation of conclusions based upon experience or observation; no real understanding is involved, and the logical element does not appear. Empirical reasoning often entails stodgy fiddling with special cases, observation of coincidences, experience

---

[2] For comments on a possible prehuman origin of mathematics see D. E. Smith [1], vol. 1, chap. 1, and H. Eves [3], Items 1°, 2°, 3°, 4°.

[3] This value for π is also found in the Bible; see I Kings 7:23, and II Chron. 4:2.

at good guessing, considerable experimentation, and flashes of intuition. Perhaps a very simple hypothetical illustration of empirical reasoning might clarify what is meant by this type of procedure.

Suppose a farmer wishes to enclose with 200 feet of fencing a rectangular field of greatest possible area along a straight river bank, no fencing being required along the river side of the field. If we designate as the *depth* of the field the dimension of the field perpendicular to the river bank and as the *length* of the field the dimension parallel to the river bank (see Figure 1.1), the farmer could soon form the following table:

Depth in feet	Length in feet	Area in square feet
10	180	1800
20	160	3200
30	140	4200
40	120	4800
50	100	5000
60	80	4800
70	60	4200
80	40	3200
90	20	1800

Examination of the table shows that the maximum area recorded occurs when the depth is 50 feet and the length is 100 feet. The interested farmer might now try various depths close to, but on each side of, 50 and would perhaps make the following additional table:

Depth	Length	Area
48	104	4992
49	102	4998
50	100	5000
51	98	4998
52	96	4992

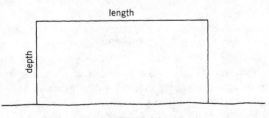

FIGURE 1.1

By now the farmer would feel quite certain that the maximum area is obtained when the depth is 50 feet and the length is 100 feet; that is, he would accept the proposition that *the maximum area occurs when the length of the field is twice the depth of the field.* A further strengthening of this belief would result from his examination of the symmetry observed in his table, and he would no doubt use his conjecture and pass it along to others as a reliable mathematical fact. Of course, the farmer's conclusion is by no means established, and no present-day student of mathematics would be permitted to "prove" the conjecture in this fashion. Shrewd guessing has taken the place of deductive logic; patience has replaced brilliance.

In spite of the empirical nature of ancient oriental mathematics, with its complete neglect of proof and the seemingly little attention paid to the difference between exact and approximate truth, one is nevertheless struck by the extent and the diversity of the problems successfully attacked. Particularly has this become evident in recent years with the scholarly deciphering of many Babylonian mathematical tablets. Apparently a great deal of elementary mathematical truth can be discovered by empirical methods when supplemented by extensive experimentation carried on patiently over a long period of time.

How were the mathematical findings of the ancient orient stated? Here we must rely on such primary sources as the Rhind, the Moscow, and other Egyptian mathematical papyri and on the approximately three hundred Babylonian mathematical tablets that have so far been deciphered.

The Rhind, or Ahmes, papyrus is a mathematical text dating from about 1650 B.C. Partaking of the nature of a practical handbook, it contains 85 problems copied by the scribe Ahmes from a still earlier work. Now possessed by the British Museum, it was originally purchased in Egypt by the Scottish antiquarian, A. Henry Rhind. This papyrus and the somewhat older Moscow papyrus, a similar mathematical text containing 25 problems, constitute our chief sources of information concerning ancient Egyptian mathematics. All of the 110 problems found in these papyri are numerical, and many of them are very easy. In general, each problem is first formulated and then followed by a step-by-step solution using the special numbers given at the beginning. Although special numbers are employed in this fashion, one feels that they are incidental and are being used merely to illustrate a general procedure. Many of the problems require nothing more than a simple linear equation, and are generally solved by the method known later in Europe as the *rule of false position.* This rule clearly reflects the empirical nature of the mathematical procedures of the time. As an example, suppose we are to solve the simple equation $x + (x/5) = 24$. Assume any convenient value for x, say, $x = 5$. Then $x + (x/5) = 6$, instead of 24. Since 6 must be multiplied by 4 to give the required 24, the correct value of x must be 4(5), or 20.

The Babylonian mathematical tablets are of two types, *table texts* and *problem texts.* There must be at least 500,000 Babylonian tablets now scattered among various museums of the world; of these only about 100 problem texts, and somewhat more than twice this number of table texts, are known to us. The table texts exhibit a wide variety of mathematical tables, such as multiplication tables, tables of reciprocals (for reducing division to multiplication), tables of squares and square roots and cubes and cube roots, tables of sums of squares and cubes (for solving certain types of cubic equations), exponential tables (for computing compound interest), and many others. The ancient Babylonians were indefatigable table makers, as one might have expected, for the construction of tables is indispensable to empirical procedure.

The problem texts also show considerable variety and are all more or less concerned with the formulation and solution of algebraic and geometric problems. A large group of the problem texts, like the Egyptian papyri considered above, formulate a problem in terms of specific numbers and then proceed with a step-by-step solution using the specific numbers. Such texts often terminate with the phrase, "such is the procedure." Again it is apparent that it is the general procedure, and not the numerical result, that is considered important. If, in a multiplication, a factor has the value 1, multiplication by this 1 will be explicitly performed, for this step is necessary in the general case. The remaining problem texts contain on a single tablet, often not as large as a page of this book, a large number of related numerical problems carefully arranged from the simplest cases up through the more complicated ones. The apparent purpose of such a text was to teach, by repetition and gradual introduction of complexities, a certain method or procedure, and the accompanying numbers serve merely as a guide to illustrate the underlying general procedure. The solution of quadratic equations, for example, both by general formula and by the method of completing the square, is explained in this way on ancient Babylonian tablets.

In summary, then, we find that pre-Hellenic mathematics was empirical. Nowhere do we find in ancient oriental mathematics a single instance of what we today call a logical demonstration. Instead of an argument we find a description of a process explained by means of specific numerical cases. In short, we are instructed to "Do thus and so." It is very interesting to note that although today confirmed students of the scientific method find this "Do thus and so" procedure highly unsatisfactory it is the procedure employed in much of our elementary teaching.

======= **1.2 Induction Versus Deduction** =======

Empirical conclusions, we have seen, are generalizations based on a limited number of observations or experiments. For example, the farmer of the previous section obtained a general rule by observing a limited number of computed areas. Another farmer may observe that unusually good crops have followed a number of winters of heavy snow, and empirically conclude that snowy winters are beneficial to crops. As a further example, a scientist may observe that particularly fine displays of the aurora borealis always occurred in his experience during periods of pronounced sun-spot activity and conclude that there must be a connection between the two phenomena. This type of reasoning, which concludes on the basis of a limited number of instances that something is always true, is known as *induction*.[4] Modern probability considerations have served to introduce refinements into inductive procedures. It is important to note,

---

[4] *Induction* should not be confused with so-called *mathematical induction*, which is considered in Section 7.3.

however, that no matter how fully the conclusions of inductive reasoning may seem warranted by the facts, these conclusions are not established beyond all possible doubt; conclusions obtained by induction are only more or less probable.

Empirical conclusions are sometimes reached by using a primitive form of induction known as reasoning by analogy. For example, if we cut off the top of a triangle by a line parallel to the base of the triangle, a trapezoid will remain, and the area of a trapezoid is given by the product of its altitude and the arithmetic average of its two bases. Now, if we cut off the top of a pyramid by a plane parallel to the base of the pyramid, a frustum will remain. By analogy one might expect the volume of a frustum of a pyramid to be given, as before, by the product of the altitude and the arithmetic average of the two bases. This is the incorrect Babylonian formula noted in the previous section. Reasoning by analogy certainly is useful, but obviously its conclusions cannot be regarded as established.

In sharp contrast to reasoning by analogy or by induction is reasoning by deduction, because the conclusions reached by deduction, provided one accepts the premises that are adopted and the system of logic that is employed, are incontestable. To illustrate deductive procedure, consider the following two statements: (1) All Canadians are North Americans; (2) Two particular men under consideration are Canadians. If we accept these two statements, we are logically compelled, following accepted principles of Aristotelian logic, to accept a third statement—namely, (3) The two men under consideration are North Americans. This is an example of deductive reasoning, which at this point may be described as those ways of deriving new statements from accepted ones that *compel* us to accept the derived statements. In the example, the first two statements are called *premises*, and the third statement the *conclusion*.

It is very important to realize that in deductive reasoning we are not concerned with the *truth* of the conclusion but rather whether the conclusion does or does not follow from the premises. If the conclusion follows from the premises, we say that our reasoning is *valid*; if it does not, we say that our reasoning is *invalid*. For example, from the two

Premises: (1) All college students are clever,
                (2) All freshmen are college students,

follows the

Conclusion: All freshmen are clever.

Now the last statement certainly is not regarded generally as true, but the reasoning leading to it is valid. *If both of the premises had been true, the conclusion also would have been true*; it is essential that one understand early in the treatment of this book this meaning of the deductive process.

A useful and easy way to test the validity of a piece of deductive reasoning, like either of the examples given above, is by a diagrammatic procedure ascribed to the eminent Swiss mathematician Leonhard Euler (1707–1783). Consider our last example. We may represent the class of all clever people by a planar region within a closed boundary, and we may do likewise for the class of all college students and for the class of all freshmen. But statement (1) insists that the class of all college students is contained in the class of all clever people, and statement (2) insists that the class of all freshmen is contained in the class of all college students. Thus our various classes must be represented by their corresponding regions as shown in Figure 1.2. Clearly, the requirements of our premises *forced* us to place the class of all freshmen entirely within the class of all clever people, which is exactly what our conclusion asserts. Hence, although the conclusion is undoubtedly false, the reasoning leading to it is valid. It cannot be over-emphasized at this point that the expert in the use of deduction is not fundamentally concerned with *truth* but with *validity*; he merely wants to be able to assert that his conclusions are implied by the premises. It would then follow that *if* the premises should happen to be true, the conclusion *must*, of necessity, also be true.

Consider, as a second example, the following:

Premises: (1) All parallelograms are polygons.
(2) All quadrilaterals are polygons.
Conclusion: All parallelograms are quadrilaterals.

Here all three statements are true, but the reasoning is invalid, for the premises do not *force* us to place the region representing the class of all parallelograms entirely within the region representing the class of all quadrilaterals; we are able to satisfy the requirements of our premises by a diagram like that shown in Figure 1.3.

As a third example, consider the following:

Premises: (1) All parallelograms are circles.
(2) All circles are polygons.
Conclusion: All parallelograms are polygons.

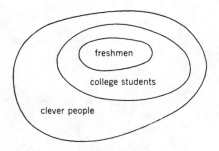

FIGURE 1.2

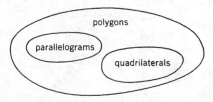

FIGURE 1.3

Here the premises are both false, the conclusion is true, and, as tested by the diagram of Figure 1.4, the reasoning is valid. Thus false assumptions may actually yield a true conclusion. True premises can yield only true conclusions when deductive logic is applied, but false premises may or may not yield true conclusions.

Finally, we shall examine the following:

Premises: (1) No quadrilaterals are triangles.
            (2) Some quadrilaterals are parallelograms.
Conclusion: Some parallelograms are not triangles.

Since, by (1), the region representing the class of all quadrilaterals and that representing the class of all triangles *cannot* overlap, and, by (2), the region representing the class of all quadrilaterals and that representing the class of all parallelograms *must* overlap, the conclusion (see Figure 1.5) certainly follows, and the reasoning is valid. Note, however, that we cannot conclude, from our premises, that *no* parallelogram is a triangle, for there is nothing that *forces* us to keep the region representing the class of all parallelograms from cutting into the region representing the class of all triangles.

Euler's diagrammatic device can be used in a great variety of situations, and it is recommended to the person unfamiliar with logical procedure.

We shall not, for the present, go beyond the above superficial study of inductive and deductive reasoning. As already indicated, deductive reasoning has the advantage that its conclusions are unquestionable if the premises are accepted, and it has the additional advantage of considerable economy: before a

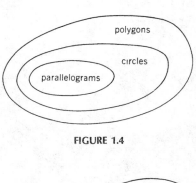

FIGURE 1.4

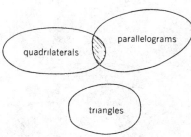

FIGURE 1.5

bridge is built and put into use, deductive reasoning can determine the outcome. But in spite of these particular advantages, deductive reasoning does not supplant the inductive approach; actually, each way of obtaining knowledge has its advantages and disadvantages. The significant thing, from the point of view of our present study, is that the ancient Greeks found in deductive reasoning the vital element of the modern mathematical method.

1.3 Early Greek Mathematics and the Introduction of Deductive Procedures

The origin of early Greek mathematics is clouded by the greatness of Euclid's *Elements*, written about 300 B.C., because this work so clearly excelled many preceding Greek writings on mathematics that the earlier works were thenceforth discarded. As the great mathematician David Hilbert (1862–1943) once remarked, one can measure the importance of a scientific work by the number of earlier publications rendered superfluous by it.

The debt of Greek mathematics to ancient oriental mathematics is difficult to evaluate, nor has the path of transmission from the one to the other yet been satisfactorily uncovered. That the debt is considerably greater than formerly believed became evident with twentieth-century researches of Babylonian and Egyptian records. Greek writers themselves expressed respect for the wisdom of the East, and this wisdom was available to anyone who could travel to Egypt and Babylonia. There are also internal evidences of a connection with the East. Early Greek mysticism in mathematics smacks strongly of oriental influence, and some Greek writings, like those of Heron and Diophantus, exhibit a Hellenic perpetuation of the more arithmetic tradition of the orient. Also, there are strong links connecting Greek and Mesopotamian astronomy.

But whatever the strength of the historical connection between Greek and ancient oriental mathematics, the Greeks transformed the subject into something vastly different from the set of empirical conclusions worked out by their predecessors. The Greeks insisted that mathematical facts must be established, not by empirical procedures, but by deductive reasoning; mathematical conclusions must be assured by logical demonstration rather than by laboratory experimentation.

This is not to say that the Greeks shunned preliminary empirical and experimental methods in mathematics, for it is probably quite true that few, if any, significant mathematical facts have ever been found without some preliminary empirical work of one form or another. Before a mathematical statement can be proved or disproved by deduction, it must first be thought of, or conjectured, and a conjecture is nothing but a guess made more or less plausible by intuition, observation, analogy, experimentation, or some other form of empirical procedure. Deduction is a convincing formal mode of exposition, but it is hardly a means of discovery. It is a set of complicated machinery that needs material to work upon, and the material is frequently furnished by empirical considerations. Even the steps of a deductive proof or disproof are not dictated to us by the deductive apparatus itself but must be arrived at by trial and error, experience, and shrewd guessing. Indeed, skill in the art of good guessing is one of the prime ingredients in the make-up of a worthy mathematician. What is important here is that the Greeks insisted that a conjectured or laboratory-obtained mathematical statement must be followed up with a rigorous proof or disproof by deduction and that no amount of verification by experiment is sufficient to *establish* the statement.

It is difficult to give a wholly adequate explanation of just why the Greeks of 600 to 400 B.C. decided to abandon empirical methods of establishing mathematical knowledge and to insist that all mathematical conclusions be established only by deductive reasoning. This completely new viewpoint on mathematical method is usually explained by the peculiar mental bias of the Greeks of classical times toward philosophical inquiries. In philosophical speculations, reasoning centers about abstract concepts and broad generalizations and is concerned with inevitable conclusions following from assumed premises. Now the empirical method affords no way of discriminating between a valid and an invalid argument and so is hardly applicable to philosophic considerations. It is deductive reasoning that philosophers find to be their indispensable tool, and so the Greeks naturally gave preference to this method when they began to consider mathematics.

Another explanation of the Greek preference for deduction stems from the Hellenic love for beauty. Appreciation of beauty is an intellectual as well as an emotional experience, and from this point of view the orderliness, consistency, completeness, and conviction found in deductive argument are very satisfying.

A still further explanation for the Greek preference for deductive procedures has been found in the nature of Greek society in classical times. Philosophers, mathematicians, and artists belonged to a social class that in general disdained manual work and practical pursuits, which were carried on by a large slave class. In Greek society the slave class ran the businesses and managed the industries, took care of households, and did both the technical and the unskilled work of the time. This slave basis naturally fostered a separation of theory from practice and led the members of the privileged class to a preference for deduction and abstraction and a disdain for experimentation and practical application.

It is disappointing that, unlike the situation with ancient Egyptian and Babylonian mathematics, there exist virtually no source materials for contemporary study that throw much light on early Greek mathematics. We are forced to rely on manuscripts and accounts that are dated several hundred years after the original treatments were written. In spite of this difficulty, however, scholars of classicism have been able to build up a rather consistent, though somewhat hypothetical, account of the history of early Greek mathematics and have even plausibly restored many of the original Greek texts. This work required amazing ingenuity and patience; it was carried through by painstaking comparisons of derived texts and by the examination of countless literary fragments and scattered remarks made by later authors, philosophers, and commentators.

Our principal source of information concerning very early Greek mathematics is the so-called *Eudemian Summary* of Proclus. This summary constitutes a few pages of Proclus's *Commentary on Euclid, Book I*, and is a very brief outline of the development of Greek geometry from the earliest times to Euclid. Although Proclus lived in the fifth century A.D., a good thousand years after the inception of Greek mathematics, he still had access to a number of historical and critical works that are now lost to us except for the fragments and allusions preserved by him and others. Among these lost works was apparently a full history of Greek geometry, covering the period before 335 B.C., written by Eudemus, a pupil of Aristotle. The *Eudemian Summary* is so named because it is based on this earlier work.

According to the *Eudemian Summary*, Greek mathematics appears to have started in an essential way with the work of Thales of Miletus in the first half of the sixth century B.C. This versatile genius, declared to be one of the "seven wise men" of antiquity, was a worthy founder of systematic mathematics and is the first known individual with whom the use of deductive methods in mathematics is associated. Thales, the summary tells us, sojourned for a time in Egypt and brought back geometry with him to Greece, where he began to apply to the subject the deductive procedures of philosophy. In particular, he is credited with the following elementary geometrical results:

1. A circle is bisected by any diameter.

2. The base angles of an isosceles triangle are equal.

3. Vertical angles formed by two intersecting straight lines are equal.

4. Two triangles are congruent if two angles and a side in one are equal respectively to two angles and the corresponding side of the other. (It is thought that Thales used this result to determine the distance of a ship from shore.)

5. An angle inscribed in a semicircle is a right angle. (The Babylonians of some 1400 years earlier were acquainted with this geometrical fact.)

We are not to measure the value of these results by their content but rather by the belief that Thales supported them with a certain amount of logical reasoning instead of intuition and experiment. For the first time a student of mathematics was committed to a form of deductive reasoning, crude and incomplete though it may have been. Moreover, the fact that the first deductive thinking was done in the field of geometry instead of algebra, for instance, inaugurated a tradition in mathematics that was maintained, as we shall see, until very recent times.

The next outstanding Greek mathematician mentioned in the *Eudemian Summary* is Pythagoras, who is claimed to have continued the purification of geometry that was begun some fifty years earlier by Thales. Pythagoras was born about 572 B.C. on the island of Samos, one of the Aegean islands near Thales's home city of Miletus, and it may be that he studied under the older man. It seems that he visited Egypt and perhaps traveled even more extensively about the orient. When, on returning home, he found Samos under the tyranny of Polycrates and Ionia under Persian dominion, he decided to migrate to the Greek seaport of Crotona in southern Italy. Here he founded the celebrated

Pythagorean school, a brotherhood knit together with secret and cabalistic rites and observances and committed to the study of philosophy, mathematics, and natural science.

The philosophy of the Pythagorean school was built on the mystical assumption that whole number is the cause of the various qualities of man and matter. This oriental outlook, perhaps acquired by Pythagoras in his eastern travels, led to the exaltation and study of number relations and to a perpetuation of numerological nonsense that has lasted even into modern times. However, in spite of the unscientific nature of much of Pythagorean study, members of the society contributed, during the two hundred or so years following the founding of their organization, a good deal of sound mathematics. They developed the properties of parallel lines and used them to prove that the sum of the angles of any triangle is equal to two right angles. They contributed in a noteworthy manner to Greek geometrical algebra; they effected the geometrical equivalent of addition, subtraction, multiplication, division, extraction of roots, and even the complete solution of the general quadratic equation insofar as it has real roots. They developed a fairly complete theory of proportion, though it was limited only to commensurable magnitudes, and used it to deduce properties of similar figures. They were aware of the existence of at least three of the regular polyhedral solids, and they discovered the incommensurability of a side and a diagonal of a square. Although much of this information was already known to the Babylonians of earlier times, the deductive aspect of mathematics is thought to have been considerably exploited and advanced in this work of the Pythagoreans. Chains of propositions in which successive propositions were derived from earlier ones in the chain began to emerge. As the chains lengthend, and some were tied to others, the bold idea of developing all of geometry in one long chain suggested itself. It is claimed in the *Eudemian Summary* that the Pythagorean, Hippocrates of Chios,[5] was the first to attempt, with at least partial success, a logical presentation of geometry in the form of a single chain of propositions based upon a few initial definitions and assumptions.

The famous Greek philosopher, Plato, was strongly influenced by the Pythagoreans, and Plato, in turn, exerted a considerable influence on the development of mathematics in Greece. Plato's influence was not due to any mathematical discoveries he made but rather to his enthusiastic conviction that the study of mathematics furnished the finest training field for the mind, and hence was essential in the cultivation of philosophers and those who should govern his ideal state. This belief explains the renowned motto over the door of his Academy, "Let no one unversed in geometry enter here." Thus, because of its logical element and the pure attitude of mind that he felt its study creates, mathematics seemed of utmost importance to Plato, and for this reason it occupied a valued place in the curriculum of the Academy. Some see in certain of Plato's dialogues what may perhaps be considered the first serious attempt at a philosophy of mathematics. Certainly mathematics in Greece at the time of Plato had advanced a long way from the empirical mathematics of ancient Egypt and Babylonia.

---

[5] He is not to be confused with Hippocrates of Cos, the eminent Greek physician of antiquity.

1.4 Material Axiomatics

Much was accomplished by the Greeks during the three hundred years between Thales in 600 B.C. and Euclid in 300 B.C. Not only did the Pythagoreans and others develop the material that ultimately was organized into the *Elements* of Euclid, but there were developed notions concerning infinitesimals and summation processes (notions that did not attain final clarification until the rigorization of the calculus in modern times) and also considerable higher geometry (the geometry of curves other than the circle and the straight line and of surfaces other than the sphere and plane). Curiously enough, much of this higher geometry originated in continued attempts to solve the three famous construction problems of antiquity—the duplication of a cube, the trisection of an arbitrary angle, and the quadrature of a circle—illustrating the principle that the growth of mathematics is stimulated by the presence of outstanding unsolved problems.

Also, some time during the first three hundred years of Greek mathematics, there developed the Greek notion of a logical discourse as a sequence of statements obtained by deductive reasoning from an accepted set of initial statements. Certainly, if one is going to present an argument by deductive procedure, any statement of the argument will have to be derived from some previous statement or statements of the argument, and such a previous statement must itself be derived from some still more previous statement or statements. Clearly this cannot be continued backward indefinitely, nor should one resort to illogical circularity by deriving statement q from statement p and then later deriving statement p from statement q. The only way out of the difficulty is to set down, toward the start of the discourse, a collection of fundamental statements whose truths are to be accepted and then to proceed by purely deductive reasoning to derive all the other statements of the discourse. Now both the initial and the derived statements of the discourse are statements about the technical matter of the discourse and hence involve special or technical terms. The meanings of these terms must be made clear to the reader, and so, the Greeks felt, the discourse should start with a list of explanations and definitions of these technical terms. After these explanations and definitions have been given, the initial statements, called *axioms* and/or *postulates* of the discourse, are to be listed. These initial statements, according to the viewpoint held by some of the Greeks, should be so carefully chosen that their truths are quite acceptable to the reader in view of the explanations and definitions already cited.

A discourse that is conducted according to the above plan is described today as a development by *material axiomatics*. Certainly the most outstanding contribution of the early Greeks to mathematics was the formulation of axiomatic procedure and the insistence that mathematics be systematized by such a procedure. Euclid's *Elements* is the earliest extensively developed example of axiomatic procedure that has come down to us; it largely follows the pattern of material axiomatics, and we shall certainly want to examine it in some detail. In more recent years, the pattern of material axiomatics has been significantly refined to yield a more abstract form of discourse known as *formal axiomatics* (see Chapter 6). For the time being we will content ourselves by summarizing the pattern of material axiomatics.

Pattern of Material Axiomatics

1. Initial explanations of certain basic technical terms of the discourse are given, the intention being to suggest to the reader what is to be meant by these basic terms.

2. Certain primary statements that concern the basic terms and that are felt to be acceptable as true on the basis of the properties suggested by the initial explanations are listed. These primary statements are called the *axioms* or *postulates* of the discourse.

3. All other technical terms of the discourse are defined by means of previously introduced terms.

4. All other statements of the discourse are logically deduced from previously accepted or established statements. These derived statements are called the *theorems* of the discourse.

To gain a feeling for the pattern of material axiomatics, let us consider an example. Suppose one is faced with the task of developing a logical discourse on carpentry. The subject of carpentry contains many special or technical terms, such as nail, spike, brad, screw, wood, hard wood, soft wood, board, strut, beam, hammer, saw, screw driver, plane, or chisel. Some of these technical terms can be defined in terms of others. For example, a spike and a brad can each be defined as a special kind of nail; hard wood and soft wood can be defined as certain special kinds of wood; board, strut, and beam can be defined as pieces of wood of certain shapes used for certain purposes; various kinds of hammers and saws can be defined in terms of the basic hammer and saw. It is certainly logical, then, to commence the discourse with some sort of explanation or description of the basic technical terms—say, nail, wood, hammer, saw, and others—and then to define further technical terms, either at the start or as needed, in terms of the basic ones. After giving these initial explanations and possible definitions, the next thing to do is to list some fundamental statements about the explained and defined terms that will be assumed so that the discourse may get under way. Now these assumed statements, from the point of view of material axiomatics, should be such that the reader is perfectly willing to accept them on account of the initial explanations of the basic terms involved. For example, one may wish to assume that *it is always possible to drive a nail with a hammer into a piece of wood*, that *it is always possible with a saw to cut a piece of wood in two by a planar cut*, etc. That two boards of desired lengths can be fastened together with nails now follows as a consequence of these assumptions, and is thus a theorem of the discourse. Probably enough has been said to illustrate the Greek notion of material axiomatics.

The theory of some simple games can be rather easily developed by material axiomatics. Consider, for example, the familiar game of tic-tac-toe. Among the technical terms of this game are the *playing board, nought, cross,* a *win,* a *draw,* and so on. These technical terms are to be explained or defined. The rules of the game are then stated as the postulates of the discourse, these rules being perfectly acceptable once one understands the basic terms of the discourse. From these rules one can then proceed to deduce the theory of the game, proving as a theorem, for example, that *with sufficiently good playing, the player who starts a game need not lose the game.*

1.5 The Origin of the Axiomatic Method

We do not know with whom the axiomatic method originated. By the account given in the *Eudemian Summary*, the method seems to have evolved with the Pythagoreans as a natural outgrowth and refinement of the early application of deductive procedures to mathematics. This is the traditional and customary account and is based principally on Proclus's summary, which, in turn, is based on the lost history of geometry written by Eudemus about 335 B.C. The account may be the true one, and, if so, we must concede to the Pythagoreans a very high place in the history of the development of mathematics.

There are some historians of ancient mathematics who find the account of the early history of Greek mathematics, as reconstructed from the *Eudemian Summary*, somewhat difficult to believe and who feel that the traditional stories about Thales and Pythagoras must be discredited as purely legendary and unhistorical in content. For example, the *Eudemian Summary* says that Thales proved that a circle is bisected by any one of its diameters. The realization that so obvious a matter as this should need demonstration seems to reflect a mathematical sophistication of a much more advanced period, when the importance and delicacy of initial assumptions had become much clearer. Eudemus may have hypothetically restored the sequence of events so that they accorded with the state of the theory of his time, as many historians do when source material is not available. Actually, we can have very little idea of the roles played in the history of mathematics by Thales and Pythagoras, and it may be much closer to reality to assume that early Greek mathematics cannot have differed greatly from the oriental type. An essential turn in the development of a subject is usually brought about by some crucial circumstance, and in mathematics such a circumstance arose some time in the fifth century B.C. with the devastating discovery of the irrationality of $\sqrt{2}$.

Let us pause a moment to consider the significance of the last statement. Since the rational numbers consist of all numbers of the form p/q, where p and q are integers with $q \neq 0$, the discovery alluded to states that there are no integers p and q such that $p/q = \sqrt{2}$; that is, $\sqrt{2}$ is *not* a rational number and hence, by definition, is an *irrational* (nonrational) number. The traditional proof of this fact, apparently known to Aristotle (384–322 B.C.), is simple and runs as follows: Suppose, on the contrary, that there are two integers p and q such that $p/q = \sqrt{2}$, where, without any loss of generality, we may assume that p and q have no common positive integral factor other than unity. Then $p^2 = 2q^2$. Since p^2 is twice an integer, we see that p^2, and hence p, must be even. So we may put $p = 2r$. Then we find $4r^2 = 2q^2$, or $2r^2 = q^2$, from which we conclude that q^2, and hence q, must be even. But this is impossible, since we assumed that p and q have no common integral factor different from unity. The supposition that $\sqrt{2}$ is rational has led to a contradictory situation, whence it follows that $\sqrt{2}$ must be irrational. This result was surprising and disturbing on several grounds. First of all, it seemed to deal a mortal blow to the Pythagorean philosophy that all depends on the integers. Next, it seemed contrary to common sense, for it was felt intuitively that any magnitude could be expressed by *some* rational number. The geometrical counterpart was equally startling, for who could doubt that for any two given line segments one is able to find some third line segment, perhaps very very small, that can be marked off a whole number of times into each of the given segments. But take as the two given segments a side s and a diagonal d of a square. Now if there exists a third segment t which can be marked off a whole number of times into s and d we would have $s = qt$ and $d = pt$, where p and q are integers. But $d = s\sqrt{2}$, whence $pt = qt\sqrt{2}$, or $\sqrt{2} = p/q$, a rational number. Contrary to intuition, then, there exist line segments having no common unit of measure. But the whole Pythagorean theory of proportion was built on the seemingly obvious assumption that any two line segments are commensurable, that is, do have some common unit of measure.

No wonder the discovery of the irrationality of $\sqrt{2}$ led to some consternation in the Pythagorean ranks. The situation must have caused a profound reaction in mathematical thinking, and must have very considerably emphasized the extreme importance of careful agreement on what can be taken for basic assumptions. A crisis, like this one of the discovery of irrational numbers, could well account for the origin of the axiomatic method, and, if so, the credit for the invention might largely go to Eudoxus, the genius of the time who finally resolved the crisis that had arisen.[6]

This second explanation of the possible origin of the axiomatic method has other points in its favor. For example, it places less stress on any peculiar mentality possessed by the Greeks of very early times, and it accounts for the relatively large number of Greek papyrus fragments containing texts after the pattern of oriental mathematics. These texts, like the similar ones from Babylonian times, probably formed the backbone of instruction in elementary mathematics. At this elementary level the highly sophisticated axiomatic method had as little influence as it has today in much of our elementary teaching. Writings of this sort, then, do not reflect any degeneration of the so-called Greek spirit in mathematics but simply exhibit the continuance, on an elementary level, of older traditions. Heron's geometry, for example, can be properly considered a Hellenic form of oriental tradition; it should not be regarded as a sign of decline in Greek mathematics just because it does not employ the refined procedures of the axiomatic method.

Perhaps it is needless to hypothesize about the origin of the axiomatic method. Certainly, by the middle of the fourth century B.C., the method had been fairly well developed, for in Aristotle's *Analytica posteriora*, we find a good deal of light thrown on some of its features. Aristotle was not a mathematician, but as the systematizer of classical logic, he found in elementary mathematics excellent models of logical reasoning, and his mathematical illustrations tell us a great deal about the principles of the axiomatic method as accepted in his time. By the turn of the century the stage was set for Euclid's magnificent and epoch-making application of the axiomatic method.

---

[6] See Appendix, Section A.6.

PROGRAMS[7] ~~PROBLEMS~~ PROBLEMS[7]

1.1.1 In the Rhind papyrus the area of a circle is taken as equal to that of a square on 8/9 of the circle's diameter. Show that this is equivalent to taking $\pi = 3.1604 \cdots$.

1.1.2 The *Śulvasūtras*, ancient Hindu religious writings dating from about 500 B.C., are of interest in the history of mathematics because they embody certain geometrical rules for the construction of altars and show an acquaintance with the Pythagorean theorem. Among the rules furnished there appear empirical solutions of the circle-squaring problem that are equivalent to taking $d = (2 + \sqrt{2})s/3$ and $s = 13d/15$, where d is the diameter of the circle and s is the side of the equal square. These formulas are equivalent to taking what values for π?

1.1.3 Show that the ancient Babylonian formula $K = (a + c)(b + d)/4$, for the area of a quadrilateral having a, b, c, d for consecutive sides, gives too large an answer for all nonrectangular quadrilaterals.

1.1.4 *(for students who have studied calculus)* Prove, by elementary differential calculus, the farmer's conjecture that the rectangular field of maximum area lying along a straight river bank and utilizing a given amount of fencing has a length that is twice the depth of the field.

1.1.5 A disc of radius R spins vertically on a horizontal axis held above the surface of a liquid. As the disc spins it cuts into the liquid. Estimate, by empirical methods, how high the axis must be above the liquid's surface so that the wetted area of the spinning disc above the surface of the liquid shall be a maximum. (This problem arose in the manufacture of fruit syrups from fruit juices, and was solved empirically by the manufacturer, who found the required height r of the axis above the surface of the liquid to be about $(3/10)R$. It is not very difficult to show by differential calculus that $r = R/(1 + \pi^2)^{1/2}$. It is interesting that the General Electric Company began studies of this evaporation method in the early 1960s in connection with the design of a diffusion still.)

1.1.6 Two ladders, 60 ft long and 40 ft long, lean from opposite sides across an alley lying between two buildings, the feet of the ladders resting against the bases of the buildings. If the ladders cross each other at a height of 10 ft above the alley, how wide is the alley? Solve this problem empirically from drawings. [An algebraic treatment of this problem requires the solution of a quartic equation. If a and b represent the lengths of the ladders, c the height at which they cross, and x the width of the alley, it can be shown that $(a^2 - x^2)^{-1/2} + (b^2 - x^2)^{-1/2} = c^{-1}$.]

1.1.7 How good is the following empirical straightedge and compass trisection of an angle of 30°? Let AOB be the given angle, with $OA = OB$. On AB as diameter draw a semicircle lying on the same side of AB as is the point O. Take D and E on the semicircle such that $AD = DE = EB$. Take F on DE such that $DF = DE/4$. Then OF is a sought trisector.

1.1.8 Solve, by the rule of false position, the following problem found in the Rhind papyrus: "A quantity, its 2/3, its 1/2, and its 1/7, added together, become 33. What is the quantity?"

1.1.9 Find the length of side BC in the quadrilateral pictured in Figure 1.6.

1.1.10 In the study of geometrical constructions there is a counterpart of the rule of false position, generally known as the *method of similitude*. The method lies in constructing a figure similar to the one desired, and then, by the use of proportion, "blowing it up" to proper size. Suppose, for example, we wish to inscribe a square

---

[7]Note that triple numbering is used in the problems. The first number is the chapter number, the second is the section number, and the third is the sequence number.

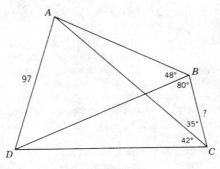

FIGURE 1.6

in a given triangle ABC so that one side of the square lies along the base BC of the triangle (see Figure 1.7). First draw a square $D'E'F'G'$ of any convenient size, as indicated. If F' falls on AC, the problem is solved. Otherwise we have solved the problem for a triangle $A'BC'$ similar to triangle ABC and having B as a center of similitude. It follows that line BF' cuts AC in the vertex F of the sought square inscribed in triangle ABC.

　　Construct, by the method of similitude, a line segment DE, where D is on side AB and E on side AC of a given triangle ABC, so that $BD = DE = EC$.

1.1.11 In the Rhind papyrus we find, "If you are asked, what is 2/3 of 1/5, take the double and the sixfold; that is 2/3 of it. One must proceed likewise for any other fraction." Interpret this and prove the general statement.

1.1.12 In the Moscow papyrus we find the following numerical example: "If you are told: A truncated pyramid of 6 for the vertical height by 4 on the base by 2 on the top. You are to square this 4, result 16. You are to double 4, result 8. You are to square 2, result 4. You are to add the 16, the 8, and the 4, result 28. You are to take one third of 6, result 2. You are to take 28 twice, result 56. See, it is 56. You will find it right." Show that this illustrates the general formula

$$V = \frac{h(a^2 + ab + b^2)}{3},$$

giving the volume of a frustum of a square pyramid in terms of the height h and the sides a and b of the bases.

1.1.13 Interpret the following, found on a Babylonian tablet dating from about 2600 B.C.:

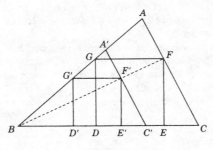

FIGURE 1.7

"60 is the circumference, 2 is the perpendicular, find the chord." "Thou, double 2 and get 4, dost thou not see? Take 4 from 20, thou gettest 16. Square 20, thou gettest 400. Square 16, thou gettest 256. Take 256 from 400, thou gettest 144. Find the square root of 144. 12, the square root, is the chord. Such is the procedure."

1.1.14 In 1936 a group of Old Babylonian tablets was lifted at Susa, about 200 miles east of Babylon. On one of the tablets the ratio of the perimeter of a regular hexagon to the circumference of the circumscribed circle is given as $57/60 + 36/3600$. Show that this leads to $3\frac{1}{8}$ as an approximation of π.

1.1.15 **(a)** A Babylonian tablet has been discovered that gives the values of $n^3 + n^2$ for $n = 1$ to 30. Make such a table for $n = 1$ to $n = 10$, and use it to find a root of the cubic equation $x^3 + 2x^2 - 3136 = 0$.

(b) A Babylonian problem of about 1800 B.C. seems to call for the solution of the simultaneous system $xyz + xy = 7/6$, $y = 2x/3$, $z = 12x$. Solve this system using the table of part (a).

1.1.16 It is known that the infinite series obtained by expanding $(a^2 + h)^{1/2}$ by the process of the binomial theorem converges to $(a^2 + h)^{1/2}$ if $-a^2 < h < a^2$.

(a) Establish the approximation formula

$$(a^2 + h)^{1/2} \approx a + \frac{h}{2a}, \qquad 0 < h < a^2.$$

(b) Take $a = 4/3$ and $h = 2/9$ in the approximation formula of part (a), and thus find a Babylonian rational approximation for $\sqrt{2}$. Find a rational approximation for $\sqrt{5}$ by taking $a = 2$, $h = 1$.

1.1.17 The Hindu mathematician, Āryabhata, wrote early in the sixth century A.D. His work is a poem of 33 couplets called the *Ganita*. Following are translations of two of the couplets: (1) The area of a triangle is the product of the altitude and half the base; half of the product of this area and the height is the volume of the solid of six edges. (2) Half the circumference multiplied by half the diameter gives the area of the circle; this area multiplied by its own square root gives the volume of the sphere. Show that, in each of these couplets, Āryabhata is correct in two dimensions but wrong in three.

1.1.18 An early Chinese work that dates probably from the second century B.C. and that had considerable influence on the development of mathematics in China was the *K'ui-ch'ang Suan-shu*, or *Arithmetic in Nine Sections*. In this work we find the empirical formula $s(c + s)/2$ for the area of a circular segment of chord c and depth s.

(a) Show how this formula might have been obtained.

(b) Obtain a correct formula.

1.1.19 **(a)** Devise an empirical procedure, using templates and a balance, for showing that the area under one arch of the cycloid curve is equal to three times the area of the generating circle. (An experiment of this nature was performed by Galileo in 1599. The first published mathematical demonstration that the area of a cycloidal arch is exactly three times that of the generating circle was furnished in 1644, by Galileo's pupil, Evangelista Torricelli.)

(b) Devise an empirical procedure, using a right circular cone, a right circular cylinder of the same radius and altitude, and some sand, for showing that the volume of a right circular cone is one third the product of its altitude and the area of its base.

(c) Devise an empirical procedure, using a circular disc, a hemisphere of the same radius, and a long piece of thick cord, for showing that the area of a sphere is equal to four times that of a great circle.

    (d) Show empirically, by folding paper, that the sum of the angles of a triangle is equal to a straight angle.

1.2.1   Criticize the following inductions:

    (a) Mr. Smith and Mr. Brown were both born in January, and both suffer from colds. They resigned themselves to their fate on the ground that all people born in January must suffer from colds.

    (b) John had never eaten yeast, and at the beginning of the year weighed 120 pounds. For the next six months he ate three yeast cakes a day, and at the end of that time weighed 150 pounds. Therefore eating yeast makes people gain weight.

1.2.2   Criticize the following inductions:

    (a) During a certain summer someone noted that the number of pounds of butter sold in New York City each month varied more or less directly with the number of inches of rainfall in New York City each month and conjectured that there must be some connection between the two.

    (b) During another summer a high degree of correlation was observed between the number of people each day at a beach resort and the corresponding number of people each day taking a boat ride on a river leading from a large city to the beach resort. From this correlation it was induced that many people travel to the beach by boat.

    (c) From an observation that students who made high grades in English also generally made high grades in mathematics it was induced that English helps mathematics.

    (d) Statistics show that over the years our principal roads have been made wider and wider, and at the same time accidents have increased. Apparently wide roads (perhaps because they cause people to drive faster) are a cause of accidents.

1.2.3   The three altitudes of a triangle are concurrent. Would you expect the four altitudes of a tetrahedron to be concurrent? (Many theorems concerning tetrahedra were first suggested by the corresponding theorems about triangles. In this case, however, the analogy leads to an incorrect result. Only for the so-called *orthocentric tetrahedra* are the four altitudes concurrent. An orthocentric tetrahedron is a tetrahedron each edge of which is perpendicular to its opposite edge.)

1.2.4   Two lines through the vertex of an angle and symmetrical with respect to the bisector of the angle are called a pair of *isogonal conjugate lines* of the angle. There is an attractive theorem about triangles that states that if three lines through the vertices of a triangle are concurrent, then the three isogonal conjugate lines through the vertices of the triangle are also concurrent. Try to construct an analogous definition and theorem for the tetrahedron.

1.2.5   List from the following statements those that are equivalent to the statement, "All parallelograms are quadrilaterals":

    (a) Every parallelogram is a quadrilateral.

    (b) If a figure is a quadrilateral, then it must be a parallelogram.

    (c) If a figure is not a quadrilateral, then it is not a parallelogram.

    (d) If a figure is a parallelogram, then it surely is not a quadrilateral.

1.2.6   List the following statements that are equivalent to the statement, "When the sunset is red, it is sure to rain the next day":

    (a) If it is raining today, then the sunset last evening must have been red.

    (b) If it does not rain today, then the sunset last evening must have been red.

    (c) If it does not rain today, then the sunset last evening must not have been red.

    (d) Whenever it rains during the day, the sunset of the previous evening was red.

1.2.7 List the following statements that are equivalent to the statement, "It never rains in June":

    (a) If it is June, it is not raining.

    (b) If it is not raining, it is not June.

    (c) In June it never rains.

    (d) Never in June does it rain.

    (e) If it is raining, it is not June.

    (f) Sometimes in June it does not rain.

1.2.8 Draw diagrams illustrating each of the following types of categorical propositions:

    (a) Universal Affirmative: All a are b.

    (b) Universal Negative: No a are b.

    (c) Particular Affirmative: Some a are b.

    (d) Particular Negative: Some a are not b.

1.2.9 Test the following arguments for validity:

    (a) Premise: All x are y.

       Conclusion: All non-x are non-y.

    (b) Premises: (1) All games played in the street are dangerous.

               (2) No bull fighting is played in the street.

       Conclusion: Bull fighting is not a dangerous game.

    (c) Premises: (1) No x are y.

               (2) Some x are z.

       Conclusion: Some z are not y.

    (d) Premises: (1) All trapezoids are quadrilaterals.

               (2) All parallelograms are quadrilaterals.

       Conclusion: All parallelograms are trapezoids.

    (e) Premises: (1) All useful books are amusing.

               (2) All books of tables are useful books.

       Conclusion: All books of tables are amusing.

    (f) Premise: All knowledge is useful.

       Conclusion: No knowledge is useless.

    (g) Premises: (1) Some doctors are not paid enough.

               (2) Some doctors are college professors.

       Conclusion: Some college professors are not paid enough.

    (h) A student must study to deserve good grades. John studied. Therefore he deserves good grades.

    (i) In a certain triangle the sum of the squares on two sides equals the square on the third. Hence the triangle is a right triangle by the Pythagorean theorem.

    (j) "He that is of God heareth God's words; ye therefore hear them not, because ye are not of God." (John 8 : 47.)

    (k) "I have tasted eggs, certainly," said Alice, . . . "but little girls eat eggs quite as much as serpents do, you know."

        "I don't believe it," said the Pigeon; "but if they do, why, then they're a kind of serpent: that's all I can say." (Lewis Carroll, *Alice in Wonderland*.)

1.2.10 Let T stand for *true*, F for *false*, V for *valid*, and I for *invalid*. Try to construct simple arguments satisfying each of the following possibilities:

Premises:	T	T	T	T	F	F	F	F
Argument:	V	V	I	I	V	V	I	I
Conclusion:	T	F	T	F	T	F	T	F

1.2.11 Consider the following four statements, called, respectively, the *direct* statement, the *converse* statement, the *inverse* statement, and the *contrapositive* statement:

1. All *a* are *b*.
2. All *b* are *a*.
3. All non-*a* are non-*b*.
4. All non-*b* are non-*a*.

(a) Show that the direct and contrapositive statements are equivalent.
(b) Show that the converse and inverse statements are equivalent.
(c) Taking "All parallelograms are quadrilaterals" as the direct statement, state the converse, inverse, and contrapositive statements.

1.2.12 (a) The categorical statement, "All *a* are *b*," may be stated in the equivalent hypothetical form, "If *w* is an *a*, then *w* is a *b*." State the corresponding converse, inverse, and contrapositive statements in hypothetical form.

(b) State the converse, inverse, and contrapositive of "If a triangle is isosceles, then the bisectors of its base angles are equal." (The direct proposition is very easily established. The converse proposition is known as the *Steiner-Lehmus theorem* and is troublesome to establish. If one can manage to establish the inverse proposition, then, of course, by Problem 1.2.11(b), the Steiner-Lehmus theorem will follow.)

1.3.1 We are told that Thales measured the distance of a ship from shore by using the fact that two triangles are congruent if two angles and the included side of one are equal to two angles and the included side of the other. Thomas L. Heath, the historian, has conjectured that this computation was probably made by an instrument consisting of two rods AC and AD, hinged together at A. The rod AD was held vertically over a point B on shore, while rod AC was pointed toward the ship P. Then, without changing the angle DAC, the instrument was revolved about AD, and point Q noted on the ground at which arm AC was directed. What distance must be measured in order to find the distance from B to the inaccessible point P?

1.3.2 The *Eudemian Summary* says that in Pythagoras's time there were three means, the *arithmetic*, the *geometric*, and the *subcontrary*, the last name being later changed to *harmonic* by Archytas and Hippasus. We may define these three means of two positive numbers *a* and *b* as

$$A = (a + b)/2, \qquad G = \sqrt{ab}, \qquad H = 2ab/(a + b),$$

respectively.

(a) Show that $A \geq G \geq H$, equality holding if and only if $a = b$.
(b) Show that H is the harmonic mean between a and b if there exists a number n such that $a = H + a/n$ and $H = b + b/n$. This was the Pythagorean definition of the harmonic mean of a and b.
(c) Since 8 is the harmonic mean of 12 and 6, Philolaus, a Pythagorean of about 425 B.C., called the cube a "geometrical harmony." Explain this.

1.3.3 Tradition is unanimous in ascribing to Pythagoras the independent discovery of the theorem on the right triangle that now universally bears his name (the square on the hypotenuse of a right triangle is equal to the sum of the squares on the two legs). This theorem was known to the Babylonians of Hammurabi's time, more than a thousand years earlier, but the first general proof of the theorem may well have been given by Pythagoras. There has been considerable conjecture regarding the proof Pythagoras might have offered; the common belief is that it probably was a dissection type of proof such as is suggested by Figure 1.8. Supply the proof. (To prove that the central piece of the second dissection is actually a square

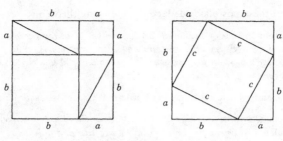

FIGURE 1.8

of side c we need to employ the fact that the sum of the angles of a right triangle is equal to two right angles. But the *Eudemian Summary* attributes this theorem for the general triangle to the Pythagoreans. Since a proof of this theorem requires, in turn, a knowledge of some properties of parallels, the early Pythagoreans are also credited with the development of that theory.)

1.3.4 State and prove the converse of the Pythagorean theorem.

1.3.5 Closely allied to the Pythagorean theorem is the problem of finding integers a, b, c to represent the legs and hypotenuse of a right triangle. Such a triple of numbers is known as a *Pythagorean triple*, and there is fairly convincing evidence that the ancient Babylonians knew how to calculate such triples.

    (a) Show that, for any *odd* integer m, the three numbers m, $(m^2 - 1)/2$, and $(m^2 + 1)/2$ yield a Pythagorean triple. (The Pythagoreans have been credited with this discovery.)

    (b) Show that for *any* integer m, the three numbers $2m$, $m^2 - 1$, and $m^2 + 1$ yield a Pythagorean triple. (This slight generalization of the above result is attributed to Plato. Neither formula yields all Pythagorean triples.)

1.3.6 Draw three unequal line segments. Label the longest one a, the medium one b, and take the shortest one as 1 unit. With straightedge and compasses construct line segments of lengths

    (a) $a + b$ and $a - b$,

    (b) ab,

    (c) a/b,

    (d) $\sqrt{a}$,

    (e) a/n, n a positive integer.

1.3.7 Show that there can be no more than five regular polyhedra.

1.4.1 Explain how the commonly heard statement, "An axiom is a self-evident truth," reflects part of the pattern of material axiomatics.

1.4.2 As a simple example of a discourse conducted by material axiomatics, consider a certain (finite and nonempty) collection S of people and certain clubs formed among these people, a club being a (nonempty) set of people organized for some common purpose. Our basic terms are thus the *collection* S *of people* and the *clubs* to which these people belong. About these people and their clubs we assume:

> **Postulate 1:** *Every person of* S *is a member of at least one club.*
>
> **Postulate 2:** *For every pair of people of* S *there is one and only one club to which both belong.*
>
> **Definition** Two clubs having no members in common are called *conjugate clubs*.
>
> **Postulate 3:** *For every club there is one and only one conjugate club.*

From these postulates deduce the following theorems:

Theorem 1: *Every person of S is a member of at least two clubs.*
Theorem 2: *Every club contains at least two members.*
Theorem 3: *S contains at least four people.*
Theorem 4: *There exist at least six clubs.*

1.4.3 Using the same basic terms as in Problem 1.4.2, let us assume:

Postulate 1: *Any two distinct clubs have one and only one member in common.*
Postulate 2: *Every person of S belongs to two and only two clubs.*
Postulate 3: *There are exactly four clubs.*

From these postulates deduce the following theorems:

Theorem 1: *There are exactly six people in S.*
Theorem 2: *There are exactly three people in each club.*
Theorem 3: *For each person in S there is exactly one other person in S not in the same club.*

1.4.4 Establish the theorem about the game of tic-tac-toe cited in the text.

1.4.5 The *cigar game* is played on a rectangular table top by two players with a large stock of cigars. The two players, taking turns, lay (at each turn) a cigar on the table top so that it does not overlap any other cigar nor protrude over the edge of the table top. The last player able to place a cigar on the table top wins the game. Prove the following theorem about this game: *With proper strategy, the player who starts the game can win the game.*

1.5.1 (a) Prove that the straight line through the points $(0, 0)$ and $(1, \sqrt{2})$ passes through no point, other than $(0, 0)$, of the coordinate lattice.
  (b) Show how the coordinate lattice may be used for finding rational approximations of $\sqrt{2}$.

1.5.2 If p is a prime number and n is an integer greater than 1, show that $\sqrt[n]{p}$ is irrational.

1.5.3 Give a purely geometric proof of the irrationality of $\sqrt{2}$.

1.5.4 The most important of Heron's geometrical works is his *Metrica*, discovered in Constantinople by R. Schöne as recently as 1896. In this work is found Heron's method of approximating the square root of a nonsquare integer, a process frequently used by computers today. If $n = ab$, then $\sqrt{n}$ is approximated by $(a + b)/2$, the approximation improving with the closeness of a to b. The method permits of successive approximations. Thus, if a_1 is a first approximation to $\sqrt{n}$, then $a_2 = (a_1 + n/a_1)/2$ is a better approximation, and $a_3 = (a_2 + n/a_2)/2$ is still better, and so on. Approximate successively, by Heron's method, $\sqrt{3}$ and $\sqrt{720}$.

1.5.5 In some problems in the Heronian collection appear the formulas

$$a, b = \frac{(r + s) \pm \{(r + s)^2 - 8rs\}^{1/2}}{2},$$

for the legs a and b of a right triangle of perimeter $2s$ and inradius r. Obtain these formulas.

1.5.6 (a) In his work *Catoptrica*, Heron proves, on the assumption that light travels by the shortest path, that the angles of incidence and reflection in a mirror are equal. Prove this.
  (b) A man wishes to go from his house to the bank of a straight river for a pail of water, which he will then carry to his barn, on the same side of the river as his

house. Find the point on the riverbank that will minimize the distance the man must travel.

1.5.7 A regular heptagon (seven-sided polygon) cannot be constructed with straight-edge and compasses. In his work *Metrica*, Heron takes, for an approximate construction, the side of the heptagon equal to the apothem of a regular hexagon having the same circumcircle. How good an approximation is this?

1.5.8 Assuming the equality of alternate interior angles formed by a transversal cutting a pair of parallel lines, prove the following:
(a) The sum of the angles of a triangle is equal to a straight angle.
(b) The sum of the interior angles of a convex polygon of n sides is equal to $n - 2$ straight angles.

1.5.9 Assuming (1) A central angle of a circle is measured by its intercepted arc, (2) The sum of the angles of a triangle is equal to a straight angle, (3) The base angles of an isosceles triangle are equal, (4) A tangent to a circle is perpendicular to the radius drawn to the point of contact, establish the following chain of theorems:
(a) An exterior angle of a triangle is equal to the sum of the two remote interior angles.
(b) An inscribed angle in a circle is measured by one half its intercepted arc.
(c) An angle formed by two intersecting chords in a circle is measured by one half the sum of the two intercepted arcs.
(d) An angle formed by two intersecting secants of a circle is measured by one half the difference of the two intercepted arcs.
(e) An angle formed by a tangent to a circle and a chord through the point of contact is measured by one half the intercepted arc.
(f) An angle formed by a tangent and an intersecting secant of a circle is measured by one half the difference of the two intercepted arcs.
(g) An angle formed by two intersecting tangents of a circle is measured by one half the difference of the two intercepted arcs.

1.5.10 Assuming the area of a rectangle is given by the product of its two dimensions, establish the following chain of theorems:
(a) The area of a parallelogram is equal to the product of its base and altitude.
(b) The area of a triangle is equal to half the product of any side and the altitude on that side.
(c) The area of a right triangle is equal to half the product of its two legs.
(d) The area of a triangle is equal to half the product of its perimeter and the radius of its inscribed circle.
(e) The area of a trapezoid is equal to the product of its altitude and half the sum of its bases.
(f) The area of a regular polygon is equal to half the product of its perimeter and the radius of its inscribed circle.
(g) The area of a circle is equal to half the product of its circumference and its radius.

EUCLID'S *ELEMENTS*

2.1 The Importance and Formal Nature of Euclid's *Elements*

The earliest extensively developed example of the use of the axiomatic method that has come down to us is the very remarkable and historically important *Elements* of Euclid. The production of this treatise is generally regarded as the first great landmark in the history of mathematical thought and organization, and its subsequent influence on scientific thinking can hardly be overstated.

Of Euclid himself, however, disappointingly little is known. It is from Proclus's *Commentary on Euclid, Book I*, that we obtain our most satisfying information about Euclid. He writes,

> Euclid, who put together the *Elements*, collected many of the theorems of
> Eudoxus. He perfected many of the theorems of Theaetetus, and also
> brought to irrefragable demonstration the things which were only somewhat
> loosely proved by his predecessors. This man lived in the time of the first
> Ptolemy, for Archimedes, who came immediately after the first Ptolemy,
> makes mention of Euclid, and furthermore, it is said that Ptolemy once asked
> him if there was in geometry any shorter way than that of the *Elements*, and
> Euclid answered that there was no royal road to geometry. It is evident, then,
> that Euclid came after the time of Plato, but preceded Eratosthenes and
> Archimedes.[1]

This statement would imply that Euclid lived about 300 B.C. Also, from other evidence, it seems quite certain that Euclid was the first professor of mathematics at the famous University of Alexandria,[2] and that he was the founder of the

---

[1] The quotations from Proclus and Aristotle that appear in this and the next chapter are adapted, by permission, from T. L. Heath, pp. 1, 115, 116, 117–118, 119, 121–122, 153–155, 202–203, 241–242.

[2] For an interesting exposition on Alexandria, see R. E. Langer.

distinguished and long-lived Alexandrian School of Mathematics. Even his birthplace is not known, but there is some reason to believe that he received his mathematical training in the Platonic School at Athens.

Although Euclid wrote at least ten treatises on mathematics, posterity has come to know him chiefly through his *Elements*, a monumental work written in thirteen books, or parts. This extraordinary work so quickly and so completely superseded all previous works of the same nature that now no copies remain of the earlier efforts. Apparently from its very first appearance it was accorded the highest respect, and the mere citation of Euclid's book and proposition numbers has been regarded ever since as sufficient to identify a particular theorem or construction. With the single exception of the Bible, no work has been more widely studied or edited. For more than two millennia it has dominated all teaching of geometry, and over a thousand editions of it have appeared since the first one printed in 1482. And, as the prototype of the axiomatic or postulational method, its impact on the development of mathematics has been enormous.

Proclus has clarified for us the meaning of the term *elements*. It seems that the elements of any demonstrative study are to be regarded as the leading, or key, theorems that are of wide and general use in the subject. Their function has been compared to that of the letters of the alphabet in relation to language; as a matter of fact, letters are called by the same name in Greek. The selection of the theorems to be taken as the elements of the subject requires the exercise of considerable judgment. As Proclus says,

> Now it is difficult, in each science, both to select and arrange in due order the elements from which all the rest is resolved. And of those who have made the attempt some were able to put together more and some less; some used shorter proofs; some extended their investigations to an indefinite length; some avoided the method of *reductio ad absurdum*; some avoided proportion; some contrived preliminary steps directed against those who reject the principles; and, in a word, many different methods have been invented by various writers of elements.
>
> It is essential that such a treatise should be rid of everything superfluous (for this is an obstacle to the acquisition of knowledge); it should select everything that embraces the subject and brings it to a point (for this is of supreme service to science); it must have great regard at once to clearness and conciseness (for their opposites trouble our understanding); it must aim at the embracing of theorems in general terms (for the piecemeal division of instruction into the more partial makes knowledge difficult to grasp). In all these ways Euclid's system of elements will be found to be superior to the rest.

And elsewhere, Proclus says,

> Starting from these elements, we shall be able to acquire knowledge of the other parts of this science as well, while without them it is impossible for us to get a grasp of so complex a subject, and knowledge of the rest is unattainable. As it is, the theorems which are most of the nature of principles, most simple, and most akin to the first hypotheses are here collected, in their appropriate order; and the proofs of all other propositions use these theorems

as thoroughly known, and start from them. Thus Archimedes in the books on the sphere and cylinder, Apollonius, and all other geometers, clearly use the theorems proved in this very treatise as constituting admitted principles.

Aristotle, in his *Metaphysics*, speaks of "elements" in the same sense when he says, "Among geometrical propositions we call those 'elements' the proofs of which are contained in the proofs of all or most of such propositions."

It is no reflection on the brilliance of Euclid's work that there had been other *Elements* anterior to his own. According to the *Eudemian Summary*, Hippocrates of Chios made the first effort along this line, and the next attempt was that of Leon, who in age fell somewhere between Plato and Eudoxus. It is said that Leon's work contained a more careful selection of propositions than did that of Hippocrates, and that these propositions were more numerous and more serviceable. The textbook of Plato's Academy was written by Theudius of Magnesia and was praised as an admirable collection of elements. The geometry of Theudius seems to have been the immediate precursor of Euclid's work and was undoubtedly available to Euclid, especially if he studied in the Platonic School. Euclid was acquainted also with the important work of Theaetetus and Eudoxus. Thus it is probable that Euclid's *Elements* is, for the most part, a highly successful compilation and systematic arrangement of works of earlier writers. No doubt Euclid had to supply a number of the proofs and to perfect many others, but the chief merit of his work lies in the skillful selection of the propositions and in their arrangement into a logical sequence presumably following from a small handful of initial assumptions.

In the thirteen books that comprise Euclid's *Elements* there is a total of 465 propositions. Contrary to popular impression, many of these propositions are concerned, not with geometry, but with number theory and with elementary (geometric) algebra. Book I contains the necessary preliminary material, together with theorems on congruence, parallel lines, and rectilinear figures. Book II is devoted to geometric algebra, Book III to circles, and Book IV to the construction of regular polygons. Books V and VI contain the Eudoxian theory of proportion and its application to geometry. Books VII, VIII, and IX, containing a total of 102 propositions, deal with elementary number theory. Book X is devoted to the study of irrationals, much of the material probably from Theaetetus. The remaining three books are concerned with solid geometry. The material of Books I, II, and IV was, in all likelihood, developed by the early Pythagoreans. The material found in current American high school plane and solid geometry texts is largely that found in Euclid's Books I, III, IV, VI, XI, and XII.

Certainly there is a good deal in the contents of Euclid's *Elements* that is of considerable interest, but in the present study our concern is with the formal nature of the *Elements* rather than with its mathematical contents. In fact, the various consequences of the formal character of this great work will constitute some of our chief avenues of investigation. At the moment, we are especially interested in Euclid's conception of the axiomatic method and in the precise manner in which he applied the method to the development of his *Elements*. We consider these matters in the two following sections.

2.2   Aristotle and Proclus on the Axiomatic Method

It is a misfortune that no copy of Euclid's *Elements* has been found that actually dates from the author's own time. Modern editions of the work are based on a revision that was prepared by the Greek commentator Theon of Alexandria, who lived almost 700 years after the time of Euclid. Theon's revision was, until the early nineteenth century, the oldest edition of the *Elements* known to us. In 1808, however, when Napoleon ordered valuable manuscripts to be taken from Italian libraries and to be sent to Paris, F. Peyrard found, in the Vatican library, a tenth-century copy of an edition of Euclid's *Elements* that predates Theon's recension. A study of this older edition and a careful sifting of citations and remarks made by early commentators indicate that the introductory material of Euclid's original treatise undoubtedly underwent some editing in the subsequent revisions, but that the propositions and their proofs, except for minor additions and deletions, have remained essentially as Euclid wrote them.

Because of our lack of a copy of Euclid's original treatise, and because of the changes and additions made by later editors, it is not certain precisely what statements Euclid assumed at the start of his work, nor even how many such statements he had. Also, unfortunately, there is no known commentary by Euclid himself on the nature of the deductive organization used so successfully in his mathematical studies. It would be valuable to have Euclid's own point of view on the meaning of proof or on the significance that he attached to such terms as *definition, axiom,* and *postulate.* Even partially to understand Euclid, therefore, we must study the ideas held by Euclid's contemporaries. Aristotle, in particular, is an important source of information. Since Aristotle studied at Plato's Academy, his scholastic background may have been quite similar to that of Euclid.

A student of mathematics would do well to study Aristotle's *Analytica posteriora.* The following passage from that work is particularly full and enlightening:

By the first principles of a subject I mean those the truth of which it is not possible to prove. What is *denoted* by the first terms and those derived from them is assumed; but, as regards their *existence,* this must be assumed for the principles but proved for the rest. Thus what a unit is, what a straight line is, or what a triangle is, must be assumed; and the existence of the unit and of magnitude must also be assumed, but the existence of the rest must be proved. Now of the premises used in demonstrative sciences some are peculiar to each science and others common to all, the latter being common by analogy, for of course they are actually useful insofar as they are applied to the subject-matter included under the particular science. Instances of first principles peculiar to a science are the assumptions that a line is of such and such a character, and similarly for a straight line; whereas it is a common principle, for instance, that if equals be subtracted from equals, the remainders are equal. But it is enough that each of the common principles is true as regards the particular subject-matter; in geometry, for instance, the effect will be the same even if the common principles be assumed to be true, not of everything, but only of magnitudes, and, in arithmetic, of numbers.

Now the things peculiar to the science, the existence of which must be assumed, are the things with reference to which the science investigates the essential attributes, for example arithmetic with reference to units, and geometry with reference to points and lines. With these things it is assumed that they exist and that they are of such and such a nature. But, with regard to their essential properties, what is assumed is only the meaning of each term employed; thus arithmetic assumes the answer to the question what is meant by "odd" or "even," "a square" or "a cube," and geometry to the question what is meant by "the irrational," or "deflection," or the so-called "verging" to a point; but that there are such things is proved by means of the common principles and of what has already been demonstrated. It is similar with astronomy. For every demonstrative science has to do with three things, (1) the things which are assumed to exist, namely the subject-matter in each case, the essential properties of which the science investigates, (2) the so-called common axioms, which are the primary source of demonstration, and (3) the properties, with regard to which all that is assumed is the meaning of the respective terms used.

This remarkable passage is almost modern in its point of view. It says that a demonstrative science must start from a set of assumptions, known as the *first principles* of the subject. These first principles constitute a sort of platform of initial agreement from which the rest of the discourse can be launched by purely deductive procedures. Of these principles, according to Aristotle, some are common to all sciences and others are peculiar to the particular science being studied. The first principles common to all sciences are called *axioms* (illustrated by, "if equals be subtracted from equals, the remainders are equal"). Among the first principles, or initial assumptions, peculiar to the science being studied, we have, first of all, statements of the *existence* of the subject matter and of the fundamental things whose properties the science intends to investigate (for example, in geometry, we must assume the *existence* of "magnitude," of "points," and of "lines"). Also among the first principles peculiar to the science being studied we have the *connotation* of the technical terms employed in the discourse. That is, we must accept certain definitions concerning manifestations or attributes of our subject matter (for example, in geometry, we must assume what is *meant* by *triangle* and by *irrational*). These definitions, however, say nothing of the existence of the things defined but must be merely understood. The existence of only the subject matter and the fundamental things is assumed; the existence of all other things defined must be proved.

In addition to the definitions, one might expect to find among the first principles that are peculiar to the particular science being studied some statements concerning properties or relationships of the technical terms of the discourse. Certainly, since we cannot prove all the statements of our discourse, we anticipate the need for some such assumed statements for the purpose of getting started. About such assumptions Aristotle, again in his *Analytica posteriora*, has the following to say:

Now anything that the teacher assumes, though it is matter of proof, without proving it himself, is a hypothesis if the thing assumed is believed by the learner, and it is moreover a hypothesis, not absolutely, but relatively to the particular pupil; but if the same thing is assumed when the learner either has no opinion on the subject or is of contrary opinion, it is a postulate. This is the difference between a hypothesis and a postulate; for a postulate is that which is rather contrary than otherwise to the opinion of the learner, or whatever is assumed and used without being proved, although matter for demonstration. Now definitions are not hypotheses, for they do not assert the existence or non-existence of anything, while hypotheses are among propositions. Definitions only require to be understood; a definition is therefore not a hypothesis, unless indeed it be asserted that any audible speech is a hypothesis. A hypothesis is that from the truth of which, if assumed, a conclusion can be established.

It must be admitted that Aristotle's notion of a postulate and of the role that a postulate plays in a demonstrative science is not too clear. His remarks imply that a postulate represents the assumption of a thing which is properly a subject of demonstration, and that the assumption is made without, perhaps, the assent of the student. In other words, a postulate may not appeal to a person's sense of what is right, but it has been adopted as basic in order that the work may proceed. From this point of view, then, a postulate is a first principle. In contradistinction to this, a hypothesis is an assumption believed in by the learner, and thus is introduced apparently in order to continue an argument. For example, once a theorem has been established, and hence is acceptable to the learner, that theorem may be taken as a hypothesis from which to deduce some later theorem. If we read further in the works of Aristotle we find other passages that are of special significance in comprehending the organization of Euclid's *Elements*. In several places we find that Aristotle regards an axiom as a universal assumption that is so self-evident that no sane person would question it; also he considers an axiom to be too fundamental ever to be regarded as matter for demonstration. We thus seem to have, according to Aristotle, the following four distinctions between an axiom and a postulate. An axiom is common to all sciences, whereas a postulate is related to a particular science; an axiom is self-evident, whereas a postulate is not; an axiom cannot be regarded as a subject for demonstration, whereas a postulate is properly such a subject; an axiom is assumed with the ready assent of the learner, whereas a postulate is assumed without, perhaps, the assent of the learner. Some of Aristotle's statements appear somewhat contradictory, but the interpretations just given seem especially appropriate in any attempt to understand Euclid's work.

Aristotle's characterizations of definitions, axioms, and postulates are further clarified by the following account given by Proclus in his *Commentary on Euclid, Book I*.[3]

> The compiler of elements in geometry must give separately the principles of the science, and, after that, the conclusions from those principles, not giving any account of the principles but only of their consequences. No science proves its own principles, or even discourses about them; they are treated as self-evident. . . . Thus, the first essential was to distinguish the principles from their consequences. Euclid carries out this plan practically in every book and, as a preliminary to the whole enquiry, sets out the common principles of this science. Then he divides the common principles themselves into *definitions*, *postulates*, and *axioms*. For all these are different from one another; an axiom,

---

[3] We have everywhere corrected a confusion that exists in the original statement caused by Proclus's consistent misuse of the term *hypothesis* for the term *definition*.

a postulate, and a definition are not the same thing, as the inspired Aristotle has somewhere pointed out. Whenever that which is assumed and ranked as a principle is both known to the learner and convincing in itself, such a thing is an *axiom*, for example the statement that things which are equal to the same thing are also equal to one another. When, on the other hand, the pupil has not the notion of what is told him which carries conviction in itself, but nevertheless lays it down and assents to its being assumed, such an assumption is a *definition*. Thus we do not preconceive by virtue of a common notion, and without being taught, that the circle is such and such a figure, but, when we are told so, we assent without demonstration. When, again, what is asserted is both unknown and assumed even without the assent of the learner, then, he says, we call this a *postulate*, for example that all right angles are equal. This view of a postulate is clearly implied by those who have made a special and systematic attempt to show, with regard to one of the postulates, that it cannot be assented to by any one straight off. According then to the teaching of Aristotle, an axiom, a postulate, and a definition are thus distinguished.

That there was no unanimity of opinion, even among the early Greek mathematicians themselves, concerning the precise nature of, and the difference between, an axiom and a postulate is borne out by remarks made by Proclus. Proclus points out the following three distinctions advocated by various parties: (1) An axiom is a self-evident assumed statement about something, and a postulate is a self-evident assumed construction of something; thus axioms and postulates bear a relation to one another much like that between theorems and construction problems. (2) An axiom is an assumption common to all sciences, whereas a postulate is an assumption peculiar to the particular science being studied. (3) An axiom is an assumption of something that is both obvious and acceptable to the learner; a postulate is an assumption of something that is neither necessarily obvious nor necessarily acceptable to the learner. (This last is essentially the Aristotelian distinction.) Further confusion is indicated by Proclus when he points out that some preferred to call them all postulates.

In summary, then, according to the Greek conception of the axiomatic method, every demonstrable science must start from assumed first principles. These first principles consist of definitions, axioms (or common notions), and postulates. The definitions describe the technical terms used in the discourse and, except in the case of a few fundamental terms, are not meant to imply the existence of the entities described. The axioms and the postulates are initial statements that must be assumed so that the discourse may proceed. Just which of these statements should be called axioms and which postulates was a matter of varying opinion.

2.3   Euclid's Definitions, Axioms, and Postulates

Adhering to the Greek conception of the axiomatic method, we find, at the very start of Book I of Euclid's *Elements*, a list of the definitions, postulates, and common notions that are to serve as the first principles of the work. Some of the succeeding books of the work commence with additional lists of definitions. It is presumed by the author that all of the 465 propositions included in the treatise are logically deduced from these principles. For reference, we now give here the complete set of first principles for Book I essentially as furnished by T. L. Heath[4] in his translation of the distinguished Heiberg text of Euclid's *Elements*.

Definitions

1. A *point* is that which has no part.

2. A *line* is length without breadth.

3. The extremities of a line are points.

4. A *straight line* is a line which lies evenly with the points on itself.

5. A *surface* is that which has only length and breadth.

6. The extremities of a surface are lines.

7. A *plane surface* is a surface which lies evenly with the straight lines on itself.

8. A *plane angle* is the inclination to one another of two lines in a plane if the lines meet and do not lie in a straight line.

9. When the lines containing the angle are straight lines, the angle is called a *rectilinear angle*.

10. When a straight line erected on a straight line makes the adjacent angles equal to one another, each of the equal angles is called a *right angle*, and the straight line standing on the other is called a *perpendicular* to that on which it stands.

11. An *obtuse angle* is an angle greater than a right angle.

12. An *acute angle* is an angle less than a right angle.

13. A *boundary* is that which is an extremity of anything.

14. A *figure* is that which is contained by any boundary or boundaries.

15. A *circle* is a plane figure contained by one line such that all the straight lines falling upon it from one particular point among those lying within the figure are equal.

16. The particular point (of Definition 15) is called the *center* of the circle.

17. A *diameter* of a circle is any straight line drawn through the center and terminated in both directions by the circumference of the circle. Such a straight line also bisects the circle.

18. A *semicircle* is the figure contained by a diameter and the circumference cut off by it. The center of the semicircle is the same as that of the circle.

19. *Rectilinear figures* are those which are contained by straight lines, *trilateral* figures being those contained by three, *quadrilateral* those contained by four, and *multilateral* those contained by more than four straight lines.

---

[4] T. L. Heath, 1, 153–155.

20. Of the trilateral figures, an *equilateral triangle* is one which has its three sides equal, an *isosceles triangle* has two of its sides equal, and a *scalene triangle* has its three sides unequal.

21. Furthermore, of the trilateral figures, a *right-angled triangle* is one which has a right angle, an *obtuse-angled triangle* has an obtuse angle, and an *acute-angled triangle* has its three angles acute.

22. Of the quadrilateral figures, a *square* is one which is both equilateral and right-angled; an *oblong* is right-angled but not equilateral; a *rhombus* is equilateral but not right-angled; and a *rhomboid* has its opposite sides and angles equal to one another but is neither equilateral nor right-angled. Quadrilaterals other than these are called *trapezia*.

23. *Parallel* straight lines are straight lines which, being in the same plane and being produced indefinitely in both directions, do not meet one another in either direction.

Postulates

Let the following be postulated:

1. A straight line can be drawn from any point to any point.
2. A finite straight line can be produced continuously in a straight line.
3. A circle may be described with any center and distance.
4. All right angles are equal to one another.
5. If a straight line falling on two straight lines makes the interior angles on the same side together less than two right angles, the two straight lines, if produced indefinitely, meet on that side on which the angles are together less than two right angles.

Common notions

1. Things which are equal to the same thing are also equal to one another.
2. If equals be added to equals, the wholes are equal.
3. If equals be subtracted from equals, the remainders are equal.
4. Things which coincide with one another are equal to one another.
5. The whole is greater than the part.

We observe that the first principles of Euclid's *Elements* fit quite well the Aristotelian account of definitions, postulates, and axioms as given in Section 2.2. It would also seem that Euclid strove to keep his list of postulates and axioms to an irreducible minimum. This economy, too, is in keeping with Aristotle's views, for in his *Analytica posteriora* he says, "other things being equal, that proof is the better which proceeds from the fewer postulates, or hypotheses, or propositions."

We shall pass over Euclid's definitions without much comment. Most of them probably were taken from earlier works, which would account for the fact that some terms, like *oblong*, *rhombus*, and *rhomboid*, are included but are never used anywhere in the work. It is curious that after having defined parallel lines Euclid does not give a formal definition of *parallelogram*. The existence of a parallelogram is established in I 33,[5] and in I 34 it is referred to as a *parallelogramic area*; then in I 35 this latter expression is shortened to *parallelogram*. We note that to the definition of a diameter of a circle (Definition 17) is appended the statement, "Such a straight line also bisects the circle." This addition is, of course, really a theorem (one of those attributed by Proclus in the *Eudemian Summary* to Thales), but its statement in Definition 17 is necessary in order to justify the definition of a semicircle that immediately follows. There are indications for believing that the definitions of a straight line and of a plane (Definitions 4 and 7) were original with Euclid. These definitions are not easy to understand but can be comprehended, at least partially, if we appeal to sight by considering an eye placed at an extremity of the line or the plane and looking, respectively, along the line or the plane. Other interpretations of these definitions have been given. A number of Euclid's definitions are vague and virtually meaningless; we shall return to this in the next section. The work of Heath previously referred to contains a full and valuable commentary on Euclid's definitions.

Some aspects of Euclid's postulates are of especial interest. The first three are postulates of construction, for they assert what we are permitted to draw. Since these postulates restrict constructions to only those that can be made in a permissible manner with straightedge and compasses, these instruments, so limited, have become known as *Euclidean tools*, although their use under these restrictions certainly predates Euclid. The construction of figures with only straightedge and compasses, viewed as a game played according to the rules set down in Euclid's first three postulates, has proved to be one of the most fascinating and absorbing games ever devised. One is surprised at the really intricate constructions that can be accomplished in the allowed manner, and accordingly it is hard to believe that certain seemingly simple construction problems, like that of trisecting a given arbitrary angle, for example, cannot also be so accomplished. The energetic efforts of early Greek geometers to solve legitimately some of the construction problems that are now known to be beyond the use of Euclidean tools profoundly influenced the development of much of the content of early geometry. For example, the invention of the conic sections, of many cubic and quartic curves, and of several transcendental curves resulted from this work. A later outgrowth was the development, in modern times, of portions of the theory of equations, of the theory of algebraic numbers, and of group theory. This whole line of mathematical development, so intimately tied to Euclid's first three postulates, has little connection with our present line of investigation and so will not be further considered here. We shall return to the subject, however, in Section A.2 of the Appendix.

Postulates 1 and 3 refer to existence. In other words, the existence of a straight line joining any two given points is assumed, as is the existence of a circle having any given center and radius. From applications that Euclid makes of

---

[5] I 33 means Proposition 33 of Book I.

Postulates 1 and 2, it appears that these postulates are meant also to imply that the straight line segment joining two points in the one case, and the produced portion in the other case, are *unique*, although it must be admitted that the postulates do not explicitly say as much. Postulate 3 may be construed as implying something in regard to the continuity and extent of the space under consideration, since the radius of the circle may be as small or as large as one desires.

Postulates 4 and 5 are quite different from the first three postulates. The meaning of Postulate 4 is certainly evident, but there has been much debate on whether it is properly classified when placed among the postulates. If it should be classified as a theorem its proof would have to be accomplished by applying one pair of adjacent right angles to another such pair, but Euclid preferred to shun, as much as possible, such *proofs by superposition*. In any event, Euclid had to place Postulate 4 before his Postulate 5, since the condition in Postulate 5 that a certain pair of interior angles be together less than two right angles would be useless unless it were first made clear that all right angles are equal.

Postulate 5, known as Euclid's *parallel postulate*, has become, as we shall see, one of the most famous statements in mathematical history. There is more evidence for the origin of this postulate with Euclid than for the origin of any of the other four. Aristotle alludes to a *petitio principii*, or a circularity in reasoning, that was involved in the theory of parallels current in his time. It is a mark of Euclid's mathematical acumen that he perceived that the only way out of the difficulty was to lay down *some* postulate as a basis for the theory of parallels that is so essential to the development of his geometry. The postulate that he formulated serves this purpose admirably and also, at the same time, furnishes a criterion for determining whether two straight lines in a figure will or will not meet if extended. This fact is an advantage of Euclid's postulate over the substitutes that were later suggested to take its place, and this advantage is actually employed in the *Elements* as early as I 44. The consequences of investigations carried on in connection with Euclid's fifth postulate proved to be very far-reaching. Not only did these investigations supply the stimulus for the development of much of the mathematics that we characterize as modern, but they led to a far deeper examination, and consequent refinement, of the axiomatic method. These investigations are therefore vital to our present study and will constitute the dramatic story of the next chapter.

Of the common notions, or axioms, there is reason to believe that the first three were given by Euclid but that the last two may have been added at a later time. Axiom 4 has been criticized on the ground that its subject matter is special rather than general and that it ought therefore to be listed as a postulate instead of as an axiom. Objections that can be raised to the method of superposition, used by Euclid with apparent reluctance to establish some of his early congruence theorems, can be at least partially met by Axiom 4. Again the student is referred to the excellent commentary given by Heath.

In conclusion, we may summarize Euclid's conception and use of the axiomatic method as follows: Every deductive system requires assumptions from which the deduction may proceed. Therefore, as initial premises, Euclid puts down five postulates, or assumed statements about his subject matter. In addition to the five postulates, Euclid lists five axioms, or common notions, that he also needs for his proofs. These axioms are not peculiar to his subject matter but are general principles valid in any field of study. Now in the postulates a number of terms occur, such as *point*, *straight line*, *right angle*, and *circle*, of which it is not certain that the reader has a precise notion. Hence some definitions are also given. These definitions are not, like the postulates, assumptions about the nature of the subject matter but are merely explanations of the meanings of the terms. Definition 10, for example, tells what a right angle is and how an angle may be identified as a right angle, but it says nothing about the existence of right angles, nor does it state what is assumed about such angles. These latter functions are left to the postulates and to deduced propositions. Thus Postulate 4 informs us that all right angles are equal, and Proposition I 11 proves that right angles exist. On the other hand, Postulate 4 gives no clue regarding the nature of a right angle, nor does it tell how the term is to be employed; it merely states a fundamental assumption about such angles. Finally, the natural order for presenting the postulates, axioms, and definitions to the student is, first, the definitions explaining the meanings of the technical terms of the discourse, next, the postulates that are so closely related to the definitions, and last, the axioms or common notions.

2.4 Some Logical Shortcomings of Euclid's Elements

It would be very surprising indeed if Euclid's *Elements*, because it is such an early and extensive application of the axiomatic methods, should be free of logical blemishes. Therefore it is no great discredit to the work that critical investigations have revealed a number of defects in its logical structure. Probably the gravest of these defects are certain tacit assumptions that are employed later in the deductions and are not granted by the first principles of the work. This danger exists in any deductive study when the subject matter is overly familiar to the author. Usually a thorough grasp of the subject matter in a field of human endeavor is regarded as an indispensable prerequisite to serious work, but in developing a deductive system such knowledge can be a definite disadvantage unless proper precautions are taken.

A deductive system differs from a mere collection of statements in that it is organized in a very special way. The key to the organization lies in the fact that all statements of the system other than the original assumptions must be deducible from these initial hypotheses, and that if any additional assumptions should creep into the work the desired organization is not realized. Now anyone formulating a deductive system knows more about his subject matter than just the initial assumptions he wishes to employ. He has before him a set of statements belonging to his subject matter, some of which he selects for postulates and the rest of which he presumably deduces from his postulates as theorems. But with a large body of information before one, it is very easy to employ in the proofs some piece of this information that is not embodied in the postulates. Any piece of information used in this way may be so apparently obvious or so seemingly elementary that it is assumed unconsciously. Such a tacit assumption, of course, spoils the rigidity of the organization of the deductive system. Moreover, should that piece of information involve some misconception, its introduction may lead to results that not only do not strictly follow from the postulates but that may actually contradict some previously established theorem. Herein, then, lies the pitfall of too great a familiarity with the subject matter of the discourse; at all times in building up a deductive system one must proceed with the appearance of being completely ignorant of the developing material. This does not mean that in building up a deductive system one refrains from making any use of one's intuitive appreciation of the significance of the axioms and of possible interpretations of the primitive terms. On the contrary one makes *full* use of these things, but only to conjecture possible theorems and possible avenues of investigation. In the actual establishment of these theorems and in the actual development of these avenues of investigation, one must be careful to proceed only in terms of the accepted assumptions.

The tacit assumption by Euclid of something that is not contained in his first principles is exemplified in the very first deduced proposition of the *Elements*. In order to examine the difficulty we shall quote Proposition I 1 verbatim from Heath's translation.[6]

> *On a given finite straight line to construct an equilateral triangle* [see Figure 2.1].
>   Let *AB* be the given finite straight line.
>   Thus it is required to construct an equilateral triangle on the straight line *AB*.
>   With center *A* and distance *AB*, let the circle *BCD* be described. [Postulate 3]
>   Again, with center *B* and distance *BA*, let the circle *ACE* be described. [Postulate 3]
>   And from the point *C*, in which the circles cut one another, to the points *A*, *B*, let the straight lines *CA*, *CB* be joined. [Postulate 1]
>   Now, since the point *A* is the center of the circle *CDB*, *AC* is equal to *AB*. [Definition 15]
>   Again, since the point *B* is the center of the circle *CAE*, *BC* is equal to *BA*. [Definition 15]

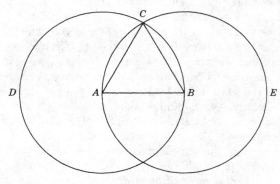

FIGURE 2.1

---

[6] T. L. Heath, 241, 242.

But *CA* was also proved equal to *AB*; therefore each of the straight lines *CA*, *CB* is equal to *AB*. And things which are equal to the same thing are also equal to one another; therefore *CA* is also equal to *CB*. [Axiom 1]

Therefore the three straight lines *CA*, *AB*, *BC* are equal to one another.

Therefore the triangle *ABC* is equilateral; and it has been constructed on the given finite straight line *AB*.

(Being) what it was required to do.

Now the construction of the two circles in this demonstration is certainly justified by Postulate 3, but there is nothing in Euclid's first principles which explicitly guarantees that the two circles shall intersect in a point *C*, and that they will not, somehow or other, slip through each other with no common point. The existence of this point, then, must be either postulated or proved, and it can be shown that Euclid's postulates are insufficient to permit the latter (see Problem 2.4.3). Only by the introduction of some additional assumption can the existence of the point *C* be established. Therefore the proposition does not follow from Euclid's first principles, and the proof of the proposition is invalid.

The fallacy here lies not in assuming something contrary to our concept of circles but in assuming something that is not implied by our accepted first principles. This is an example where the tacit assumption is so evident and elementary that there does not appear to be any assumption. The fallacy is a subtle one, but had Euclid known nothing more about circles than what his first principles say of them, he certainly could not have fallen into this error.

What is needed here is some additional postulate that will guarantee that the two circles concerned will intersect. Postulate 5 gives a condition under which two straight lines will intersect. We need similar postulates telling when two circles will intersect and when a circle and a straight line will intersect. What is essentially involved here is the continuity of circles and straight lines, and in modern treatments of geometry the existence of the desired points of intersection is taken care of by some sort of continuity postulate.

Another tacit assumption made by Euclid is that the straight line is of infinite extent. Although Postulate 2 asserts that a straight line may be produced indefinitely, it does not necessarily imply that a straight line is infinite in extent but merely that it is endless, or boundless. The arc of a great circle joining two points on a sphere may be produced indefinitely along the great circle, making the prolonged arc endless, but certainly it is not infinite in extent. Now it is conceivable that a straight line may behave similarly, and that after a finite prolongation it, too, may return on itself. It was the great German mathematician Bernhard Riemann (1826–1866) who, in his famous probationary lecture, *Über die Hypothesen welche der Geometrie zu Grunde liegen*, of 1854, distinguished between the boundlessness and the infinitude of straight lines. There are numerous occasions where Euclid unconsciously assumes the infinitude of a straight line. Let us briefly consider, for example, Proposition I 16:

In any triangle, if one of the sides be produced, the exterior angle is greater than either of the interior and opposite angles.

A précis of Euclid's proof runs as follows. Let *ABC* (Figure 2.2) be the given triangle, with *BC* produced to *D*. Let *E* be the midpoint of *AC*. Draw *BE* and extend it its own length to *F*. Draw *CF*. Then triangles *BEA* and *FEC* can easily

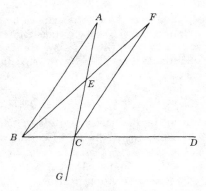

FIGURE 2.2

be shown to be congruent, whence $\angle FCE = \angle BAC$. But $\angle ACD > \angle FCE$, whence $\angle ACD > \angle BAC$. By producing AC to G, we may similarly show that $\angle BCG$, which is equal to $\angle ACD$, is also greater than $\angle ABC$.

Now if a straight line should return on itself, like the great circle arc considered here, BF may be so long that F will coincide with B or lie on the segment BE. Should this be the case, the proof would certainly fail. The author has been misled by his visual reference to the figure rather than to the principles that should be the basis of his argument. Clearly, then, to make the proof universally valid we must either prove or postulate the infinitude of straight lines.

One can point out many other tacit assumptions that, like the preceding one, were unconsciously made by Euclid and that vitiate the true deductive character of his work. For example, in Proposition I 21, Euclid unconsciously assumes that if a straight line enters a triangle at a vertex it must, if sufficiently produced, intersect the opposite side. It was Moritz Pasch (1843–1930) who recognized the necessity of a postulate to take care of this situation. Again, Euclid makes no provision for *linear order*, and his concept of "betweenness" is without any postulational foundation, with the result that paradoxes are possible. We have already pointed out that Postulate 1, which guarantees the existence of at least one straight line joining two points A and B, probably was meant to imply uniqueness of this line, but the postulate fails to assert so much. Also, the objections that can be raised against the principle of superposition, employed in many early popular textbooks, are only partially met by Euclid's Axiom 4.

In short, the truth of the matter is that Euclid's first principles are simply not sufficient for the derivation of all of the 465 propositions of the *Elements*. In particular, the set of postulates needs to be considerably amplified. The work of perfecting Euclid's initial assumptions, so that all of his geometry can rigorously follow, occupied mathematicians for more than two thousand years. Not until the end of the nineteenth century and the early part of the twentieth century, after the foundations of geometry had been subjected to an intensive study, were satisfactory sets of postulates supplied for Euclidean plane and solid geometry. The history of this struggle is of major concern in our present study, and in following it we shall encounter the device that mathematicians contrived for avoiding the pitfall, into which Euclid so often fell, of overfamiliarity with the subject matter.

Not only is Euclid's work marred by numerous tacit assumptions, but some of his preliminary definitions are also open to criticism. Euclid, following the Greek pattern of material axiomatics, makes some sort of attempt to define, or at least to explain, all the terms of his discourse. Actually, it is as impossible to define *explicitly* all of the terms of a discourse as it is to prove all of the statements of the discourse, for a term must be defined by means of other terms, and these other terms by means of still others, and so on. In order to get started, and to avoid circularity of definition where term x is defined by means of term y, and then later term y by means of term x, one is forced to set down at the very start of the discourse a collection of primitive, or basic, terms whose meanings are not to be questioned. All subsequent terms of the discourse must be defined, ultimately, by means of these initial primitive ones. The postulates of the discourse are, then, in final analysis, assumed statements about the primitive terms. From this point of view, the primitive terms may be regarded as defined *implicitly*, in the sense that they are any things or concepts that satisfy the postulates, and this implicit definition is the only kind of definition that the primitive terms can receive.

In Euclid's development of geometry the terms *point* and *line*, for example, could well have been included in a set of primitive terms for the discourse. At any rate, Euclid's definition of a point as "that which has no part" and of a line as "length without breadth" are easily seen to be circular and therefore, from a logical viewpoint, woefully inadequate. One distinction between the Greek conception and the modern conception of the axiomatic method lies in this matter of primitive terms; in the Greek conception there is no listing of the primitive terms. The excuse for the Greeks is that to them geometry was not just an abstract study but was an attempted logical analysis of idealized physical space. Points and lines were, to the Greeks, idealizations of very small particles and of very thin threads. It is this idealization that Euclid attempts to express in his two initial definitions.

Other differences between the Greek and the modern views of the axiomatic method will be discussed in a later chapter.

═══   ## 2.5   The End of the Greek Period and the Transition   ═══
to Modern Times[7]

Very little in the further development of the axiomatic method took place after Euclid until relatively modern times. We must mention, however, the brilliant exploitation of the method by Archimedes (*ca.* 287–212 B.C.), one of the greatest mathematicians of all time, and certainly *the* greatest of antiquity. Although Archimedes lived most of his long life in the Greek city of Syracuse, on the island of Sicily, it seems that he studied for a time at the University of Alexandria. He was thoroughly schooled in the Euclidean tradition, and he left deep imprints on

---

[7] This section is largely skimmed from the appropriate places in H. Eves [1].

both geometry and mechanics. Archimedes' works are masterpieces of mathematical exposition and resemble to a remarkable extent, because of their high finish, economy of presentation, and rigor in demonstration, the articles found in present-day research journals. It is interesting that Archimedes employed the axiomatic method in his writings on theoretical mechanics, as well as in his purely geometrical studies, always laying down the first principles of the work and then deducing a sequence of propositions. Thus, in his treatise *On Plane Equilibriums*, Archimedes establishes twenty-five theorems of mechanics on the basis of three simple postulates suggested by common experience. The postulates are as follows:

1. *Equal weights at equal distances balance; equal weights at unequal distances do not balance but incline toward the weight that is at the greater distance.*

2. *If, when weights at certain distances balance, something is added to one of the weights, equilibrium will not be maintained, but there will be inclination on the side of the weight to which the addition was made; similarly, if anything is taken away from one of the weights, there will be inclination on the side of that weight from which nothing was taken.*

3. *When equal and similar plane figures coincide if placed on one another, their centroids similarly coincide; and in figures that are unequal but similar, the centroids will be similarly situated.*

From these simple postulates Archimedes locates, for example, the centroid of any parabolic segment and of any portion of a parabola lying between two parallel chords. Problems of this sort would today be worked out by means of the integral calculus.

Again, in his work *On Floating Bodies*, Archimedes rests the establishment of the nineteen propositions of the work on two fundamental postulates. This treatise is the first recorded application of mathematics to hydrostatics, and it begins by developing those familiar laws of hydrostatics that nowadays are encountered in an elementary physics course. The treatise then goes on to consider several rather difficult problems, culminating with a remarkable investigation of the positions of rest and of stability of a right segment of a paraboloid of revolution floating in a fluid. Not until the sixteenth-century work of Simon Stevin did the science of statics and the theory of hydrodynamics appreciably advance beyond the points reached by Archimedes. It is worthy of note that these early researches in theoretical physics were developed by the use of the axiomatic method.

A geometrical assumption explicitly stated by Archimedes in his work *On the Sphere and Cylinder* deserves special mention; it is one of the five postulates assumed at the start of Book I of the work and it has become known as the *postulate of Archimedes*. A simple statement of the postulate is as follows: *Given two unequal linear segments, there is always some finite multiple of the shorter one which is longer than the other.* In some modern treatments of geometry this postulate serves as part of the postulational basis for introducing the concept of continuity. It is a matter of interest that in the nineteenth and twentieth centuries geometric systems were constructed that denied the Archimedean postulate, thus giving rise to so-called non-Archimedean geometries. Although named after Archimedes, this postulate had been considered earlier by Eudoxus.

There were other able Greek mathematicians in ancient times after Euclid besides Archimedes—for example, Apollonius, Eratosthenes, Menelaus, Claudius Ptolemy, Heron, Diophantus, and Pappus—but these men did little to advance the development of the axiomatic method and so have slight connection with our present study. After Pappus, who flourished toward the end of the third century A.D., Greek mathematics practically ceased as a living study, and thenceforth merely its memory was perpetuated by minor writers and commentators, such as Theon and Proclus. This closing period of ancient times was dominated by Rome. One Greek center after another had fallen before the power of the Roman armies; in 146 B.C. Greece had become a province of the Roman Empire, although Mesopotamia was not conquered until 65 B.C., and Egypt held out until 30 B.C. The economic structure of the empire was based essentially on agriculture and an increasing use of slave labor. Conditions proved more and more stifling to original scientific work, and a gradual decline in creative thinking set in. The eventual collapse of the slave market, with its disastrous effect on Roman economy, found science reduced to a mediocre level. The famous Alexandrian school gradually faded with the breakup of ancient society, and finally, in A.D. 641, Alexandria was taken by the Arabs, who put the torch to what the Christians had left. The long and glorious era of Greek mathematics was over.

The period starting with the fall of the Roman Empire in the middle of the fifth century and extending into the eleventh century is known as Europe's Dark Ages, for during this period civilization in western Europe reached a very low ebb. Schooling became almost nonexistent, Greek learning all but disappeared, and many of the arts and crafts bequeathed by the ancient world were forgotten. Only the monks of the Christian monasteries, and a few cultured laymen, preserved a slender thread of Greek and Latin learning. The period was marked by great physical violence and intense religious faith. The old social order gave way, and society became feudal and ecclesiastical.

The Romans had never taken to abstract mathematics but had contented themselves with merely a few practical aspects of the subject that were associated with commerce and civil engineering. With the fall of the Roman Empire and the subsequent closing of much of east–west trade and the abandonment of state engineering projects, even these interests waned, and it is no exaggeration to say that very little in mathematics, beyond the development of the Christian calendar, was accomplished in the West during the whole of the half millennium covered by the Dark Ages.

During this bleak period of learning the people of the east, especially the Hindus and the Arabs, became the major custodians of mathematics. However, the Greek concept of rigorous thinking—in fact, the very idea of proof—seemed distasteful to the Hindu way of doing things. Although the Hindus excelled in computation, contributed to the devices of algebra, and played an important role in developing our present positional numeral system, they produced nothing of importance as far as basic methodology is concerned. Hindu mathematics of this period is largely empirical and lacks those outstanding Greek characteristics of clarity and logicality in presentation and of insistence on rigorous demonstration.

The spectacular episode of the rise and decline of the Arabian empire occurred during the period of Europe's Dark Ages. Within a decade following Mohammed's flight from Mecca to Medina in A.D. 622, the scattered and disunited tribes of the Arabian peninsula were consolidated by a strong religious fervor into a powerful nation. Within a century, force of arms had extended the Moslem rule and influence over a territory reaching from India, through Persia, Mesopotamia, northern Africa, and into Spain. Of considerable importance for the preservation of much of world culture was the manner in which the Arabs seized on Greek and Hindu erudition. The Baghdad caliphs not only governed wisely and well but many became patrons of learning and invited distinguished scholars to their courts. Numerous Hindu and Greek works in astronomy, medicine, and mathematics were industriously translated into the Arabic tongue and thus were saved until later European scholars were able to retranslate them into Latin and other languages. But for the work of the Arabian scholars a great part of Greek and Hindu science would have been irretrievably lost over the long period of the Dark Ages.

Not until the latter part of the eleventh century did Greek classics in science and mathematics begin once again to filter into Europe. There followed a period of transmission during which the ancient learning preserved by Moslem culture was passed on to the western Europeans through Latin translations made by Christian scholars traveling to Moslem centers of learning, and through the opening of western European commercial relations with the Levant and the Arabian world. The loss of Toledo by the Moors to the Christians in 1085 was followed by an influx of Christian scholars to that city to acquire Moslem learning. Other Moorish centers in Spain were infiltrated, and the twelfth century became, in the history of mathematics, a century of translators. One of the most industrious translators of the period was Gherardo of Cremona, who translated into Latin more than ninety Arabian works, among which were Ptolemy's *Almagest* and Euclid's *Elements*. At the same time Italian merchants came in close contact with eastern civilization, thereby picking up useful arithmetical and algebraical information. These merchants played an important part in the European dissemination of the Hindu-Arabic system of numeration.

The thirteenth century saw the rise of the universities at Paris, Oxford, Cambridge, Padua, and Naples. Universities were to become potent factors in the development of mathematics, since many mathematicians associated themselves with one or more such institutions. During this century Campanus made a Latin translation of Euclid's *Elements*, which later, in 1482, became the first printed version of Euclid's great work.

The fourteenth century was a mathematically barren one. It was the century of the Black Death, which swept away more than a third of the population of Europe; and during this century the Hundred Years' War, with its political and economic upheavals in northern Europe, got well under way.

The fifteenth century witnessed the beginning of the European Renaissance in art and learning. With the collapse of the Byzantine Empire, culminating in the fall of Constantinople to the Turks in 1453, refugees flowed into Italy, bringing with them treasures of Greek civilization. Many Greek classics, up to that time known only through the often inadequate Arabic translations, could now be studied from original sources. Also, the middle of the century witnessed the invention of printing, which revolutionized the book trade and enabled knowledge to be disseminated at an unprecedented rate. Mathematical activity in this century was largely centered in the Italian cities and in the central European cities of Nuremberg, Vienna, and Prague, and it concentrated on arithmetic, algebra, and trigonometry, under the practical influence of trade, navigation, astronomy, and surveying.

In the sixteenth century the development of arithmetic and algebra continued, the most spectacular mathematical achievement of the century being the discovery, by Italian mathematicians, of the algebraic solution of cubic and quartic equations. In 1572 Commandino made a very important Latin translation of Euclid's *Elements* from the Greek. This translation served as a basis for many subsequent translations, including a very influential work by Robert Simson, from which, in turn, so many English editions were derived.

The seventeenth century proved to be particularly outstanding in the history of mathematics. Early in the century Napier revealed his invention of logarithms, Harriot and Oughtred contributed to the notation and codification of algebra, Galileo founded the science of dynamics, and Kepler announced his laws of planetary motion. Later in the century Desargues and Pascal opened a new field of pure geometry, Descartes launched modern analytic geometry, Fermat laid the foundations of modern number theory, and Huygens made distinguished contributions to the theory of probability and other fields. Then, toward the end of the century, after many mathematicians had prepared the way, the epoch-making creation of the calculus was made by Newton and Leibniz. Thus, during the seventeenth century, many new and vast fields were opened for mathematical investigation. The dawn of modern mathematics was at hand, and it was perhaps inevitable that sooner or later some aspect of the axiomatic method itself should once again claim the attention of researchers.

PROBLEMS

2.1.1   Which of the following two theorems should more likely appear among the "elements" of a course in plane geometry, and why? (1) The three altitudes of a triangle, produced if necessary, meet in a point. (2) The sum of the three angles of a triangle is equal to two right angles.

2.1.2   A mathematics instructor is going to present the subject of geometric progressions to his college algebra class. After defining this type of progression, what theorems about geometric progressions should the instructor offer as the "elements" of the subject?

2.1.3   Imagine yourself building up an elementary treatment of trigonometric identities. Which identities would you select for the "elements" of your treatment, and in what order would you arrange them?

2.1.4   As an illustration of nongeometrical material found in Euclid's *Elements*, let us consider the *Euclidean algorithm*, or process, for finding the greatest common integral divisor (g.c.d.) of two positive integers. The process is found at the start of Euclid's Book VII, although perhaps it was known before Euclid's time. This algorithm is basic to several developments in modern mathematics. Stated in the form of a rule, the process is this: *Divide the larger of the two positive integers by the smaller one. Then divide the divisor by the remainder. Continue this process, of .*

dividing the last divisor by the last remainder, until the division is exact. The final divisor is the sought g.c.d. of the two original positive integers.

(a) Find, by the Euclidean algorithm, the g.c.d. of 5913 and 7592.

(b) Find, by the Euclidean algorithm, the g.c.d. of 1827, 2523, and 3248.

(c) Prove that the Euclidean algorithm does lead to the g.c.d.

(d) Let h be the g.c.d. of the positive integers a and b. Show that there exist integers p and q (not necessarily positive) such that $pa + qb = h$.

(e) Find p and q for the integers of part (a).

(f) Prove that a and b are relatively prime if and only if there exist integers p and q such that $pa + qb = 1$.

2.1.5 (a) Prove, using Problem 2.1.4 (f), that if p is a prime and divides the product uv then either p divides u or p divides v.

(b) Prove, from part (a), the "fundamental theorem of arithmetic": *Every integer greater than 1 can be uniquely factored into a product of primes.* This is essentially Proposition IX 14 of Euclid's *Elements*.

2.1.6 The fundamental theorem of arithmetic says that, for any given positive integer a, there are unique non-negative integers $a_1, a_2, a_3, \ldots$, only a finite number of which are different from zero, such that

$$a = 2^{a_1} 3^{a_2} 5^{a_3} \ldots ,$$

where $2, 3, 5, \ldots$ are the consecutive primes. This suggests a useful notation. We shall write

$$a = (a_1, a_2, a_3 \ldots, a_n),$$

where a_n is the last nonzero exponent. Thus we have $12 = (2, 1)$, $14 = (1, 0, 0, 1)$, $27 = (0, 3)$, and $360 = (3, 2, 1)$.

Prove the following theorems:

(a) $ab = (a_1 + b_1, a_2 + b_2, \ldots)$.

(b) b is a divisor of a if and only if $b_i \leqq a_i$ for each i.

(c) The number of divisors of a is $(a_1 + 1)(a_2 + 1) \cdots (a_n + 1)$.

(d) A necessary and sufficient condition for a number n to be a perfect square is that the number of divisors of n be odd.

(e) Set g_i equal to the smaller of a_i and b_i if $a_i \neq b_i$ and equal to either a_i or b_i if $a_i = b_i$. Then $g = (g_1, g_2, \ldots)$ is the g.c.d. of a and b.

(f) If a and b are relatively prime and b divides ac, then b divides c.

(g) If a and b are relatively prime and if a divides c and b divides c, then ab divides c.

(h) Show that $\sqrt{2}$ and $\sqrt{3}$ are irrational.

2.1.7 Prove the famous Proposition IX 20 of Euclid's *Elements: The number of prime numbers is infinite.*

2.1.8 A number is said to be perfect if it is the sum of its proper divisors. For example, 6 is a perfect number, since $6 = 1 + 2 + 3$. The last proposition of the ninth book of Euclid's *Elements* proves that *if $2^n - 1$ is a prime number, then $2^{n-1}(2^n - 1)$ is a perfect number.* The perfect numbers given by Euclid's formula are even numbers, and it has been shown that every even perfect number must be of this form. The existence or nonexistence of odd perfect numbers is one of the celebrated unsolved problems in number theory. There is no number of this type having less than 100 digits.

(a) Show that in Euclid's formula for perfect numbers, n must be prime.

(b) What are the first four perfect numbers given by Euclid's formula?

(c) Prove that the sum of the reciprocals of *all* the divisors of a perfect number is equal to 2.

2.2.1 Discuss Euclid's axioms and postulates, as listed in Section 2.3, in relation to the three distinctions that, according to Proclus, were advocated by various early Greeks.

2.3.1 How does the modern definition of a circle differ from Euclid's definition?

2.3.2 "Prove" Euclid's Postulate 4 by the method of superposition.

2.3.3 One should understand precisely the intention of Euclid's Postulate 3. When Euclid says that "a circle may be described with any center and distance," he means that *a circle may be described with any point as center and having any straight line segment radiating from this center as a radius.* It follows that the Euclidean compasses differ from our modern compasses, for with the modern compasses we are permitted to draw a circle having any point A as center and any segment BC as radius. In other words, we are permitted to transfer the distance BC to the center A, using the compasses as dividers. The Euclidean compasses, on the other hand, may be supposed to collapse if either leg is lifted from the paper.

A student reading Euclid's *Elements* for the first time might experience surprise at the opening propositions of Book I. The first three propositions are the construction problems:

1. *To describe an equilateral triangle upon a given finite straight line.*

2. *From a given point to draw a straight line equal to a given straight line.*

3. *From the greater of two given straight lines to cut off a part equal to the lesser.*

These three constructions are trivial with straightedge and *modern* compasses but require some ingenuity with straightedge and *Euclidean* compasses.

(a) Solve Proposition 1 of Book I with Euclidean tools.

(b) Solve Proposition 2 of Book I with Euclidean tools.

(c) Solve Proposition 3 of Book I with Euclidean tools.

(d) Show that Proposition 2 of Book I proves that the straightedge and *Euclidean* compasses are equivalent to the straightedge and *modern* compasses.

2.4.1 If an assumption tacitly made in a deductive development should involve a misconception, its introduction may lead not only to a result that does not follow from the postulates of the deductive system but to one that may actually contradict some previously established theorem of the system. From this point of view, criticize the following three geometrical paradoxes:

(a) *To prove that any triangle is isosceles.*

Let ABC be any triangle (see Figure 2.3). Draw the bisector of $\angle C$ and the perpendicular bisector of side AB. From their point of intersection E, drop perpendiculars EF and EG on AC and BC, respectively, and draw EA and EB. Now right triangles CFE and CGE are congruent, since each has CE as hypotenuse and since $\angle FCE = \angle GCE$. Therefore $CF = CG$. Again, right triangles EFA and EGB are congruent, since leg EF of one equals leg EG of the other (any point E on the bisector of an angle C is equidistant from the sides of the angle) and since hypotenuse EA of one equals hypotenuse EB of the other (any point E on the perpendicular bisector of a line segment AB is equidistant from the extremities of that line segment). Therefore $FA = GB$. It now follows that $CF + FA = CG + GB$, or $CA = CB$, and the triangle is isosceles.

(b) *To prove that a right angle is equal to an obtuse angle.*

Let $ABCD$ be any rectangle (see Figure 2.4). Draw BE outside the rectangle and equal in length to BC, and hence to AD. Draw the perpendicular bisectors of DE and AB; since they are perpendicular to

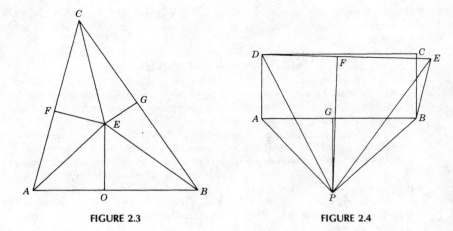

FIGURE 2.3　　　　　　　　FIGURE 2.4

nonparallel lines, they must intersect in a point P. Draw AP, BP, DP, EP. Then $PA = PB$ and $PD = PE$ (any point on the perpendicular bisector of a line segment is equidistant from the extremities of the line segment). Also, by construction, $AD = BE$. Therefore triangles APD and BPE are congruent, since the three sides of one are equal to the three sides of the other. Hence $\angle DAP = \angle EBP$. But $\angle BAP = \angle ABP$, since these angles are base angles of the isosceles triangle APB. By subtraction it now follows that right angle $DAG =$ obtuse angle EBA.

(c)　*To prove that there are two perpendiculars from a point to a line.*

Let two circles intersect in A and B (see Figure 2.5). Draw the diameters AC and AD, and let the join of C and D cut the respective circles in M and N. Then angles AMC and AND are right angles, since each is inscribed in a semicircle. Hence AM and AN are two perpendiculars to CD.

2.4.2　To guarantee the existence of certain points of intersection (of line with circle and circle with circle) Richard Dedekind (1831–1916) introduced into geometry the following continuity postulate:

If all points of a horizontal straight line fall into two classes, such that every point of the first class lies to the left of every point of the second class, then

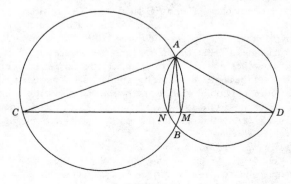

FIGURE 2.5

*there exists one and only one point that produces this division of all points into
two classes—that is, this severing of the straight line into two portions.*

(a) Complete the details of the following indicated proof of the theorem:

> *The straight line segment joining a point A inside a circle to a point B
> outside the circle has a point in common with the circle.*

Let O be the center and r the radius of the given circle (see Figure 2.6),
and let C be the foot of the perpendicular from O on the line determined by A
and B. The points of the segment AB can be divided into two classes: those
points P for which $OP < r$ and those points Q for which $OQ \geqq r$. It can be
shown that, in every case, $CP < CQ$. Hence, by Dedekind's postulate, there
exists a point R of AB such that all points that precede it belong to one class
and all that follow it belong to the other class. Now $OR \not< r$, for otherwise we
could choose S on AB, between R and B, such that $RS < r - OR$. But since
$OS < OR + RS$, this would imply the absurdity that $OS < r$. Similarly, it
can be shown that $OR \not> r$. Hence we must have $OR = r$, and the theorem is
established.

(b) How might Dedekind's postulate be extended to cover angles?

(c) How might Dedekind's postulate be extended to cover circular arcs?

2.4.3 Let us, for convenience, restate Euclid's first three postulates in the following
equivalent forms:

1. *Any two distinct points determine a straight line.*

2. *A straight line is boundless.*

3. *There exists a circle having any given point as center and passing through
   any second given point.*

Show that Euclid's postulates, partially restated above, hold if the points of
the plane are restricted to those whose rectangular Cartesian coordinates for some
fixed frame of reference are rational numbers. Show, however, that under this
restriction a circle and a line through its center need not intersect each other.

2.4.4 Show that Euclid's postulates (as partially restated in Problem 2.4.3) hold if we
interpret the plane as the surface of a sphere, straight lines as great circles on the
sphere, and points as points on the sphere. Show, however, that in this
interpretation the following are true:

(a) Parallel lines do not exist.

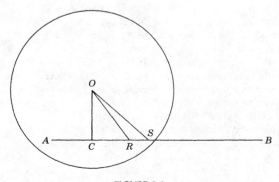

FIGURE 2.6

(b) All perpendiculars to a given line erected on one side of the line intersect in a point.

(c) It is possible to have two distinct lines joining the same two points.

(d) The sum of the angles of a triangle exceeds two right angles.

(e) There exist triangles having all three angles right angles.

(f) An exterior angle of a triangle is not always greater than either of the two remote interior angles.

(g) The sum of two sides of a triangle can be less than the third side.

(h) A triangle with a pair of equal angles may have the sides opposite them unequal.

(i) The greatest side of a triangle does not necessarily lie opposite the greatest angle of the triangle.

2.4.5 In 1882 Moritz Pasch formulated the following postulate:

Let A, B, C *be three points not lying in the same straight line, and let* m *be a straight line lying in the plane of* ABC *and not passing through any of the points* A, B, C. *Then, if the line* m *passes through a point of the segment* AB, *it will also pass through a point of the segment* BC *or a point of the segment* AC.

This postulate is one of those assumptions classified by modern geometers as a *postulate of order,* and it assists in bringing out the idea of "betweenness."

(a) Prove, as a consequence of Pasch's postulate, that *if a line enters a triangle at a vertex, it must cut the opposite side.*

(b) Show that Pasch's postulate does not always hold for a spherical triangle cut by a great circle.

2.4.6 New terms defined by means of more primitive terms are not essential to a deductive system but are convenient in the drawing of inferences, for the new terms serve as shorthand for complex and unmanageable phrases involving the more primitive terms. To illustrate the cumbersomeness that would result if, in Euclid's *Elements,* we should dispense with, say, the terms *point* and *line,* we would have to describe a straight line as "a breadthless length which lies evenly with all the entities on itself which have no parts." State Euclid's first two postulates without using the terms *point* and *line.*

2.5.1 Let W_1 and W_2 be two weights at distances d_1 and d_2, respectively, from a fulcrum. On the basis of the first two postulates of Archimedes' treatise *On Plane Equilibriums,* establish the following theorems:

(a) If we have equilibrium and $W_1 = W_2$, then $d_1 = d_2$.

(b) If we have equilibrium and $d_1 = d_2$, then $W_1 = W_2$.

(c) If we have equilibrium and $W_1 \neq W_2$, then $d_1 \neq d_2$.

(d) If, when weights at certain distances balance, one of the distances should be increased, equilibrium will not be maintained, but there will be inclination on the side of the distance which was increased.

(e) Try to show that we have equilibrium if and only if $W_1 d_1 = W_2 d_2$.

2.5.2 (a) Assuming the existence of the usual one-to-one correspondence between real numbers and points on an *x*-axis, show that an arithmetized form of the postulate of Archimedes is, "If *a* and *b* are any two positive real numbers, there exists a positive integer *n* such that $na > b$."

(b) State the postulate of Archimedes for angles, and indicate how it might be deduced from the arithmetized form of the postulate.

3

NON-EUCLIDEAN GEOMETRY

3.1 Euclid's Fifth Postulate

Postulate 5 of Euclid's *Elements* has been described as "perhaps the most famous single utterance in the history of science."[1] Certainly it has been the source of much controversy, and the dissatisfaction of mathematical scholars with its statement as a postulate is indicated by the fact that many reputable geometers attempted over a period of some twenty centuries either to prove it as a theorem or to replace it by a more acceptable equivalent. As we shall soon see, this concern over Euclid's fifth postulate furnished the stimulus for the development of a great deal of modern mathematics and also led to deep and revealing inquiries into the logical and philosophical foundations of the subject.

A rigorous development of the theory of parallels apparently gave the early Greeks considerable trouble. Euclid met the difficulties by defining parallel lines as coplanar straight lines that do not meet one another however far they may be produced in either direction, and by adopting as an assumption his now famous fifth postulate. Proclus tells us that this postulate was attacked from the very start. Even a cursory reading of Euclid's five postulates discloses a very noticeable difference between the fifth postulate and the other four; the fifth postulate lacks the terseness and the simple comprehensibility possessed by the other four, and it certainly does not have that quality of ready acceptance demanded by material axiomatics. A more studied examination reveals that the fifth postulate is actually the converse of Proposition I 17.[2] It is not surprising that it seemed more like a proposition than a postulate. Moreover, Euclid himself made no use of it until he reached Proposition I 29. It was very natural to wonder whether the postulate was really needed at all and to think that perhaps it could be derived as a theorem from the remaining nine "axioms" and "postulates" or,

---

[1] C. J. Keyser [1], p. 113.

[2] See Appendix, Section A.1, for the statements of the first twenty-eight propositions of Euclid's Book I.

at least, that it could be replaced by a more acceptable equivalent. Proclus, who was under the illusion that he possessed a proof of the postulate, favored deleting the postulate from the first principles. A quotation from Proclus might be of interest:

> This ought to be struck out of the postulates altogether; for it is a theorem involving many difficulties. Ptolemy, in a certain book, set himself to solve it, and it requires for its demonstration a number of definitions as well as theorems. Moreover, its converse is actually proved by Euclid himself as a theorem. It may be that some persons would be deceived, and would think it proper to place the assumption in question among the postulates as affording ground for an instantaneous belief that the straight lines converge and meet when the two angles are made less than two right angles. To such persons, Geminus correctly replied that we have learned from the very pioneers of this science not to have any regard for mere plausible imaginings when it is really a question of the reasonings to be included in our geometrical doctrine. Aristotle says that it is as justifiable to ask scientific proofs of a rhetorician as to accept mere plausibilities from a geometer, and Simmias is made by Plato to say that he recognizes as quacks those who fashion for themselves proofs from probabilities. So in this case, when the two right angles are lessened, the fact that the straight lines converge is true and necessary; but the statement that they will meet sometime, since they converge more and more as they are produced, is plausible, but it is not necessary in the absence of some argument showing that this is true. It is a known fact that some lines exist which approach each other indefinitely, but yet remain nonintersecting; this seems improbable and paradoxical, but nevertheless it is true and fully ascertained with regard to other species of lines. May not the same thing which happens in the case of the lines referred to be possible in the case of straight lines? Indeed, until the statement in the postulate is clinched by proof, the facts shown in the case of other lines may direct our imagination the opposite way. Though the controversial arguments against the meeting of the straight lines should contain much that is surprising, is that not all the more reason why we should expel from our body of doctrine this merely plausible and unreasoned hypothesis?

There were many attempts to "prove" the fifth, or parallel, postulate and many substitutes devised for its replacement. Of the various substitutes, the one most commonly favored is that made well known in modern times by the Scottish physicist and mathematician, John Playfair (1748–1819), although this particular alternative had been used by others and had even been stated as early as the fifth century by Proclus. This substitute is the one most often encountered in present-day high school geometry texts—namely, *Through a given point not on a given line can be drawn only one line parallel to the given line.*[3] Some other alternatives for the parallel postulate that have been either proposed or tacitly assumed over the years are these:

1. (Posidonius and Geminus) *There exists a pair of coplanar straight lines everywhere equally distant from one another.*

---

[3] Propositions I 27 and I 28 guarantee, under the tacit assumption of the infinitude of straight lines, the existence of at least *one* parallel.

2. (Wallis, Saccheri, Carnot, and Laplace) *There exists a pair of similar noncongruent triangles.*

3. (Saccheri) *If in a quadrilateral a pair of opposite sides are equal and if the angles adjacent to a third side are right angles, then the other two angles are also right angles.*

4. (Lambert and Clairaut) *If in a quadrilateral three angles are right angles, the fourth angle is also a right angle.*

5. (Legendre) *There exists at least one triangle having the sum of its three angles equal to two right angles.*

6. (Legendre) *Through any point within an angle less than 60° there can always be drawn a straight line intersecting both sides of the angle.*

7. (Legendre and W. Bolyai) *A circle can be passed through any three noncollinear points.*

8. (Gauss) *There is no upper limit to the area of a triangle.*

It constitutes an interesting and challenging collection of exercises for the student to try to show the equivalence of these alternatives to the original postulate stated by Euclid. To show the equivalence of Euclid's postulate and a particular one of the alternatives, one must show that the alternative follows as a theorem from Euclid's assumptions and also that Euclid's postulate follows as a theorem from Euclid's system of assumptions with the parallel postulate replaced by the considered alternative.

It would be difficult to estimate the number of attempts that have been made, throughout the centuries, to deduce Euclid's fifth postulate as a consequence of the other Euclidean assumptions, either explicitly stated or tacitly implied. All these attempts ended unsuccessfully, and most of them were sooner or later shown to rest on an assumption equivalent to the postulate itself. The earliest effort, of which we are today aware, to prove the postulate was made by Claudius Ptolemy (*ca.* A.D. 150), alluded to by Proclus in the quotation given above. Claudius Ptolemy was the author of the famous and very influential *Almagest*, the great definitive Greek work on astronomy. Proclus exposed the fallacy in Ptolemy's attempt by showing that Ptolemy had unwittingly assumed that through a point only one parallel can be drawn to a given line; this assumption is the Playfair equivalent of Euclid's postulate. Proclus submitted an attempt of his own, but his "proof" rests on the assumption that parallel lines are always a bounded distance apart, and this assumption can be shown to imply Euclid's fifth postulate. Among the more noteworthy attempts of somewhat later times is one made in the thirteenth century by Nasir-ed-din (1201–1274), a Persian astronomer and mathematician who compiled, from an earlier Arabic translation, an improved edition of the *Elements* and who wrote a treatise on Euclid's postulates, but his attempt, too, involves a tacit assumption equivalent to the postulate being "proved."

An important stimulus to the development of geometry in western Europe after the Renaissance was a renewal of the criticism of Euclid's fifth postulate. Hardly any critical comments are to be found in the early printed editions of the *Elements* made at the end of the fifteenth century and at the beginning of the sixteenth century. However, after the translation, in 1533, of Proclus's *Commentary on Euclid, Book I*, many men once again embarked upon a critical analysis of the fifth postulate. For example, John Wallis (1616–1703), while lecturing at Oxford University, became interested in the work of Nasir-ed-din and in 1663 offered his own "proof" of the parallel postulate, but this attempt involves the equivalent assumption that similar noncongruent triangles exist. So it was with all the many attempts to derive Euclid's postulate as a theorem; each attempt involved the vitiating circularity of assuming something equivalent to the thing being established or else committed some other form of fallacious reasoning. Most of this vast amount of work is of little real importance in the actual evolution of mathematical thought until we come to the remarkable investigation of the parallel postulate made by Girolamo Saccheri in 1733.

3.2   Saccheri and the *Reductio ad Absurdum* Method

Every student of elementary geometry has encountered the so-called *indirect*, or *reductio ad absurdum*,[4] method of proof. It is a powerful, and at times seemingly indispensable, method that is employed frequently by Euclid in his *Elements*. The method, it will be recalled, consists of assuming, by way of hypothesis, that a proposition that is to be established is false; if an absurdity follows, one concludes that the hypothesis is untenable and that the original proposition must then be true. It was this method of proof that we employed in Section 1.5 to show that $\sqrt{2}$ is irrational.

To illustrate further the *reductio ad absurdum* method, let us briefly consider Euclid's Proposition I 6, the first proposition in the *Elements* established by this type of proof. We wish to prove the theorem: *If in a triangle two angles are equal to each other, then the two sides opposite these angles are also equal to each other.* Let *ABC* (Figure 3.1) be the triangle, and suppose $\angle ABC = \angle ACB$. We wish to show that side AB = side AC. Suppose the sides AB and AC are not equal to each other. Then one of them—say, AB—is greater than the other, and we may mark off on BA a segment BD equal to the lesser side AC. Now in triangles ABC and DCB we have $CB = BC$, $CA = BD$, $\angle BCA = \angle CBD$. It follows that the triangles are congruent. But this conclusion is impossible, since triangle DCB is only a part of triangle ABC. Our hypothesis that $AB \neq AC$ has led to an absurd situation and hence is untenable. We must conclude, therefore, that $AB = AC$, and our theorem is established.

The *reductio ad absurdum* method rests on two cardinal principles of classical logic—namely, the *law of contradiction* and the *law of the excluded middle*. Somewhat loosely described, the law of contradiction says that *if* S *is any statement, then* S *and a contradiction* (that is, the denial) *of* S *cannot both hold*, and the law of the excluded middle says that *either* S *or the denial of* S *must hold* (that

---

[4] In a more refined treatment one distinguishes several slight variations in the indirect method of proof, and then it is customary to assign the technical terminology, *reductio ad absurdum*, to a particular one of these variations. We do not make this refinement here.

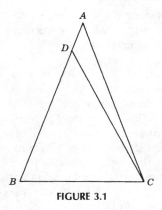

FIGURE 3.1

is, there is no third, or middle, possibility). As an illustration, suppose S is the statement, "Archimedes was born in 287 B.C." The law of contradiction then asserts that Archimedes cannot have been born in both 287 B.C. and A.D. 400, for example, and the law of the excluded middle asserts that Archimedes either was born in 287 B.C. or he was not born in 287 B.C. Now let S be the statement of any proposition to be established by the *reductio ad absurdum* method—for example, the statement of Proposition I 6 above. By the method, we set about and show that the denial of S implies the denial of some previously assumed or established statement T. By the law of contradiction, T and the denial of T cannot both be true (that is, cannot both follow from the postulates). Since T is true, the denial of T is then false, from which, since a true statement can never imply a false one, it follows that the denial of S must also be false. By the law of the excluded middle, however, either S is true or the denial of S is true. Since the denial of S is false, it follows that S is true, and our proposition is established.

The law of contradiction and the law of the excluded middle have settled so deeply into the warp and woof of human thinking that it is difficult to conceive of questioning their validity. We shall see later, however, that although these laws are usually pronounced as universally true, some sort of limitation must be made concerning their applicability. Indeed, since 1912 some mathematicians have felt that we must drastically restrict the free use of the law of the excluded middle as part of the logical machinery used in deducing theorems from postulates. But more of this in its proper place. For the time being we shall accept these laws, particularly insofar as they apply to the *reductio ad absurdum* method of proof.

One concluding remark about the *reductio ad absurdum* method seems appropriate here. In the game of chess a *gambit* is one of various possible openings in which a pawn or a piece is risked in order to obtain an advantageous attack. The eminent English mathematician, G. H. Hardy (1877–1947), delightfully pointed out that *reductio ad absurdum* "is a far finer gambit than any chess gambit: a chess player may offer the sacrifice of a pawn or even a piece, but a mathematician offers *the game*."[5] *Reductio ad absurdum* emerges as the most stupendous gambit conceivable.

---

[5] G. H. Hardy [2], p. 34.

Reductio ad absurdum certainly constitutes one of the finest weapons in a mathematician's armory of attack; with this weapon Girolamo Saccheri in 1733 made the first really scientific assault on the problem of Euclid's parallel postulate.

Little is known of Saccheri's life. He was born in San Remo in 1667, showed marked precocity as a youngster, completed his novitiate for the Jesuit Order at the age of twenty-three, and then spent the rest of his life filling a succession of university teaching posts. While instructing rhetoric, philosophy, and theology at a Jesuit College in Milan, Saccheri read Euclid's *Elements* and became enamored with the powerful method of *reductio ad absurdum*. Later, while teaching philosophy at Turin, Saccheri published his *Logica demonstrativa*, in which the chief innovation is the application of the method of *reductio ad absurdum* to the treatment of formal logic. Some years after, while a professor of mathematics at the University of Pavia, it occurred to Saccheri to apply his favorite method of *reductio ad absurdum* to a study of Euclid's parallel postulate. He was well prepared for the task, having dealt ably in his earlier work on logic with such matters as definitions and postulates. Also, he was acquainted with the work of others regarding the parallel postulate and had succeeded in pointing out the fallacies in the attempts of Nasir-ed-din and Wallis.

Saccheri's effort to establish Euclid's parallel postulate by attempting to institute a *reductio ₐad absurdum* was apparently the first time anyone had conceived the idea of denying the postulate and of studying the consequences of a contradiction of the famous assumption. The result of these researches was a little book entitled *Euclides ab omni naevo vindicatus* (Euclid Freed of Every Flaw), which was printed in Milan in 1733, only a few months before the author's death. In this work Saccheri accepts the first twenty-eight propositions of Euclid's *Elements*, which, as we have previously stated, do not require the fifth postulate for their proof. With the aid of these theorems he then proceeds to study the *isosceles birectangle*—that is, a quadrilateral *ABDC* in which (see Figure 3.2) *AC* = *BD* and the angles at *A* and *B* are right angles. By drawing the diagonals *AD* and *BC* and then using simple congruence theorems (which are found among Euclid's first 28 propositions), Saccheri easily shows that the angles at *C* and *D* are equal to each other. But nothing can be ascertained in regard to the magnitude of these angles. Of course, as a consequence of Euclid's fifth postulate, it follows that these angles are both right angles, but the assumption of this postulate is not to be employed. As a result, the two angles might both be right angles, obtuse angles, or acute angles. Here Saccheri

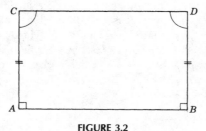

FIGURE 3.2

maintains an open mind and names the three possibilities: the *hypothesis of the right angle*, the *hypothesis of the obtuse angle*, and the *hypothesis of the acute angle*. The plan of the work is to rule out the last two possibilities by showing that their respective assumptions lead to absurdities, thus leaving, by the *reductio ad absurdum* method, the first hypothesis. But this hypothesis can be shown to be equivalent to Euclid's fifth postulate. In this way the parallel postulate is to be established and the blemish of its assumption by Euclid removed.

The task of eliminating the hypothesis of the obtuse angle and the hypothesis of the acute angle turns out to be rather arduous. With real geometrical skill and fine logical penetration, Saccheri establishes a number of theorems, of which the following are among the more important:

1. *If one of the hypotheses is true for a single isosceles birectangular quadrilateral, it is true for every such quadrilateral.*

2. *On the hypothesis of the right angle, the obtuse angle, or the acute angle, the sum of the angles of a triangle is respectively equal to, greater than, or less than two right angles.*

3. *If there exists a single triangle for which the sum of the angles is equal to, greater than, or less than two right angles, there follows the truth of the hypothesis of the right angle, the obtuse angle, or the acute angle.*

4. *On the hypothesis of the right angle two distinct straight lines intersect, except in the one case in which a transversal cuts them under equal corresponding angles. On the hypothesis of the obtuse angle two straight lines always intersect. On the hypothesis of the acute angle there is an infinitude of straight lines through a given point not on a given straight line and which do not meet the given straight line.*

5. *The locus of the extremity of a perpendicular of constant length that moves with its other end on a fixed straight line is a straight line on the hypothesis of the right angle, a curve convex to the fixed line on the hypothesis of the obtuse angle, and a curve concave to the fixed line on the hypothesis of the acute angle.*

After establishing a chain of thirteen propositions, Saccheri manages to dispose of the hypothesis of the obtuse angle, but in so doing he makes the same tacit assumption that Euclid made concerning the infinitude of the straight line. With this tacit assumption (introduced by using Euclid's Proposition I 18, which depends on I 16) Saccheri shows that the hypothesis of the obtuse angle implies Euclid's fifth postulate, which, in turn, implies that the sum of the angles of a triangle is equal to two right angles. But this second implication contradicts the theorem that, on the hypothesis of the obtuse angle, the sum of the angles of a triangle is greater than two right angles.

The case of the hypothesis of the acute angle proves to be even more stubborn, and Saccheri requires nearly twenty more propositions before he feels he can dispose of it. After obtaining many of the theorems that were later to become classical in non-Euclidean geometry, Saccheri weakly forces into his development an unconvincing contradiction involving vague concepts about elements at infinity. The contradiction that he reaches is that there exist two straight lines that when produced to infinity merge into one another and there have a common perpendicular. Coming after the careful work that has been presented up to this point, it is difficult to believe that Saccheri himself was really convinced by his lame ending. Indeed, in a second part to his work, he attempted, with no greater success, a second attack on the hypothesis of the acute angle. Had Saccheri not been so eager to exhibit a contradiction but rather had boldly admitted his inability to find one, he would today unquestionably be credited with the discovery of non-Euclidean geometry.

It is difficult to evaluate the influence that Saccheri's work may have had on later researches connected with the parallel postulate, for subsequently his little publication was lost for a long time. It was dramatically resurrected in 1889 by Saccheri's compatriot, Eugenio Beltrami (1835–1900), a mathematician who, as we shall soon see, made notable contributions of his own to the subject of non-Euclidean geometry. The first part of Saccheri's work has been translated into English[6] and can be easily read by any student of elementary plane geometry.

3.3 The Work of Lambert and Legendre

In 1766, thirty-three years after Saccheri's publication, the German mathematician Johann Heinrich Lambert (1728–1777) wrote an investigation of the parallel postulate entitled *Die Theorie der Parallellinien*, which, however, was not published until eleven years after the author's death. Lambert's treatise is in three parts. The first part considers whether Euclid's fifth postulate can be proved from Euclid's other assumptions or only if additional assumptions are made. The second part is concerned with the reduction of the parallel postulate to various equivalent propositions. It is the last part of the study that closely resembles the earlier work by Saccheri. Here Lambert chooses as a fundamental figure the *trirectangle*, or quadrilateral containing three right angles, which can be regarded as the half of a Saccheri isosceles birectangle formed by joining the midpoints of the latter's bases. As with Saccheri, three hypotheses arise, according to whether the fourth angle of the trirectangle is right, obtuse, or acute.

Lambert went considerably beyond Saccheri in deducing propositions under the hypotheses of the obtuse and acute angles. Thus, not only did he show that for the three hypotheses the sum of the angles of a triangle is equal to, greater than, or less than two right angles, respectively, but in addition he showed that the excess above two right angles in the hypothesis of the obtuse angle, or the deficiency below two right angles in the hypothesis of the acute angle, is proportional to the area of the triangle. This result led him to observe the resemblance to spherical geometry of the geometry following from the hypothesis of the obtuse angle (in spherical geometry the area of a triangle is proportional to its spherical excess), and he conjectured that the geometry following from the hypothesis of the acute angle could perhaps be verified on a sphere of imaginary radius.

---

[6] See G. B. Halsted [2] or D. E. Smith [2], pp. 351–359.

Another notable discovery made by Lambert concerns the measurement of lengths in the two geometries that follow from the obtuse-angle and acute-angle hypotheses. In Euclidean geometry, because similar noncongruent figures exist, lengths can be measured only in terms of some arbitrary unit that has no structural connection with the geometry. Angles, on the other hand, possess a natural unit of measure, such as the right angle or the radian, which is capable of geometrical definition. This is what is meant when mathematicians say that in Euclidean geometry lengths are *relative* but angles are *absolute*. Lambert discovered that under the hypotheses of the obtuse and acute angles, angles are still absolute, but lengths are absolute also! In fact, it can be shown for these geometries that for every angle there is a corresponding line segment, so that to a natural unit of measure for angles there corresponds a natural unit of measure for lengths.

Lambert eliminated the hypothesis of the obtuse angle by making the same tacit assumption that Saccheri made, but his conclusions with regard to the hypothesis of the acute angle were indefinite and unsatisfactory. Indeed, it was this incomplete and unsettled state of affairs with regard to the acute hypothesis that held Lambert from publishing his work, with the result that it did not appear until friends finally put it through the press after his death.

Lambert was a mathematician of high quality. As the son of a poor tailor he was largely self-taught. He possessed a fine imagination and established his results with great attention to rigor. In fact, Lambert was the first to prove rigorously that the number π is irrational. He showed that if x is rational but not zero, then tan x cannot be rational; since tan $\pi/4 = 1$, it follows that $\pi/4$, or π, cannot be rational. We also owe to Lambert the first systematic development of the theory of hyperbolic functions and, indeed, our present notation for these functions. Lambert was a many-sided scholar who contributed to the mathematics of numerous other topics, such as descriptive geometry, the determination of comet orbits, and the theory of projections employed in the making of maps.

A third distinguished effort to establish Euclid's parallel postulate by the *reductio ad absurdum* method was essayed, over a long period of years, by the eminent French analyst Adrien-Marie Legendre (1752–1833). He began anew and considered three hypotheses according to whether the sum of the angles of a triangle is equal to, greater than, or less than two right angles. Tacitly assuming the infinitude of a straight line, he was able to eliminate the second hypothesis, but although he made several attempts, he could not dispose of the third hypothesis. These various endeavors appear in the successive editions of his very popular *Éléments de géométrie*,[7] which ran from a first edition in 1794 to a twelfth in 1823. Legendre's first effort is vitiated by the assumption that the choice of a unit of length will not affect the correctness of his propositions, but this, of

---

[7] This work is an attempted pedagogical improvement of Euclid's *Elements* made by considerably rearranging and simplifying the propositions. The work won high regard in continental Europe and was so favorably received in the United States that it became the prototype of the elementary geometry textbooks in this country. The first English translation was made in the United States in 1819 by John Farrar of Harvard University. The next English translation was made in 1822 by the famous Scottish litterateur, Thomas Carlyle, who early in life was a teacher of mathematics. Carlyle's translation ran through thirty-three American editions.

course, is equivalent to assuming the existence of similar noncongruent figures. The next attempt is vitiated by assuming the existence of a circle through any three noncollinear points. Later Legendre independently observed the fact already discovered by Lambert that, under the third hypothesis, the deficiency of the sum of the angles of a triangle below two right angles is proportional to the area of the triangle. Hence, Legendre reasoned, if by starting with any given triangle one could obtain another triangle containing the given triangle at least twice, then the deficiency for this new triangle would be at least twice the deficiency for the given triangle. By repeating the operation a sufficient number of times, one could finally end with a triangle whose angle sum has become negative, a situation that is absurd. But in order to solve the problem of obtaining a triangle containing a given triangle at least twice, Legendre found he had to assume that through any point within a given angle less than 60° there can always be drawn a straight line intersecting both sides of the angle, and this, as we have pointed out, is equivalent to Euclid's fifth postulate. Legendre gave an elegant proof of the theorem: *If there exists a single triangle having the sum of its angles equal to two right angles, then the sum of the angles of every triangle is equal to two right angles.* Although this theorem is contained in the results given by Saccheri, it is generally referred to as *Legendre's second theorem. Legendre's first theorem* is: *The sum of the three angles of a triangle cannot be greater than two right angles.* Of course, in proving this theorem, Legendre tacitly assumed the infinitude of straight lines. In fact, in proving both his first and second theorems, Legendre assumed the postulate of Archimedes. Max Dehn (1878–1952) has shown that this assumption is unavoidable in proving the first theorem but not necessary in proving the second.

Legendre's last paper on parallels, essentially a collection of his earlier efforts, was published in 1833, the year of his death. He perhaps holds the record for persistence in attempting to prove the famous postulate. The simple and straightforward style of his proofs, widely circulated because of their appearance in his *Éléments,* and his high eminence in the world of mathematics, created marked popular interest in the parallel postulate. Actually, however, Legendre had scarcely progressed as far as had Saccheri a hundred years earlier. Moreover, even before the appearance of his last paper, a Russian mathematician, separated from the rest of the scientific world by barriers of distance and language, had taken a most significant step, the boldness and importance of which were far to transcend anything Legendre had done on the subject.

3.4   The Discovery[8] of Non-Euclidean Geometry

We have seen that, in spite of considerable effort exerted over a long period of time, no one was able to find a contradiction under the hypothesis of the acute angle. It is no wonder that no contradiction was found under this hypothesis, for it is now known that the geometry developed from a certain basic set of

---

[8]We are not here concerned with any philosophical distinction between *discovery* and *invention.*

assumptions plus the acute angle hypothesis is as consistent as the Euclidean geometry developed from the same basic set of assumptions plus the hypothesis of the right angle. In other words, it is now known that the parallel postulate *cannot* be deduced as a theorem from the other assumptions of Euclidean geometry but is independent of those other assumptions. It took unusual imagination to entertain such a possibility, for the human mind had for two millennia been bound by the prejudice of tradition to the firm belief that Euclid's geometry was most certainly the only possible one and that any contrary geometric system simply could not be consistent.

The first to suspect the independence of the parallel postulate were Carl Friedrich Gauss (1777–1855) of Germany, Johann Bolyai (1802–1860) of Hungary, and Nicolai Ivanovitch Lobachevsky (1793–1856) of Russia. These men independently approached the subject through the Playfair form of the parallel postulate by considering the three possibilities: Through a given point not on a given straight line can be drawn *just one* line, *no* line, or *more than one* line parallel (in Euclid's sense) to the given line. These three situations are equivalent, respectively, to the hypotheses of the right, the obtuse, and the acute angle. Assuming, as did their predecessors, the infinitude of a straight line, the second case was easily eliminated. Inability to find a contradiction in the third case, however, led each of the three mathematicians to suspect, in time, a consistent geometry under that hypothesis, and each, unaware of the work of the other two, carried out, for its own intrinsic interest, an extensive development of the resulting new geometry.

Gauss was perhaps the first person really to anticipate a non-Euclidean geometry. Although he meditated a good deal on the matter from very early youth on, probably not until his late twenties did he begin to suspect the parallel postulate to be independent of Euclid's other assumptions. Unfortunately, Gauss failed, throughout his life, to publish anything on the subject, and his advanced conclusions are known to us only through copies of letters to interested friends, a couple of published reviews of works of others, and some notes found among his papers after his death. Although he refrained from publishing his own findings, he strove to encourage others to persist in similar investigations, and he called the new geometry *non-Euclidean*.

Apparently the next person to anticipate a non-Euclidean geometry was Johann Bolyai, who was a Hungarian officer in the Austrian army and the son of the mathematician Wolfgang Bolyai, a long-time personal friend of Gauss. The younger Bolyai undoubtedly received considerable stimulus for his study from his father, who had earlier shown an interest in the problem of the parallel postulate. As early as 1823 Johann Bolyai began to understand the real nature of the problem that faced him, and a letter written during that year to his father shows the enthusiasm he held for his work. In this letter he discloses a resolution to publish a tract on the theory of parallels as soon as he can find the time and opportunity to put the material in order, and exclaims, "Out of nothing I have created a strange new universe." The father urged that the proposed tract be published as an appendix to his own large two-volume semiphilosophical work on elementary mathematics. The expansion and arrangement of ideas proceeded more slowly than Johann had anticipated, but finally, in 1829, he submitted the finished manuscript to his father, and three years later, in 1832, the tract appeared as a twenty-six–page appendix to the first volume of his father's work.[9] Johann Bolyai never published anything further, although he did leave behind a great pile of manuscript pages. His chief interest was in what he called "the absolute science of space," by which he meant the collection of those propositions which are independent of the parallel postulate and which consequently hold in both the Euclidean geometry and the new geometry. For example, the familiar law of sines for triangle ABC,

$$a:b:c = \sin A:\sin B:\sin C,$$

holds only in the Euclidean geometry, but if modified to read

$$O(a):O(b):O(c) = \sin A:\sin B:\sin C,$$

where $O(r)$ denotes the circumference of a circle of radius r, then the modified law holds in both of the geometries. This, then, is the form that the law of sines takes in Bolyai's work. It is not difficult to show that this same form of the sine law also holds in the geometry of triangles on a sphere.

Although Gauss and Johann Bolyai are acknowledged to be the first to conceive a non-Euclidean geometry, actually the Russian mathematician Lobachevsky published the first really systematic development of the subject. Lobachevsky spent the greater part of his life at the University of Kasan, first as a student, later as a professor of mathematics, and finally as rector, and his earliest paper on non-Euclidean geometry was published in 1829–1830 in the *Kasan Bulletin*, two to three years before Bolyai's work appeared in print. This memoir attracted only slight attention in Russia, and, because it was written in Russian, practically no attention elsewhere. Lobachevsky followed this initial effort with other presentations. For example, in the hope of reaching a wider group of readers, he published in 1840 a little book written in German entitled *Geometrische Untersuchungen zur Theorie der Parallellinien* (Geometrical Researches on the Theory of Parallels),[10] and then still later, in 1855, a year before his death and after he had become blind, he published in French a final and more condensed treatment entitled *Pangéométrie* (Pangeometry).[11] So slowly did information of new discoveries spread in those days that Gauss probably did not hear of Lobachevsky's work until the appearance of the German publication in 1840, and Johann Bolyai was unaware of it until 1848. Lobachevsky himself did not live to see his work accorded any wide recognition, but the non-Euclidean geometry which he developed is nowadays frequently referred to as *Lobachevskian geometry*.

The characterizing postulate of Lobachevskian geometry, which replaces Euclid's parallel postulate, is that *through a given point* P, *not on a given line* m, *more than one line can be drawn lying in the plane of* P *and* m *and not intersecting* m. On the basis of this postulate, together with the other assumptions of Euclidean geometry, it is not difficult to show (see Figure 3.3) that there are always two lines through P that do not intersect m, that make equal acute angles α with the

---

[9] For a translation of this appendix, see R. Bonola, or D. E. Smith [2], pp. 375–388.
[10] N. Lobachevsky.
[11] D. E. Smith [2], pp. 360–374.

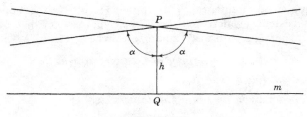

FIGURE 3.3

perpendicular PQ from P to m, and that are such that any line through P lying within the angle formed by the two lines and containing the perpendicular PQ intersects m, while any other line through P does not intersect m. The two lines through P are, then, boundary lines separating all lines through P into two classes, those that cut m and those that do not. From the viewpoint of Euclid, all the lines through P that do not cut m should be called parallels to m. Lobachevsky, however, uses this term more reservedly and refers to only the two boundary lines as being *parallel* to m. Other lines passing through P but not cutting m may be said to be *hyperparallel* to m. The acute angle α is called the *angle of parallelism*, and it plays an important role in Lobachevsky's development. He shows that the size of α depends on the length h of the perpendicular PQ, and emphasizes this by denoting α by the functional symbol $\Pi(h)$. In fact, he shows that if the unit of length is chosen as the distance that corresponds to the particular angle of parallelism

$$\alpha = 2 \operatorname{Arc} \tan e^{-1},$$

where e is the base for natural logarithms, then

$$\Pi(h) = 2 \operatorname{Arc} \tan e^{-h} \quad \text{and} \quad h = \ln \cot \frac{\Pi(h)}{2}.$$

We note that the angle of parallelism, $\Pi(h)$, increases from 0 to $\pi/2$ as h decreases from ∞ to 0, so that, "in the small," Lobachevskian geometry approximates Euclidean geometry. Also, since to each angle $\Pi(h)$ is associated a definite distance h, we see why distances, as well as angles, are absolute in Lobachevskian geometry. It further turns out that the trigonometrical formulas in Lobachevskian geometry are nothing but the familiar formulas of spherical trigonometry when the sides a, b, c of the triangle are replaced by a/i, b/i, c/i, and we are reminded of Lambert's suggestion about an imaginary sphere, mentioned in Section 3.3.

It is not the purpose of our study to go deeply into the Lobachevskian non-Euclidean geometry resulting from the hypothesis of the acute angle, and perhaps we have already indicated a sufficient number of propositions in the geometry to give the reader some idea of its content. We have seen that the hypothesis of the obtuse angle was discarded by all who did research in this subject because it contradicted the assumption that a straight line is infinite in length. Recognition of a second non-Euclidean geometry, based on the hypothesis of the obtuse angle, was not fully achieved until some years later, when Bernhard Riemann, in his probationary lecture of 1854, discussed the concepts of boundlessness and infiniteness. With these concepts clarified, one can realize an equally consistent geometry satisfying the hypothesis of the obtuse angle if Euclid's Postulates 1, 2, and 5 are modified to read:

(1') *Two distinct points determine at least one straight line.*

(2') *A straight line is boundless.*

(5') *Any two straight lines in a plane intersect.*

Much in this new non-Euclidean geometry is interesting. For example (see Figure 3.4), it can be shown, without great difficulty, that all the perpendiculars erected on the same side of a given straight line m are concurrent in a point O, and that the lengths along these perpendiculars from O to the line m are all equal to one another. Moreover, this common length, which we shall denote by q, is independent of which straight line in the plane is chosen for m. It can also be shown that, if A, B, P are any three points on line m, then

$$AP:AB = \angle\, AOP : \angle\, AOB,$$

and that if AB is taken equal in length to q, then $\angle\, AOB = \pi/2$. It now follows that all straight lines are finite and of the same constant length $4q$, for we observe that OP coincides with OA when $\angle\, AOP = 2\pi$, so that, under such circumstances, AP becomes the total length of the line m. But now

$$AP:AB = 2\pi : \angle\, AOB,$$

from which, by taking $AB = q$ and therefore $\angle\, AOB = \pi/2$, we find that

$$AP = 4q.$$

Thus straight lines, though boundless, are finite in length. Also, as in the case of Lobachevskian geometry, lengths, as well as angles, are absolute.

Riemann's celebrated lecture[12] of 1854 is not detailed or specific in its development but is extraordinarily rich in the depth and generality of its concepts and in the originality of its powerful new points of view. It would be difficult to point out another paper that has so greatly influenced modern geometrical research. This paper inaugurated a second period in the development of non-Euclidean geometry, a period characterized by the employment of

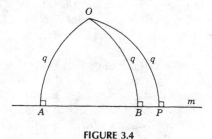

FIGURE 3.4

---

[12] The lecture was published in 1866, shortly after Riemann's death. For an English translation, see D. E. Smith [2], pp. 411–425.

the methods of differential geometry rather than the previously used methods of elementary synthetic geometry. To this paper we owe a considerable generalization of the concept of space that has led, in more recent times, to the extensive and important theory of abstract spaces; some of this theory has found application in the physical theory of relativity. Literally volumes of modern mathematical research can be traced to ideas advanced in this remarkable paper.

3.5   The Consistency and Significance of Non-Euclidean Geometry

It was some years after the appearance of the work of Lobachevsky and Bolyai that the mathematical world in general paid much attention to the subject of non-Euclidean geometry, and several decades elapsed before the full implication of the invention was appreciated. Most of the development of the subject beyond the historical point to which we have carried it is of too advanced a nature to be adequately considered here. One important matter, however, in this later development must be at least briefly touched on. Although Lobachevsky and Bolyai encountered no contradiction in their extensive investigations of the non-Euclidean geometry based on the hypothesis of the acute angle, and although they even felt confident that no contradiction would arise, the possibility still remained that such a contradiction or inconsistency might appear if the investigations should be sufficiently continued. To Beltrami goes the credit for the first proof of the consistency of this non-Euclidean geometry. In a brilliant paper,[13] published in 1868, Beltrami showed that the plane non-Euclidean geometry of Lobachevsky and Bolyai can be represented, with certain restrictions, on a surface of so-called constant negative curvature. It can be similarly shown that the plane non-Euclidean geometry of Riemann can be represented on a surface of constant positive curvature. Although Beltrami's methods are those of differential geometry and cannot be fully appreciated without an understanding of that field of mathematics, we can rather simply explain the gist of his idea.

Of the surfaces of constant positive curvature, the simplest is the sphere. Now the *geodesics* on the sphere—that is, the curves of shortest length lying on the sphere and joining pairs of points on the sphere—are the great circles of the sphere. If we should interpret the plane of the non-Euclidean geometry of Riemann as the surface of a sphere, and the straight lines of that non-Euclidean geometry as the great circles on the sphere, then it is a very simple matter to show that the postulates of the non-Euclidean geometry hold in our interpretation. For example:

(1′) *Two distinct points on the sphere determine at least one great circle on the sphere. (In fact, the great circle is unique, unless the points on the sphere happen to be diametrically opposite to each other, in which case any number of great circles may be passed through the two points.)*

---

[13] E. Beltrami.

(2′) *A great circle on the sphere is boundless. (A great circle is not infinite in length, however; in fact, all great circles on the sphere have the same finite length.)*

(3′) *With any point on the sphere as center and any great circle arc as polar distance a circle can be drawn on the sphere.*

(4′) *All right angles on the sphere are equal to one another.*

(5′) *Any two great circles on the sphere intersect.*

In view of our success in finding on the surface of a sphere a representation of Riemann's non-Euclidean geometry, it now follows that the plane non-Euclidean geometry of Riemann is consistent if Euclidean geometry is consistent, for if a deduced inconsistency were in this plane non-Euclidean geometry, there would be a corresponding deduced inconsistency in the ordinary geometry of great circles on a sphere, and this geometry is a part of the Euclidean geometry of space. What we have shown to be true of the great circles on a sphere can, by the methods of differential geometry, be shown to be true of the geodesics on any surface of constant positive curvature.

Just as the plane non-Euclidean geometry of Riemann can be realized on a surface of constant positive curvature, so also can the plane non-Euclidean geometry of Lobachevsky and Bolyai be similarly realized on a surface of constant negative curvature. Perhaps the simplest surface of constant negative curvature is the *pseudosphere*, or *tractoid*. To define this surface we first define a plane curve known as the *tractrix*. The tractrix may be generated as follows: Imagine a piece of inextensible cord lying along the positive *y*-axis (see Figure 3.5), one end of the cord lying at the origin, and the other end having attached to it a small heavy pellet. If the end lying at the origin is now pulled along the *x*-axis, the pellet will trace a kind of curve of pursuit; this curve is the tractrix, as shown in Figure 3.5. The curve is symmetrical in the *y*-axis and has the *x*-axis for an asymptote. Now the pseudosphere is the surface of revolution obtained by revolving the tractrix about its asymptote as an axis of rotation (see Figure 3.6). It can be shown that the geometry of the geodesics on this surface satisfies the postulates of the non-Euclidean geometry of Lobachevsky and Bolyai, after there has been suitable particularization of terms, but the proof here is not as simple as in the previous case of the sphere and can perhaps best be

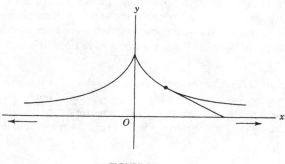

FIGURE 3.5

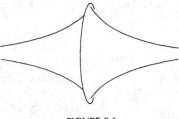

FIGURE 3.6

accomplished by the differential geometry methods employed by Beltrami for the general surface of constant negative curvature. This model, or representation, in Euclidean space, of the plane non-Euclidean geometry of Lobachevsky and Bolyai shows that this plane non-Euclidean geometry, too, is consistent if Euclidean geometry is consistent; for, once more, any inconsistency in the plane non-Euclidean geometry would imply a corresponding inconsistency in the Euclidean geometry of geodesics on the pseudosphere.

The pseudosphere and its geodesics, considered as a representation of the plane non-Euclidean geometry of Lobachevsky and Bolyai, is not as satisfactory as the sphere and its geodesics, considered as a representation of the plane non-Euclidean geometry of Riemann, for the pseudosphere represents only a limited part of the one non-Euclidean plane, whereas the sphere represents the whole of the other non-Euclidean plane. Beltrami conjectured, and it has since been proved, that no surface of constant negative curvature can represent the entire plane in the first case. Also, it is to be noted that neither of our representations takes into account any solid non-Euclidean geometry.

After the discovery of the above representations of the two classical non-Euclidean geometries on surfaces of constant curvature, many other, and in some ways more satisfying, representations of a different nature were devised. In Chapter 4 we shall examine an elementary representation of Lobachevskian non-Euclidean geometry that was devised by the great French mathematician Henri Poincaré (1854–1912). This method of models, or representations, however, does not establish absolute consistency but merely relative consistency. All we can assert from such models is that the two classical non-Euclidean geometries are consistent if Euclidean geometry is consistent. The possibility of an absolute test of consistency for a postulate set will be considered in a later chapter.

One consequence of the consistency of the non-Euclidean geometries is, of course, the final settlement of the ages-old problem of the parallel postulate. The consistency established the fact that the parallel postulate is independent of the other assumptions of Euclidean geometry and proved the impossibility of deducing the postulate as a theorem from those other assumptions, for if the parallel postulate could be so deduced there would have to be an inconsistency in the non-Euclidean systems.

But some consequences of the consistency of the non-Euclidean geometries are much more far-reaching than the settlement of the parallel postulate problem. One of the chief of these is the liberation of geometry from its traditional mold. The postulates of geometry become, for the mathematician, mere hypotheses whose physical truth or falsity need not concern him; the mathematician may take his postulates to suit his pleasure, as long as they are consistent with one another. A postulate, as the word is employed by the mathematician, has nothing to do with "self-evidence" or "truth." With the possibility of inventing such purely "artificial" geometries it became apparent that physical space must be viewed as an empirical concept derived from our external experiences, and that the postulates of a geometry designed to describe physical space are simply expressions of this experience, like the laws of a physical science. Euclid's parallel postulate, for example, insofar as it tries to interpret actual space, appears to have the same type of validity as Galileo's law of falling bodies; that is, they are both laws of observation that are capable of verification within the limits of experimental error. This point of view, that geometry when applied to actual space is an experimental science, or a branch of applied mathematics, is in striking contrast to the Kantian theory of space that dominated philosophical thinking at the time of the discovery of the non-Euclidean geometries. The Kantian theory claimed that space is a framework already existing intuitively in the human mind, that the axioms and postulates of Euclidean geometry are *a priori* judgments imposed on the mind, and that without these axioms and postulates no consistent reasoning about space can be possible. That this viewpoint is untenable was incontestably demonstrated by the invention of the non-Euclidean geometries.

Indeed, the consistency of the non-Euclidean geometries not only liberated geometry but had a similar effect on mathematics as a whole. Mathematics emerged as an arbitrary creation of the human mind, and not as something essentially dictated to us of necessity by the world in which we live. The matter is very neatly put in the following words of E. T. Bell:

> In precisely the same way that a novelist invents characters, dialogues, and situations of which he is both author and master, the mathematician devises at will the postulates upon which he bases his mathematical systems. Both the novelist and the mathematician may be conditioned by their environments in the choice and treatment of their material; but neither is compelled by any extrahuman, eternal necessity to create certain characters or invent certain systems.[14]

The invention of the non-Euclidean geometries, by puncturing a traditional belief and breaking a centuries-long habit of thought, dealt a severe blow to the *absolute truth* viewpoint of mathematics. In the words of Georg Cantor, "The essence of mathematics lies in its freedom."

Since we have a number of geometries of space—the Euclidean and the two classical non-Euclidean geometries—the question is often asked, "Which is the true geometry?" This question is, of course, quite meaningless when geometry is considered a branch of mathematics, because all we can say about truth with respect to a branch of mathematics is that if the postulates are true then the theorems are true. If, on the other hand, geometry is considered a branch of physics, then the question becomes more meaningful. But even here we cannot give a simple and definite answer. When it comes to the applications of several

---

[14] Quoted by permission from E. T. Bell [3], p. 330.

mathematical theories to a given physical situation, we are interested in that mathematical theory that best explains, or most closely agrees with, the observed facts of the physical situation and that will stand the kinds of tests customarily placed on hypotheses in any field of scientific inquiry. In the present case, then, we are interested in which of the Euclidean and non-Euclidean systems of geometry most closely agrees with the observed facts of physical space. It is not difficult to show that all three geometries under consideration fit our very limited portion of physical space equally well, and so it would seem we must be content with an indeterminate answer until some crucial experimental test on a great scale can be devised to settle the matter. Such a crucial test would appear to be the measurement of the sum of the three angles of a large physical triangle. To date no deviation, exceeding expected errors in measurement, from 180° has been found in the sum of the angles of any physical triangle. But, we recall, the discrepancy of this sum from 180° in the two non-Euclidean geometries is proportional to the area of the triangle, and the area of any triangle so far measured may be so small that any existing discrepancy is swallowed by the allowed errors in measurement. There are even some reasons for believing that physical experiments will never be able to resolve the matter anyway. In this event, then, we would do better to ask not which is the *true* geometry but which is the *most convenient* geometry, and this convenience may depend on the application at hand. Certainly, for drafting, for terrestrial surveying, and for the construction of ordinary buildings and bridges, Euclidean geometry is probably the most convenient simply because it is the easiest with which to work.

There are physical studies where geometries other than the Euclidean have been found to be more acceptable. For example, Einstein found in his study of the general theory of relativity that none of the three geometries that we have been considering is, in itself, adequate, and he adopted a suitable generalization of the Riemannian non-Euclidean geometry wherein the curvature of space may vary from point to point of the space. Again, a recent study[15] of *visual space* (the space psychologically observed by persons of normal vision) came to the conclusion that such a space can best be described by Lobachevskian non-Euclidean geometry. Other examples can be given.

Though it may be logical to call any geometry whose postulate system is not equivalent to a postulate system of Euclidean geometry a non-Euclidean geometry, custom has reserved this term only for the two geometries that result from the hypotheses of the acute and obtuse angle. Many other geometries other than these two, and that differ from Euclidean geometry, have been devised. Riemann was the originator of a whole class of these other geometries, usually referred to as *Riemannian geometries*, of which the Riemannian non-Euclidean geometry is a particular example. One of the accomplishments of the twentieth century was the development of general *non-Riemannian geometries*. Another geometry different than that of Euclid, invented through a deliberate application of the postulational method, is one by Max Dehn (1878–1952) in which the postulate of Archimedes is denied; such a geometry is referred to as a *non-Archimedean geometry*. The creation of these new geometries considerably

---

[15] R. K. Luneburg.

modified former conceptions of mathematics and led, as we shall see in the following chapters, to a profound study of the philosophy and foundations of mathematics and to the further development and understanding of the axiomatic method.

PROBLEMS

3.1.1 Give examples illustrating Proclus's statement, "It is a known fact that some lines exist which approach each other indefinitely, but yet remain nonintersecting."

3.1.2 Show that Playfair's postulate and Euclid's fifth postulate are equivalent. (One may use any of Euclid's first twenty-eight propositions; see Appendix, Section A.1.)

3.1.3 Prove that each of the following statements is equivalent to Playfair's postulate:
   (a) If a straight line intersects one of two parallel lines, it will intersect the other also.
   (b) Straight lines that are parallel to the same straight line are parallel to one another.

3.1.4 Replacing Euclid's fifth postulate by the assumption that the sum of the angles of a triangle is always equal to two right angles, establish the following consequences:
   (a) An exterior angle of a triangle is equal to the sum of the two opposite interior angles.
   (b) Through a given point P there can be drawn a line making, with a given line m, not passing through P, an angle less than any given angle α, however small.
   (c) If PA and QB are perpendicular to the line segment PQ at P and Q, respectively, and lie on the same side of PQ, and if PR is any transversal through P cutting QB in R, then $\angle APR = \angle QRP$.

3.1.5 Show that Playfair's postulate and the statement, "The sum of the angles of a triangle is always equal to two right angles," are equivalent.

3.1.6 Replacing Euclid's fifth postulate by the assumption, "If two angles of one triangle are equal to two angles of another triangle, then the third angles are also equal," show that the sum of the angles of a triangle is always equal to two right angles.

3.1.7 Find the fallacy in the following "proof," given by B. F. Thibaut (1809), of Euclid's fifth postulate: Let a straightedge be placed with its edge coinciding with side CA of triangle ABC. Rotate the straightedge successively about the three vertices A, B, C, in the direction ABC, so that it coincides in turn with AB, BC, CA. When the straightedge returns to its original position it must have rotated through four right angles. But the whole rotation is made up of three rotations equal to the exterior angles of the triangle. It now follows that the sum of the angles of the triangle must be equal to two right angles, and from this follows Euclid's parallel postulate.

3.1.8 Find the fallacy in the following "proof," given by J. D. Gergonne (1812), of Euclid's fifth postulate: Let PA and QB, lying in the same plane and on the same side of PQ, be perpendicular to PQ. Then PA and QB are parallel. Let PG be the last ray through P, and lying within angle QPA, which intersects QB. Produce QB to a point K beyond the point of intersection of PG with QB, and draw PK. It follows that PG is *not* the last ray through P that meets QB, and therefore all rays through P and lying within angle QPA must meet QB. Thus through P there is only one line parallel to line QB, and Euclid's fifth postulate follows.

3.1.9 Find the fallacy in the following "proof," given by J. K. F. Hauff (1819), of

Euclid's fifth postulate: Let AD, BE, CF be the altitudes of an equilateral triangle ABC, and let O be the point of concurrency of these altitudes. In right triangle ADC, acute angle CAD equals one-half acute angle ACD. Therefore, in right triangle AEO, acute angle OAE equals one-half acute angle AOE. A similar statement holds for each of the six small right triangles of which AEO is typical. It now follows that the sum of the angles of triangle ABC is equal to one-half the sum of the angles about O—that is, equal to two right angles. But it is known that the existence of a single triangle having the sum of its angles equal to two right angles is enough to guarantee Euclid's fifth postulate.

3.2.1 The *reductio ad absurdum* method is often useful in establishing the converse of a known theorem. Apply this procedure to establish the converses of the following theorems:

(a) *Ceva's theorem*: If D, E, F are three points on the sides BC, CA, AB of a triangle ABC such that AD, BE, CF are concurrent, then $(BD/DC)(CE/EA)$ $(AF/FB) = 1$.

(b) In a triangle the greater angle lies opposite the greater side.

3.2.2 **(a)** On the earth, on opposite sides of the international date line, it can be both Monday and Tuesday at the same time. Discuss this in relation to the *law of contradiction* (which asserts that "today cannot be *both* Monday and Tuesday"), and the *law of the excluded middle* (which asserts that "either today is Monday or today is not Monday").

(b) Does the law of the excluded middle hold for the following statements? (1) All circles are blue. (2) The king of the United States is more than six feet tall.

This exercise shows that, although the laws of contradiction and the exluded middle are usually considered universally valid, *some* sort of limitations must be made concerning their applicability.

3.2.3 Prove, by simple congruence theorems (which do not require the parallel postulate), the following theorems about isosceles birectangles:

(a) The summit angles of an isosceles birectangle are equal to each other.

(b) The line joining the midpoints of the base and summit of an isosceles birectangle is perpendicular to both the base and the summit.

(c) If perpendiculars are drawn from the extremities of the base of a triangle on the line passing through the midpoints of the two sides, an isosceles birectangle is formed.

(d) The line joining the midpoints of the equal sides of an isosceles birectangle is perpendicular to the line joining the midpoints of the base and summit.

3.2.4 The *hypothesis of the acute angle* assumes that the equal summit angles of an isosceles birectangle are acute, or that the fourth angle of a trirectangle (a quadrilateral containing three right angles) is acute. In the following we shall assume the hypothesis of the acute angle:

(a) Let ABC be any right triangle, and let M be the midpoint of the hypotenuse AB. At A construct angle BAD = angle ABC. From M draw MP perpendicular to CB. On AD mark off $AQ = PB$, and draw MQ. Prove triangles AQM and BPM congruent, thus showing that angle AQM is a right angle and that points Q, M, P are collinear. Then $ACPQ$ is a trirectangle with acute angle at A. Now show that, *under the hypothesis of the acute angle, the sum of the angles of any right triangle is less than two right angles.*

(b) Let angle A of triangle ABC be not smaller than either angle B or angle C. Draw the altitude through A, and show, by part (a), that *under the hypothesis of the acute angle, the sum of the angles of any triangle is less than two right angles.* The difference between two right angles and the sum of the angles of a triangle is known as the *defect* of the triangle.

(c) Consider two triangles, ABC and $A'B'C'$, in which corresponding angles are equal. If $A'B' = AB$, then these triangles are congruent. Suppose $A'B' < AB$. On AB mark off $AD = A'B'$, and on AC mark off $AE = A'C'$. Then triangles ADE and $A'B'C'$ are congruent. Show that E cannot fall on C, since then angle BCA would be greater than angle DEA. Show also that E cannot *fall* on AC produced, since then DE would cut BC in a point F and the sum of the angles of triangle FCE would exceed two right angles. Therefore E lies between A and C and $BCED$ is a convex quadrilateral. Show that the sum of the angles of this quadrilateral is equal to four right angles. But this is impossible under the hypothesis of the acute angle. It thus follows that we cannot have $A'B' < AB$ and that, *under the hypothesis of the acute angle, two triangles are congruent if the three angles of one are equal to the three angles of the other.* In other words, in the geometry resulting from the acute angle hypothesis similar figures of different sizes do not exist.

(d) A line segment joining a vertex of a triangle to a point on the opposite side is called a *cevian*. A cevian divides a triangle into two subtriangles, each of which may be similarly divided, and so on. Show that if a triangle is partitioned by cevians into a finite number of subtriangles, the defect of the original triangle is equal to the sum of the defects of the triangles in the partition.

3.3.1 Show that a trirectangle can be regarded as half of an isosceles birectangle.

3.3.2 A *spherical degree* for a given sphere is defined to be the spherical area which is equivalent to $(1/720)$th of the entire surface of the sphere. The *spherical excess* of a spherical triangle is defined as the excess, measured in degrees of angle, of the sum of the angles of the triangle above $180°$.

(a) Show that the area of a lune whose angle is $n°$ is equal to $2n$ spherical degrees.

(b) Show that the area of a spherical triangle, in spherical degrees, is equal to the spherical excess of the triangle.

(c) Show that the area A of a spherical triangle of spherical excess $E°$ is given by

$$A = \frac{\pi r^2 E°}{180°},$$

where r is the radius of the sphere. This shows that, for a given sphere, the area of a spherical triangle is proportional to its spherical excess.

3.3.3 Indicate how one might set up a one-to-one correspondence on a sphere between spherical angles and great circle arcs, thus showing that in the geometry on a sphere lengths, as well as angles, are *absolute*.

3.3.4 Fill in the details of the following proof of *Legendre's first theorem*: "The sum of the three angles of a triangle cannot be greater than two right angles." Show that the proof assumes the infinitude of the straight line.

Suppose that the sum of the angles of triangle ABC is $180° + \theta$ and that angle CAB is not greater than either of the other angles. Join A to D, the midpoint of BC, and produce AD its own length to E. Show that triangles BDA and CDE are congruent; hence that the sum of the angles of triangle AEC is also equal to $180° + \theta$. One of the angles CAE and CEA is not greater than $(1/2) \measuredangle CAB$. Apply the same process to triangle AEC, obtaining a third triangle whose angle-sum is $180° + \theta$ and one of whose angles is not greater than $(1/2^2) \measuredangle CAB$. By applying the construction n times, a triangle is reached whose angle-sum is $180° + \theta$ and one of whose angles is not greater than $(1/2^n) \measuredangle CAB$. By the postulate of Archimedes there exists an integer k such that $k\theta > \measuredangle CAB$. Choose n so large that $2^n > k$. Then $\theta > (1/2^n) \measuredangle CAB$, and the sum of two of the angles of the last triangle must be greater than $180°$. But this conclusion contradicts Proposition I 17.

3.3.5 In one effort to eliminate the hypothesis of the acute angle, Legendre tried to obtain, under this hypothesis, a triangle containing a given triangle at least twice. He proceeded as follows. Let ABC be any triangle such that angle A is not greater than either of the other two angles. Construct on side BC a triangle DCB congruent to triangle ABC, with angle DCB equal to angle B and angle DBC equal to angle C. Through D draw any line cutting AB and AC produced in E and F, respectively. Then triangle AEF contains triangle ABC at least twice.

Show that this construction assumes that through a point within a given angle less than 60° there can be drawn a straight line intersecting both sides of the angle.

3.3.6 Assuming Legendre's first theorem (see Problem 3.3.4), prove the following sequence of theorems credited to Legendre:

(a) If the sum of the angles of a triangle is equal to two right angles, then the same is true of any triangle obtained from the given triangle by drawing a cevian [see Problem 3.2.4(d)] through one of its vertices.

(b) If there exists a triangle with the sum of its angles equal to two right angles, then one can construct an isosceles right triangle having the sum of its angles equal to two right angles and its legs greater in length than any given line segment.

(c) *Legendre's second theorem*: If there exists a single triangle having the sum of its angles equal to two right angles, then the sum of the angles of every triangle will be equal to two right angles.

(d) If there exists a single triangle having the sum of its angles less than two right angles, then the sum of the angles of every triangle is less than two right angles.

3.4.1 **(a)** If ABC is a spherical right triangle, right-angled at C, show that $\sin A = \sin a / \sin c$.

(b) If ABC is any spherical triangle, show that $\sin a : \sin b : \sin c = \sin A : \sin B : \sin C$.

This is the *law of sines* for spherical triangles.

(c) If ABC is any spherical triangle, show that

$$O(a) : O(b) : O(c) = \sin A : \sin B : \sin C,$$

where $O(r)$ denotes the circumference of a circle, on the sphere, whose polar distance is r.

3.4.2 Prove, by using Dedekind's postulate of continuity (see Problem 2.4.2), the following theorem in Lobachevskian geometry: If m is a line and P a point not on m, then there are two lines through P that do not intersect m, that make equal acute angles α with the perpendicular PQ from P to m, and that are such that any line through P lying within the angle formed by the two lines and containing the perpendicular PQ intersects m, while any other line through P does not intersect m.

3.4.3 Given $\alpha = 2 \operatorname{Arc} \tan e^{-h}$, solve for h in terms of α.

3.4.4 (*for students who have studied calculus*) Define, for any complex number z,

$$e^z = 1 + z + \frac{z^2}{2!} + \cdots + \frac{z^{n-1}}{(n-1)!} + \cdots,$$

$$\sin z = z - \frac{z^3}{3!} + \cdots + (-1)^{n+1} \frac{z^{2n-1}}{(2n-1)!} + \cdots,$$

$$\cos z = 1 - \frac{z^2}{2!} + \cdots + (-1)^{n+1} \frac{z^{2n-2}}{(2n-2)!} + \cdots,$$

$$\sinh z = \frac{e^z - e^{-z}}{2}, \quad \cosh z = \frac{e^z + e^{-z}}{2}.$$

It can be shown that each of the series converges for all complex numbers z.

(a) Show that, if x is real,

$$e^{xi} = \cos x + i \sin x, \quad e^{-xi} = \cos x - i \sin x,$$

from which

$$\sin x = \frac{e^{xi} - e^{-xi}}{2i}, \quad \cos x = \frac{e^{xi} + e^{-xi}}{2}.$$

(b) In Lobachevskian geometry the parts a, c, A of a right triangle ABC right-angled at C can be shown to be related by

$$\sin A = \frac{\sinh a}{\sinh c}.$$

Replace a by a/i and c by c/i, and obtain the corresponding formula [see Problem 3.4.1(a)]

$$\sin A = \frac{\sin a}{\sin c}$$

of spherical geometry.

3.4.5 Prove that, in plane Riemannian non-Euclidean geometry, all the perpendiculars erected on the same side of a given straight line m are concurrent in a point O, the lengths along these perpendiculars from O to the line m are all equal to one another, and this common length is independent of which straight line in the plane is chosen for m.

3.5.1 Verify, in spherical geometry, the following theorems of plane Riemannian non-Euclidean geometry:

(a) All the perpendiculars erected on the same side of a given straight line m are concurrent in a point O, the lengths along these perpendiculars from O to the line m are all equal to one another, and this common length (call it q) is independent of which straight line in the plane is chosen for m.

(b) If A, B, P are any three points on the line m, then $AP : AB = \angle AOP : \angle AOB$.

(c) All straight lines are finite and of the same constant length $4q$.

(d) Two triangles are congruent if the three angles of one are equal to the three angles of the other.

3.5.2 (*for students who have studied calculus*) Consider the sections of a surface S made by planes containing the normal to S at a point P on S. Of these sections there is one having a maximum curvature k at P, and one having a minimum curvature k' at P. These two sections of maximum and minimum curvature are generally at right angles to each other, and their curvatures at P are called the *principal curvatures* of S at P. The product $K = kk'$ is called the (Gaussian, or total) *curvature* of the surface S at P. If the two principal curvatures are of the same sense, then K is positive; if the two principal curvatures are of opposite senses, then K is negative; if one principal curvature is zero, then K is zero. Gauss, who essentially created the differential geometry of surfaces, discovered the remarkable theorem that *if a surface is bent* (without stretching, creasing, or tearing) *the curvature of the surface at each point remains unaltered.*

(a) Is there a quadric surface whose curvature is everywhere positive? everywhere negative? everywhere zero? in some places positive and in others negative?

(b) If a surface can be bent so as to coincide with another surface, then the two surfaces are said to be *applicable* to each other. Show that a pair of applicable surfaces have their points in one-to-one correspondence such that at pairs of corresponding points the curvatures of the two surfaces are equal.

(c) Show that when one surface is bent into another surface, the geodesics of the first surface go into geodesics of the second surface.

(d) Show that a sphere of radius r has constant positive curvature equal to $1/r^2$.

(e) Show that a plane has constant zero curvature.

(f) Show that a cylindrical surface has constant zero curvature. Is a cylindrical surface applicable to a plane?

(g) Show that if a surface is applicable on itself in all positions, its curvature must be constant.

(h) Show that the only surfaces on which free mobility of figures is possible are those of constant curvature.

(i) Show that a sphere is not applicable to a plane. (This is why, in terrestrial map making, some sort of distortion in the map is necessary.)

3.5.3 (*for students who have studied calculus*) The *hyperbolic sine* and *hyperbolic cosine* functions may be defined (see Problem 3.4.4) by $\sinh x = (e^x - e^{-x})/2$ and $\cosh x = (e^x + e^{-x})/2$.

(a) Show that $\cosh^2 x - \sinh^2 x = 1$.

(b) Show that $d(\cosh u)/dx = \sinh u \, du/dx$, $d(\sinh u)/dx = \cosh u \, du/dx$.

3.5.4 (*for students who have studied calculus*) The graph of $y = k \cosh (x/k)$ is a *catenary*, the form assumed by a perfectly flexible inextensible chain of uniform density hanging from two supports not in the same vertical line. Let this catenary (see Figure 3.7) cut the y-axis in A; let P be any point on the curve, and let F be the foot of the ordinate through P; let the tangent to the curve at P cut the x-axis in T, and let Q be the foot of the perpendicular from F on PT.

(a) With simple calculus, show that QF is constant and equal to k.

(b) With the aid of integral calculus, show that QP is equal to the length of the arc AP.

(c) Show that if a string AP is unwound from the catenary, the tracing end A will describe a curve AQ having the property that the length of the tangent QF is constant and equal to k. In other words, the locus of Q, which is an involute of the catenary, is a tractrix.

(d) It can be shown that for a surface of revolution the principal curvatures (see Problem 3.5.2) at a point Q on the surface are the curvature of the meridian through Q and the curvature of the section through Q, which is normal to the

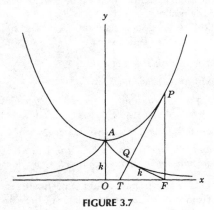

FIGURE 3.7

meridian through Q. If the normal to the surface at Q meets the axis of revolution of the surface in T, then the latter curvature is known to be equal to $1/QT$. Show that the principal curvatures at Q of the tractoid obtained by revolving the tractrix of part (c) about the x-axis are given by $1/QP$ and $1/QT$.

(e) Show that the curvature (see Problem 3.5.2) of the tractoid of part (d) is constant, and everywhere equal to $-1/k^2$.

3.5.5 (*for students who have studied calculus*) Show that the tractrix is an orthogonal trajectory of a family of circles of constant radius k with centers lying on a straight line.

3.5.6 (*for students who have studied calculus*) Show that the geometry of geodesics on a surface of constant zero curvature (for example, on a cylindrical surface) is Euclidean. (See Problem 3.5.2.)

3.5.7 Take a fixed circle, Σ, in the Euclidean plane, and interpret the Lobachevskian plane as the interior of Σ, a "point" of the Lobachevskian plane as a Euclidean point within Σ, and a "line" of the Lobachevskian plane as that part of a Euclidean line that is contained within Σ. Verify, in this model, the following statements:

(a) Two "points" determine one and only one "line."

(b) Two distinct "lines" intersect in at most one "point."

(c) Through a "point" P not on a "line" m can be passed infinitely many "lines" not meeting "line" m.

(d) Let the Euclidean line determined by the two "points" P and Q intersect Σ in S and T, in the order S, P, Q, T. Then we interpret the Lobachevskian "distance" from P to Q as $\log [(QS)(PT)/(PS)(QT)]$. If P, Q, R are three "points" on a "line," show that

$$\text{"distance" } PQ + \text{"distance" } QR = \text{"distance" } PR.$$

(e) Let "point" P be fixed and let "point" Q move along a fixed "line" through P toward T. Show that "distance" $PQ \to \infty$.

This model was devised by Felix Klein (1849–1925). With the above interpretations, and with a suitable interpretation of "angle" between two "lines" it can be shown that all the assumptions necessary for Euclidean geometry, except the parallel postulate, are true propositions in the geometry of the model. We have seen, in part (c), that the Euclidean parallel postulate is not such a proposition but that the Lobachevskian parallel postulate holds instead. The model thus proves that the Euclidean parallel postulate cannot be deduced from the other assumptions of Euclidean geometry, for if it were implied by the other assumptions it would have to be a true proposition in the geometry of the model.

3.5.8 Consider the following set of postulates about certain objects called *dabbas* and certain collections of dabbas called *abbas*:

P1. *Every abba is a collection of dabbas.*

P2. *There exist at least two dabbas.*

P3. *If* p *and* q *are two dabbas, then there exists one and only one abba containing both* p *and* q.

P4. *If* L *is an abba, then there exists a dabba not in* L.

P5. *If* L *is an abba, and* p *is a dabba not in* L, *then there exists one and only one abba containing* p *and not containing any dabba that is in* L.

(a) Devise a model (or interpretation) of the postulate set to show that P3 cannot be deduced from the remaining postulates of the set.

(b) Devise a model (or interpretation) of the postulate set to show that P5 cannot be deduced from the remaining postulates of the set.

(c) Restate the postulates by interpreting *abba* as "straight line" and *dabba* as "point." Note that P5 is now Playfair's postulate.

3.5.9 Legend has it that in an effort to determine whether physical space is Euclidean or non-Euclidean, Gauss measured the sum of the angles of a large triangle whose vertices were three mountain peaks. He found no deviation from 180°, beyond the expected error of measurement, and thus was unable to conclude anything except that perhaps the triangle was too small. Assuming such a test to be workable (in this connection see Problem 3.5.10), perhaps a crucial experiment could be devised in connection with the measurement of the parallax of stars.

In Figure 3.8 let S represent a star and E_1 and E_2 two positions of the earth at opposite ends of a diameter of its orbit. Let C represent the sun, and suppose CS is perpendicular to $E_1 E_2$. Angle $E_1 SC$ is then called the *parallax* of star S. There are two methods of determining this parallax, the *direct method* and the *differential method*. In the direct method one measures angle $SE_1 C$ and then calculates the parallax angle. In the differential method one measures angle $S'E_1 S$, where S' is a "neighboring" star whose distance is known to be so much greater than that of S that it is taken as infinite. Set

$$\theta = 90° - \angle SE_1 C - \angle S'E_1 S.$$

(a) Assuming Euclidean geometry, show that $\theta = 0$.

(b) Assuming Lobachevskian geometry, show that $\theta = 90° - \Pi(R) > 0$, where R is the radius of the earth's orbit.

(c) Assuming either geometry, show that $\angle E_1 SC > \theta$ and that it then follows that if a positive lower bound can be found for θ, the geometry of physical space would be Lobachevskian rather than Euclidean.

3.5.10 Because of the apparently inextricable entanglement of space and matter it may be impossible to determine by astronomical methods whether physical space is Euclidean or non-Euclidean. Since all measurements involve both physical and geometrical assumptions, an observed result can be explained in many different ways by merely making suitable compensatory changes in our assumed qualities of space and matter. For example, it is quite possible that a discrepancy observed in the angle-sum of a triangle could be explained by preserving the assumptions of Euclidean geometry but at the same time modifying some physical law, such as some law of optics. Again, the absence of any such discrepancy might be compatible with the assumptions of a non-Euclidean geometry, together with some suitable adjustments in our assumptions about matter. On these grounds Henri Poincaré maintained the impropriety of asking which geometry is the true one. To clarify this viewpoint, Poincaré devised an imaginary universe Σ

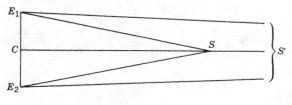

FIGURE 3.8

occupying the interior of a sphere of radius R in which he assumed the following physical laws to hold:

1. *At any point* P *of* Σ *the absolute temperature* T *is given by* $T = k(R^2 - r^2)$, *where* r *is the distance of* P *from the center of* Σ *and* k *is a constant.*

2. *The linear dimensions of a material body vary directly with the absolute temperature of the body's locality.*

3. *All material bodies in* Σ *immediately assume the temperatures of their localities.*

(a) Show that it is possible for an inhabitant of Σ to be quite unaware of the above three physical laws holding in his universe.

(b) Show that an inhabitant of Σ would feel that his universe is infinite in extent on the grounds that he would never reach a boundary after taking a finite number N of steps, no matter how large N may be chosen.

(c) Show that geodesics in Σ are curves bending toward the center of Σ. As a matter of fact, it can be shown that the geodesic through two points A and B of Σ is the arc of a circle or straight line through A and B which cuts the bounding sphere orthogonally.

(d) Let us impose one further physical law on the universe Σ by supposing that light travels along the geodesics of Σ. This condition can be physically realized by filling Σ with a gas having the proper index of refraction at each point of Σ. Show, now, that the geodesics of Σ will "look straight" to an inhabitant of Σ.

(e) Show that in the geometry of geodesics in Σ the Lobachevskian parallel postulate holds, so that an inhabitant of Σ would believe that he lives in a non-Euclidean world. Here we have a piece of ordinary, and supposedly Euclidean, space, that, because of different physical laws, appears to be non-Euclidean.

HILBERT'S *GRUNDLAGEN*

4.1 The Work of Pasch, Peano, and Pieri

The modern postulational method in mathematics, with the present-day deliberate drive toward more and ever more generalization and abstraction· in the subject, can be traced directly to two sources of approximately simultaneous origin—the bold creation by Lobachevsky and Bolyai of non-Euclidean geometry and the penetrating discovery by British mathematicians of abstract algebraic structure. We have discussed the first of these sources in the preceding chapter, and we shall deal with the second one in the succeeding chapter. In the present chapter we consider a bit further some of the historical consequences of the first of the two fountainheads.

After the discovery of non-Euclidean geometry, a need was felt for a truly satisfactory postulational treatment of Euclidean geometry. All hidden, or tacit, assumptions had to be ferreted out, and a logically acceptable set of underlying postulates for the subject had to be clearly and unequivocally put forth. Such an organization of Euclidean geometry was first accomplished in 1882 by the German mathematician Moritz Pasch.

In his treatment of Euclidean geometry, Pasch recognized the important distinction between *explicit* and *implicit* definition. Most people are familiar with the concept of explicit definition, inasmuch as this is the type of definition most frequently employed. In an explicit definition a new term is expressed by means of terms that are already accepted in the vocabulary. In a technical sense, then, a new term introduced in such a manner serves merely as an abbreviation for a complex combination of terms already present. Thus a new term introduced by explicit definition is really arbitrary, though convenient, and may be entirely dispensed with, although then the discourse in which the vocabulary is to be employed would immediately increase in complexity.

Implicit definition, on the other hand, is relatively unfamiliar to most people, though such a notion is indispensable in logical theory. The necessity for implicit definition is due to the fact that it is impossible to define all terms explicitly if we wish to avoid circularity. It is impossible, for example, to define word A in terms of word B, then word B in terms of word C, and so on indefinitely, for such a procedure would imply an infinite number of words in the vocabulary. Of course the dictionary makes an attempt to define all words explicitly through the admitted use of circularity, but it is hoped that people using the dictionary have developed an adequate vocabulary so that the words in terms of which some unknown word is defined are already familiar to them.

Most people have tried to introduce new words into their vocabularies by observing how these words are used by others. This idea of defining a word through the medium of the context in which it occurs is the basic idea of implicit definition. In a logical discourse, since not all technical terms can be explicitly defined, some must have meanings that can be realized only by observing the context in which they are employed. A logical discourse, in other words, must accept a relatively small number of primitive technical terms that can be used for explicitly defining all the other technical terms that occur in the discourse, these primitive terms of the discourse receiving no definitions except those given to them implicitly by their presence in the adopted postulates of the discourse.

Whereas Euclid attempted a kind of explicit definition of the terms *point*, *line*, and *plane*, for example, Pasch accepted these as primitive, or irreducible, terms in his development of Euclidean geometry; he considered them only implicitly defined by the basic propositions that he assumed as postulates in his treatment. These assumed basic propositions were described by Pasch as *nuclear*. Although the origin of the nuclear propositions might be found in empirical considerations, Pasch emphasized that they are to be enunciated without regard to any empirical significance. He declared that the creation of a truly deductive science demands that all logical deductions must be independent of any meanings that might be attached to the various concepts. In fact, if it becomes necessary at any point in the construction of a proof to refer to certain interpretations of the basic terms, then that is sufficient evidence that the proof is logically inadequate. On the other hand, by keeping all of the work purely formal, various applications of the discourse may be obtained by assigning different suitable meanings to the primitive terms employed. From this point of view, Euclidean geometry is essentially a symbolic system whose validity and possibility for further development do not depend on any specific meanings given to the basic terms employed in the postulates of the geometry; Euclidean geometry is reduced to a pure exercise in logical syntax. Where Euclid appears to have been guided by visual imagery and thus subjected to the making of tacit assumptions, Pasch attempted to avoid this pitfall by deliberately considering geometry as a purely hypothetico-deductive system. Pasch profoundly influenced postulational thinking in geometry, and later works in the field attempted to maintain the standards of rigor that he had introduced.

Following Pasch, the Italian mathematician Giuseppe Peano (1858–1932) gave, in 1889, a new postulational development of Euclidean geometry. Like Pasch, Peano based his treatment on certain primitive terms, among which are an entity called a "point" and a relation among points designated by "betweenness." From many points of view Peano's work is largely a translation of Pasch's treatise into the notation of a symbolic logic that Peano introduced to the mathematical world. In Peano's version no empiricism is found; his geometry is purely formalistic by virtue of the fact that it is constructed as a calculus of relations between variables. Here we have the mathematician's ultimate cloak of protection from the pitfall of overfamiliarity with his subject matter. We have seen that Euclid, working with visual diagrams in a field of study with which he was very familiar, unconsciously made numerous hidden assumptions that were not guaranteed to him by his axioms and postulates. To protect himself from similar prejudice, Peano conceived the idea of symbolizing his primitive terms and his logical processes of thought. Clearly, if one says, "Two x's determine a y," instead of, "Two points determine a straight line," one is not likely to be biased by preconceived notions about "points" and "straight lines," and if a symbolic logic is employed in the reasoning, one is not likely to fall into fallacies stemming from slippery intuition and other modes of loose reasoning. The derivation of theorems becomes an algebraic process in which only symbols and formulas are employed, and geometry is reduced to a strictly formal process that is entirely independent of any interpretations of the symbols involved.

We shall hear more of Peano in subsequent chapters, for his interest in formal systems extended beyond geometry. Today Peano is recognized not only as one of the truly original geniuses in mathematics, but also as one of the creators of modern mathematical logic. He continued the work in the field of symbolic logic that had been commenced by George Boole and his successors, essentially devising a kind of universal shorthand for the mathematics of his time. The important subject of symbolic, or mathematical, logic is considered more fully in a later chapter. For the present, we wish to point out that Peano's logical analysis of geometry rendered even clearer the concept of geometry as an abstract hypothetico-deductive system with no intrinsic content beyond that implied by the postulates.

Another Italian mathematician, Mario Pieri (1860–1904), employed, in 1899, in his studies of Euclidean geometry, a quite different approach from that of his predecessors. He considered the subject of his study to be an aggregate of undefined elements called "points" and an undefined concept of "motion." Pieri's first five postulates will indicate the important role assigned to the concept of motion. They are

1. *There is given an aggregate S of points containing at least two distinct members.*

2. *A motion establishes a pairing of the points of S such that to each point P of S there corresponds some point P' of S. For any motion that establishes a correspondence between the points P of S and the points P' of S, there is an inverse motion that establishes a correspondence between the points P' of S and the points P of S.*

3. *The resultant of two motions performed successively is equivalent to a single motion.*

As a consequence of Postulates 2 and 3, the motion equivalent to some motion followed by its inverse is a motion that makes each point of S correspond to itself, or, in other words, leaves each point of S fixed. This motion is called the *identical motion*. A motion that is not the identical motion is called an *effective motion*.

4. *For any two distinct points A and B, there exists an effective motion that*

leaves A *and* B *fixed. Such a motion may be referred to as a* rotation motion *about the two points* A *and* B.

5. *If there is an effective motion that leaves three points* A, B, C *fixed, then every motion that leaves* A *and* B *fixed also leaves* C *fixed.*

As a result of Postulate 5, it is now possible to define the straight line determined by two points A and B: "The *straight line AB* is the aggregate of points that remain fixed under any effective motion that leaves A and B fixed."

It can be seen from the brief sample of Pieri's postulate system given above that the concern is with the results of a given motion rather than with the motion's actual nature. The modern mathematician would speak of *transformations*, or *mappings*, instead of motions. It should be noted that Pieri's idea of motion, which is essentially that of rigid displacement, is adaptable to the Euclidean superposition proofs.

Although Pieri's treatment of Euclidean geometry received no wide acceptance, the development of certain modern notions is apparent in his work. Pieri was considering Euclidean geometry as a study of the properties and relations of configurations of points that remain invariant under the set of all rigid displacements. We say more about this idea later on, for the general concept of building a geometry as the invariant theory of a group of transformations pointed the way to important studies in the late nineteenth and early twentieth centuries.

4.2 Hilbert's *Grundlagen der Geometrie*

The eminent German mathematician, Professor David Hilbert (1862–1943), gave a course of lectures on the foundations of Euclidean geometry at the University of Göttingen during the 1898–1899 winter term. These lectures, which concerned themselves with a postulational discussion of Euclidean geometry, were rearranged and published in a slender volume in June 1899 under the title *Grundlagen der Geometrie (Foundations of Geometry)*. This work, in its various improved revisions, is today a classic in its field; it has done far more than any other single work since the discovery of non-Euclidean geometry to promote the modern axiomatic method and to shape the character of a good deal of present-day mathematics. The influence of the book was immediate. A French edition appeared soon after the German publication, and an English version, translated by E. J. Townsend, appeared in 1902.[1] The work went through seven German editions during the author's lifetime, the seventh edition appearing in 1930. An eighth German edition, a revision and enlargement by Paul Bernays, appeared in 1956; a ninth edition, also by Bernays, appeared in 1962; and a tenth edition, again by Bernays, appeared in 1968.

By developing a postulate set for plane and solid Euclidean geometry that does not depart too greatly in spirit from Euclid's own, and by employing a minimum of symbolism, Hilbert succeeded in convincing mathematicians, to a

---

[1]D. Hilbert [1].

far greater extent than had Pasch and Peano, of the purely hypothetico-deductive nature of geometry. But the influence of Hilbert's work went far beyond this, for backed by the author's great mathematical authority, it firmly implanted the postulational method, not only in the field of geometry but also in essentially every other branch of mathematics of the twentieth century. The stimulus to the development of the foundations of mathematics provided by Hilbert's little book is difficult to overestimate. Lacking the strange symbolism of the works of Pasch and Peano, Hilbert's work can be read, in great part, by any intelligent student of high school geometry.

Whereas Euclid made a distinction between "axioms" and "postulates," modern mathematicians consider these two terms synonymous and designate all the assumed propositions of a logical discourse by either term. From this point of view, Hilbert's treatment of plane and solid Euclidean geometry rests on twenty-one axioms or postulates, and these involve six primitive, or undefined, terms. For simplicity we here consider only those postulates of Hilbert's set that apply to *plane* geometry. Under this limitation there are only fourteen postulates and five primitive terms.

The primitive terms in Hilbert's treatment of plane Euclidean geometry are *point, line* (meaning *straight line*), *on* (a relation between a point and a line), *between* (a relation between a point and a pair of points), and *congruent* (a relation between configurations called *segments* and between configurations called *angles*, which are explicitly defined in the treatment). For convenience of language, the phrase "point A is on line m" frequently is stated alternatively by the equivalent phrases, "line m passes through point A" or "line m contains point A."

Hilbert's fourteen postulates for plane geometry are provided here, interspersed with occasional definitions when needed. The statements of the postulates are taken, with some slight modifications for the sake of clarity, from the tenth (1968) edition of Hilbert's *Grundlagen der Geometrie*.[2] The postulates are presented in certain related groups.

Group I: Postulates of Connection

I-1. *There is one and only one line passing through any two given distinct points.*

I-2. *Every line contains at least two distinct points, and for any given line there is at least one point not on the line.*

Group II: Postulates of Order

II-1. *If point C is between points A and B, then A, B, C are three distinct points on the same line, and C is between B and A, and B is not between C and A, and A is not between C and B.*

II-2. *For any two distinct points A and B there is a point D such that B is between A and D.*

---

[2] D. Hilbert [8].

Definition　If A and B are distinct points, then by *segment AB* is meant the points A and B and all points that are between A and B. Points A and B are called the *end points* of the segment. Point C is said to be *on* the segment AB if it is A or B or some point between A and B.

Definition　Two lines, a line and a segment, or two segments, are said to *intersect* if there is a point that is on both of them.

Definition　Let A, B, C be three points not on the same line. Then by the *triangle ABC* is meant the three segments AB, BC, CA. The segments AB, BC, CA are called the *sides* of the triangle, and the points A, B, C are called the *vertices* of the triangle.

II-3.　(Pasch's postulate) *A line that intersects one side of a triangle but does not pass through any of the vertices of the triangle must also intersect another side of the triangle.*

Group III: Postulates of Congruence

III-1.　*If A and B are distinct points and if A′ is a point on a line m, then there are two and only two distinct points B′ and B″ on m such that segment A′B′ is congruent to segment AB and segment A′B″ is congruent to segment AB; moreover, A′ is between B′ and B″.*

III-2.　*If two segments are congruent to the same segment, then they are congruent to each other.*

III-3.　*If point C is between points A and B and point C′ is between points A′ and B′, and if segment AC is congruent to segment A′C′ and segment CB is congruent to segment C′B′, then segment AB is congruent to segment A′B′.*

Definition　By the *ray AB* is meant the set of all points consisting of those that are between A and B, the point B itself, and all points C such that B is between A and C. The ray AB is said to *emanate from* point A.

Theorem　If $B′$ is any point on the ray AB, then the rays $AB′$ and AB are identical.

Definition　By an *angle* is meant a point (called the *vertex* of the angle) and two rays (called the *sides* of the angle) emanating from the point. By virtue of the above theorem, if the vertex of the angle is point A and if B and C are any two points other than A on the two sides of the angle, we may unambiguously speak of the angle BAC (or CAB).

Definition　If ABC is a triangle, then the three angles BAC, CBA, ACB are called the *angles* of the triangle. Angle BAC is said to be *included* by the sides AB and AC of the triangle.

III-4.　*If BAC is an angle whose sides do not lie in the same line, and if A′ and B′ are two distinct points, then there are two and only two distinct rays, A′C′ and A′C″, such that angle B′A′C′ is congruent to angle BAC and angle B′A′C″ is congruent to angle BAC; moreover, if D′ is any point on the ray A′C′ and D″ is any point on the ray A′C″, then the segment D′D″ intersects the line determined by A′ and B′.*

III-5. *Every angle is congruent to itself.*

III-6. *If two sides and the included angle of one triangle are congruent, respectively, to two sides and the included angle of another triangle, then each of the remaining angles of the first triangle is congruent to the corresponding angle of the second triangle.*

Group IV: Postulate of Parallels

IV-1. (Playfair's postulate) *Through a given point A not on a given line* m *there passes at most one line which does not intersect* m.

Group V: Postulates of Continuity

V-1. (Postulate of Archimedes) *If* A, B, C, D *are four distinct points, then there is, on the ray* AB, *a finite set of distinct points* $A_1, A_2, \ldots, A_n$ *such that* (1) *each of the segments* $AA_1, A_1A_2, A_2A_3, \ldots, A_{n-1}A_n$ *is congruent to the segment* CD, *and* (2) B *is between* A *and* A_n.

V-2. (Postulate of completeness) *The points of a line constitute a system of points such that no new points can be assigned to the line without causing the line to violate at least one of the eight postulates* I-1, I-2, II-1, II-2, II-3, III-1, III-2, V-1.

On these fourteen postulates rests the entire extensive subject of plane Euclidean geometry. To develop the geometry appreciably from these postulates is too long a task for us to undertake here, but we shall add a few words concerning the significance of some of the postulates.

The postulates of the first group define implicitly the idea expressed by the primitive term *on*, and they establish a connection between the two primitive entities, *points* and *lines*.

The postulates of the second group were first studied by Pasch, and they define implicitly the idea expressed by the primitive term *between*. In particular, they assure us of the existence of an infinite number of points on a line and that a line is not terminated at any point, and they guarantee that the order of points on a line is serial rather than cyclical. Postulate II-3 (Pasch's postulate) differs from the other postulates of the group, for, since it involves points not all on the same line, it gives information about the plane as a whole. The postulates of order are of historical interest inasmuch as Euclid completely failed to recognize any of them. It is this serious omission on Euclid's part that permits one, using only Euclid's list of assumptions, to derive paradoxes by applying apparently sound reasoning to misconceived figures.

The postulates of the third group define implicitly the idea expressed by the primitive term *congruence* as applied to segments and to angles. These postulates are included in order to circumvent the necessity of dealing with the concept of motion. For example, it is interesting to note how, in Postulate III-6, Hilbert introduces the congruency of triangles without employing Euclid's method of superposition, still found in some high school textbooks.

The Playfair parallel postulate appears as the only postulate in Group IV; it is, of course, equivalent to Euclid's parallel postulate. Using the postulates of the first three groups one can prove that at least one line passes through the given point A and does not intersect the given line m.

The first postulate of the last group (the postulate of Archimedes) corresponds to the familiar process of estimating the distance from one point of a line to another by the use of a measuring stick; it guarantees that if we start at the one point and lay off toward the second point a succession of equal distances (equal to the length of the measuring stick) we will ultimately pass the second point. On this postulate can be made to depend the entire theory of measurement and, in particular, Euclid's theory of proportion. The final postulate (the postulate of completeness) is not required for the derivation of the theorems of Euclidean geometry, but it makes possible the establishment of a one-to-one correspondence between the points on any line and the set of all real numbers, and is necessary for the free use of the real number system in analytic, or coordinate, geometry. It can be shown that, in the presence of the other twelve postulates, these last two postulates are equivalent to the postulate of Dedekind, and therefore, if we wish, they can be replaced by this postulate. In other words, the postulates of Group V can be replaced by the following Alternate Group V.

Alternate Group V

Definitions   Consider a segment AB. Let us call one end point, say A, the *origin* of the segment, and the other end point, B, the *extremity* of the segment. Given two distinct points M and N of AB, we say that M *precedes* N (or N *follows* M) if M coincides with the origin A or lies between A and N. A segment AB, considered in this way, is called an *ordered segment*.

V'-1.   (Dedekind's postulate) *If the points of an ordered segment of origin* A *and extremity* B *are separated into two classes in such a way that*
   1.   *each point of* AB *belongs to one and only one of the classes,*
   2.   *the points* A *and* B *belong to different classes (which we shall respectively call the* first *class and the* second *class),*
   3.   *each point of the first class precedes each point of the second class, then there exists a point* C *on* AB *such that every point of* AB *that precedes* C *belongs to the first class and every point of* AB *that follows* C *belongs to the second class.*

Considerably more was accomplished by Hilbert in his *Grundlagen der Geometrie* than just the establishment of a satisfactory set of postulates for Euclidean geometry. In showing the logical consistency and partial independence of his postulates, Hilbert had to devise many interesting models, or interpretations, for different subsets of the postulates. This amounted to introducing various new systems of geometry and to creating a number of unusual algebras of segments. The significance of several important postulates and theorems in the development of Euclidean geometry is clearly shown in the work, and examples of various kinds of nontraditional geometries are illustrated. For example, to show the independence of the postulate of Archimedes from the other postulates of the treatment, an example of a non-Archimedean system is offered in which all the postulates except the postulate of Archimedes are shown to hold. There is also developed in this work a theory of proportion and a theory of areas that are independent of the postulate of Archimedes. These accompanying investigations by Hilbert virtually inaugurated the twentieth-century study of abstract geometry and successfully convinced many mathematicians of the hypothetico-deductive nature of mathematics. By implanting the postulational method in nearly all of mathematics since 1900, Hilbert's *Grundlagen der Geometrie* represents a definite landmark in the history of mathematical thought.

Other postulational treatments of Euclidean geometry followed Hilbert's effort. In 1904 the American mathematician Oswald Veblen (1880–1960) furnished a new postulate set in which he replaced the primitive notion of *betweenness*, as used by Peano and Hilbert, by a more pervasive primitive relation of *order*.[3] With this new primitive relation, the terms *line*, *plane*, *on*, and *congruent* can receive explicit definition, and thus the list of primitive terms be reduced to just two—namely, *point* and *order*. The feeling among mathematicians is that the smaller one can make the number of primitive terms in a postulational development, the more aesthetically pleasing is that postulational development—a principle that was emphasized by Peano. A second study of the foundations of Euclidean geometry was made by Veblen in 1911, in which his original treatment was slightly revised to accord with some ideas put forth by R. L. Moore.[4]

A very satisfying combination of the postulates of Hilbert and Veblen has been employed by Gilbert de B. Robinson.[5] Robinson's postulates are essentially Veblen's postulates of order combined with Hilbert's postulates of congruence and continuity. In some ways Robinson's postulate set offers improvements over the two previous sets, and many students may prefer the study of this system to that of either Hilbert or Veblen.

After the construction of Veblen's system of postulates for Euclidean geometry, the study became increasingly abstract and formal. An interesting example of this tendency may be found in a paper written by E. V. Huntington (1874–1952) in 1913.[6] In this paper, Huntington offers a treatment of three-dimensional Euclidean geometry based on *sphere* and *inclusion* (one sphere lying within another) as primitive terms. This unusual approach exemplifies the fact that it is possible to characterize Euclidean geometry by systems of postulates that are superficially very different from one another. Huntington did a great deal of work with postulate systems, and we meet his name again when we consider postulational developments in algebra and arithmetic.

An excellent and detailed abstract postulational examination of Euclidean geometry appeared in 1927 in a work by Henry George Forder.[7] Here we find many alternative postulate sets compared with one another. For example, Forder considers nine different parallel postulates, which vary in the strengths of their assumptions. By adopting a strong parallel postulate and by using Dedekind's

---

[3] O. Veblen [1].

[4] O. Veblen [2].

[5] G. de B. Robinson.

[6] E. V. Huntington [3]; also see G. de B. Robinson, Appendix, pp. 157–160.

[7] H. G. Forder [1].

postulate as a postulate of continuity, Forder gives a postulate set for Euclidean geometry based on only the two primitive terms *point* and *order*. He also gives an abstract treatment of a postulate set of Pieri's and based on the two primitive terms *point* and *congruence*.

A singularly elegant and concise postulate set for plane Euclidean geometry, based on the two primitive terms *point* and *distance*, was published by L. M. Blumenthal in 1961.[8] The postulate set characterizes the Euclidean plane as a special kind of metric space (a concept that is considered in Chapter 8) and is easily modified to give a postulate set for *n*-dimensional Euclidean geometry and for plane Lobachevskian geometry.

Whether or not it is wise to attempt a postulational treatment of Euclidean geometry at the high school level is a matter of pedagogical opinion. George Bruce Halsted made an unsuccessful effort in 1904, when he published an elementary geometry textbook based on Hilbert's postulate set.[9] More successful, and certainly worthy of examination, is an attempt made in 1940 by Professors George David Birkhoff and Ralph Beatley of Harvard University.[10] Here a teachable high school course in plane Euclidean geometry is evolved from five postulates based on an ability to measure line segments and angles. Although for pedagogical reasons certain subtler mathematical and logical details are either ignored or slurred over, the work does stem from a rigorous mathematical presentation that had been made earlier by Birkhoff.[11] Since about the middle of the twentieth century a number of authors and writing groups have attempted the task of producing textual materials for the high school geometry class wherein geometry is developed rigorously from a postulational basis. In these attempts usually either the Hilbert postulate set or the Birkhoff postulate set (sometimes slightly altered or augmented) is adopted.

4.3   Poincaré's Model and the Consistency of Lobachevskian Geometry

A satisfactory postulate set for plane Lobachevskian non-Euclidean geometry may be obtained from the Hilbert postulate set given in the last section by simply replacing the postulate of parallels (Postulate IV-1) by

> IV'-1. *Through a given point* A *not on a given line* m *there pass at least two lines which do not intersect line* m.

Developments of plane Lobachevskian geometry have been made using such a postulate set as a foundation.[12] Our purpose, in the present section, is to show that plane Lobachevskian geometry is as consistent as plane Euclidean geometry.

---

[8] L. M. Blumenthal [2].

[9] G. B. Halsted [1]. Certain logical criticisms of the text were met by Halsted in 1907 in a thoroughly revised second edition; this edition has been translated into French.

[10] G. D. Birkhoff and R. Beatley.

[11] G. D. Birkhoff [1].

[12] See, for example, G. Verriest.

To accomplish this we employ a model, or representation, of plane Lobachevskian geometry that was devised by Henri Poincaré. Our aim will be accomplished when we show that this model is a system of geometrical elements and relations that, when substituted for the primitive terms in the postulates for plane Lobachevskian geometry, convert these postulates into theorems in Euclidean geometry.

The Poincaré model that we shall employ may be described as follows. Let a fixed circle Σ (see Figure 4.1) be selected and called the *fundamental circle*. A point of the Lobachevskian plane is represented in the model by a point in the interior of Σ, and will hereafter be designated, by means of boldface type, as a **point**. A line of the Lobachevskian plane is represented in the model by the arc interior to Σ of any circle, or straight line, that cuts Σ perpendicularly, and will hereafter be designated as a **line**. Boldface type will be similarly employed whenever we wish to understand that the concept considered is a Lobachevskian, rather than a Euclidean, concept. The relationships of a **point on a line**, and of a **point between two points**, have, in the model, the obvious interpretations. To interpret suitably **congruence of segments** and **congruence of angles** two definitions are made. The (positive) **length** of a **segment** AB is defined as

$$\mathbf{AB} = \log{(AB, TS)},$$

where S and T are the points in which the circle or straight line containing the **segment** AB meets Σ, A lying between S and B, and (AB, TS) denotes the so-called *cross ratio* $(AT/BT)(BS/AS)$. And the **measure** of the **angle** between two intersecting **lines** is defined as the ordinary Euclidean measure of the angle between the two circles or straight lines on which the **lines** lie. Two **segments** are **congruent** if and only if they have equal **lengths**, and two **angles** are **congruent** if and only if they have equal **measures**.

We shall now show that, with the above interpretations, the Hilbert postulates for plane Lobachevskian geometry become theorems in plane Euclidean geometry.

We verify Postulate I-1 by proving that there is one and only one circle (or possibly straight line) orthogonal to a given circle Σ and passing through two given points A and B lying inside Σ.

Let O (see Figure 4.2) be the center of Σ, and let A' be the point on OA

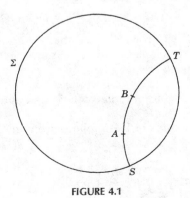

FIGURE 4.1

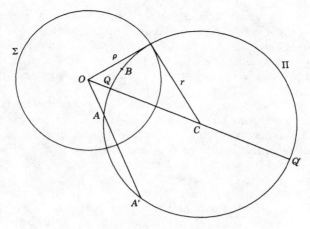

FIGURE 4.2

produced such that $OA \cdot OA' = \rho^2$, where ρ is the radius of Σ. We shall now show that the circle Π determined by the three points A, B, A' is orthogonal (that is, perpendicular) to Σ. (If these three points are collinear, circle Π will degenerate into a diametral line of Σ, which is certainly orthogonal to Σ.) Let C be the center and r the radius of Π, and let OC cut Π in Q and Q' as indicated in the figure. Then

$$\rho^2 = OA \cdot OA' = OQ \cdot OQ' = (OC - r)(OC + r) = OC^2 - r^2,$$

or

$$r^2 + \rho^2 = OC^2,$$

and circles Σ and Π cut each other orthogonally. Thus at least one circle passes through points A and B and cuts circle Σ orthogonally. To show that there is only one such circle, let Π now represent any circle through A and B and orthogonal to Σ, and let OA produced cut Π in point A'. Then, since

$$r^2 + \rho^2 = OC^2,$$

we have

$$\rho^2 = OC^2 - r^2 = (OC - r)(OC + r) = OQ \cdot OQ' = OA \cdot OA',$$

and A' is the point on OA produced such that $OA \cdot OA' = \rho^2$. Therefore the circle Π coincides with the circle considered earlier, and there is only one circle through A and B orthogonal to Σ.

The reader can easily verify that Postulates I-2, II-1, II-2, II-3, III-1, and III-2 are obviously satisfied by our interpretations of **point**, **line**, **on**, **between**, and **congruence of segments**.

The verification of Postulate III-3 follows from the fact that

$$\mathbf{AB} = \log(AB, TS)$$

$$= \log\left[\left(\frac{AT}{BT}\right)\left(\frac{BS}{AS}\right) \right]$$

$$= \log\left[\left(\frac{AT}{CT}\right) \left(\frac{CS}{AS}\right) \left(\frac{CT}{BT}\right) \left(\frac{BS}{CS}\right) \right]$$

$$= \log\left[\left(\frac{AT}{CT}\right) \left(\frac{CS}{AS}\right) \right] + \log\left[\left(\frac{CT}{BT}\right) \left(\frac{BS}{CS}\right) \right]$$

$$= \log(AC, TS) + \log(CB, TS)$$

$$= \mathbf{AC} + \mathbf{CB}.$$

Similarly, $\mathbf{A'B'} = \mathbf{A'C'} + \mathbf{C'B'}$, whence, since $\mathbf{A'C'} = \mathbf{AC}$ and $\mathbf{C'B'} = \mathbf{CB}$, we have $\mathbf{A'B'} = \mathbf{AB}$, and the **segments** $A'B'$ and AB are **congruent**.

Postulate III-4 becomes obvious once we show that there is a unique circle Π orthogonal to Σ, passing through a given point A' (other than O) within Σ, and tangent at A' to a ray $A'R$ emanating from A'. This demonstration is easily accomplished, for, by virtue of our work in connection with Postulate I-1, circle Π (see Figure 4.3) must pass through point A'' on OA' produced and such that $OA' \cdot OA'' = \rho^2$. But there is a unique circle passing through A' and A'' and tangent to $A'R$ at A'; its center lies at the intersection of the perpendicular to $A'R$ at A' and the perpendicular bisector of segment $A'A''$.

Postulates III-5 and V-1 are easily seen to be satisfied by our interpretations, and Postulate V-2 is not needed for a noncoordinate development of the subject. The only postulates still remaining to be verified are, then, Postulate III-6 and Postulate IV'-1. The task of verifying Postulate III-6 is not so easy, as it requires some knowledge of elementary college geometry and in particular of the transformation of inversion. We accordingly relegate this piece of verification to the problems at the end of the chapter, where the required theory from college geometry is developed.

We conclude, then, with a verification of the Lobachevskian parallel postulate, Postulate IV'-1. In Figure 4.4 let A be a given **point** and let m be a given **line** not passing through A. Let S and T be the points where the circle (or

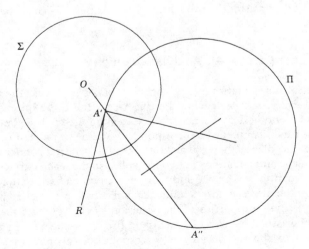

FIGURE 4.3

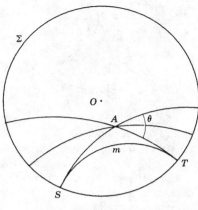

FIGURE 4.4

straight line) *m* cuts Σ. Now through *S* and *A* passes a unique circle (or straight line) cutting Σ orthogonally, and through *T* and *A* passes a unique circle (or straight line) cutting Σ orthogonally. These loci are tangent, at *S* and *T*, respectively, to circle (or straight line) *m*, and are distinct from one another. Clearly, any circle (or straight line) orthogonal to Σ, passing through *A*, and lying within the angle *θ* indicated in the figure, fails to cut **line** *m*.

We may now consider that the purpose of this section (to show that plane Lobachevskian geometry is as consistent as plane Euclidean geometry) has been accomplished. For, should there be any inconsistency in plane Lobachevskian geometry, there would have to be a corresponding inconsistency in the plane Euclidean geometry of the Poincaré model.

By replacing the fundamental circle Σ by a *fundamental sphere* and by considering spherical surfaces (and planes) cutting this fundamental sphere orthogonally, the Poincaré model can be extended to show that solid Lobachevskian geometry is consistent if solid Euclidean geometry is consistent.

4.4 Analytic Geometry

Every student of mathematics meets, early in his college work, the remarkable subject called *analytic geometry*, and he can hardly fail to be impressed by the powerful idea behind it. The essence of the idea as applied to the plane, it will be recalled, is the establishment of a correspondence between pairs of real numbers and points in the plane, thereby making possible a correspondence between curves in the plane and equations in two variables, so that each curve in the plane has a definite equation $f(x, y) = 0$ and each such equation has a definite curve, or set of points, in the plane. A correspondence is similarly established between the algebraic and analytic properties of the equation $f(x, y) = 0$ and the geometric properties of the associated curve. The task of proving a theorem in geometry is cleverly shifted to that of proving a corresponding theorem in algebra and analysis.

There is no unanimity of opinion among historians of mathematics concerning who invented analytic geometry, nor even concerning what age should be credited with the invention. Much of this difference of opinion is caused by a lack of agreement regarding just what constitutes analytic geometry. Those who favor antiquity as the era of the invention point out the well-known fact that the concept of fixing the position of a point by means of suitable coordinates was employed in the ancient world by the Egyptians and the Romans in surveying and by the Greeks in map making. If analytic geometry implies not only the use of coordinates but also the geometric interpretation of relations among coordinates, then particularly strong in the favor of the Greeks is the fact that Apollonius derived the bulk of his geometry of the conic sections from the geometrical equivalents of certain Cartesian equations of these curves, an idea that seems to have originated with Menaechmus about 350 B.C. Others claim that the invention of analytic geometry should be credited to Nicole Oresme, who was born in Normandy about 1323 and who died in 1382 after a career that carried him from a mathematics professorship to a bishopric. Oresme, in one of his mathematical tracts, anticipated another aspect of analytic geometry when he represented certain laws by graphing the dependent variable against the independent one, as the latter variable was permitted to take on small increments. Advocates for Oresme as the inventor of analytic geometry see in his work such accomplishments as the first explicit introduction of the equation of a straight line and the extension of some of the notions of the subject from two-dimensional space to three-, and even four-dimensional spaces.[13] A century after Oresme's tract was written, it enjoyed several printings and in this way may possibly have influenced later mathematicians. However, before analytic geometry could assume its present highly practical form, it had to await the development of algebraic symbolism, and accordingly it may be more nearly correct to agree with the majority of historians, who regard the decisive contributions made in the seventeenth century by the two French mathematicians, René Descartes (1596–1650) and Pierre de Fermat (1601?–1665), as the essential origin of at least the modern spirit of the subject. Not until after the great impetus given to the subject by these two men do we find analytic geometry in a form with which we are familiar.

In a history of mathematics a good deal of space would be devoted to Descartes and Fermat, for these men left very deep imprints on the subject. Also, in a history of mathematics, much would be said about the importance of analytic geometry, not only for the development of geometry and for the theory of curves and surfaces in particular, but as an indispensable force in the development of the calculus and as an influential power in molding our ideas of such far-reaching concepts as those of "function" and "dimension." In the present section, however, we restrict ourselves to the relations of analytic geometry with the axiomatic, or postulational, method; these relations prove to be both interesting and enlightening, and they may clear away some of the confusion regarding just what constitutes analytic geometry.

In the usual college textbooks of plane analytic geometry, use is made of

---

[13] See P. Duhem.

various definitions and theorems of Euclidean geometry in order to set up a coordinate system in which a point is represented by an ordered pair of real numbers. Similar use of some definitions and theorems of Euclidean geometry is made in proving that a straight line is represented by a linear equation in two variables, in deriving a formula for the distance between two points, and so on. In such a development, where the geometry itself is employed in setting up a coordinate machinery for the study of the geometry, a student might well wonder whether the methods of plane analytic geometry are adequate for the establishment of any theorem that is implied by the postulates of plane Euclidean geometry. The answer to this question will be in the affirmative, and the blemish of assuming tidbits of plane Euclidean geometry in order to set up the coordinate machinery will be removed, if we can obtain, for example, an algebraic interpretation of the Hilbert postulate set for plane geometry given in Section 4.2. We now do this.

Our task is to assign algebraic meanings to the five primitive terms, *point*, *line*, *on*, *between*, and *congruent* (as applied to segments and to angles), that convert each of Hilbert's postulates into a theorem of algebra. To accomplish this we make the following assignments:

1. By a *point* we mean any ordered pair of real numbers; these numbers will be called the *coordinates* of the point.

2. By a *line* we mean any equation in the two variables x and y of the form $ax + by + c = 0$, where a, b, c are real numbers and a and b are not both 0. Two such linear equations in x and y, whose left members differ at most by only a constant nonzero factor, will represent the same line, and either equation will be called *an equation* of the line.

3. We say a point is *on* a line if and only if the coordinates of the point satisfy an equation of the line.

4. We say the point (x, y) is *between* the points (x_1, y_1) and (x_2, y_2) if and only if there is a real number t, greater than 0 and less than 1, such that

$$x = (1 - t)x_1 + tx_2 \quad \text{and} \quad y = (1 - t)y_1 + ty_2.$$

5. We say the segment, denoted by $(x_1, y_1)(x_2, y_2)$, is *congruent* to the segment, denoted by $(x_3, y_3)(x_4, y_4)$, if and only if

$$(x_2 - x_1)^2 + (y_2 - y_1)^2 = (x_4 - x_3)^2 + (y_4 - y_3)^2.$$

We shall call the left member of this equation the square of the *distance* between the points (x_1, y_1) and (x_2, y_2).

6. Finally, we say the angle, denoted by $(x_2, y_2)(x_1, y_1)(x_3, y_3)$, is *congruent* to the angle, denoted by $(x'_2, y'_2)(x'_1, y'_1)(x'_3, y'_3)$, if and only if

$$\frac{(x_2 - x_1)(x_3 - x_1) + (y_2 - y_1)(y_3 - y_1)}{\sqrt{(x_2 - x_1)^2 + (y_2 - y_1)^2}\sqrt{(x_3 - x_1)^2 + (y_3 - y_1)^2}}$$
$$= \frac{(x'_2 - x'_1)(x'_3 - x'_1) + (y'_2 - y'_1)(y'_3 - y'_1)}{\sqrt{(x'_2 - x'_1)^2 + (y'_2 - y'_1)^2}\sqrt{(x'_3 - x'_1)^2 + (y'_3 - y'_1)^2}}.$$

We call the left-hand member of this equation the *cosine* of the angle $(x_2, y_2)(x_1, y_1)(x_3, y_3)$.

It should be kept in mind that the names *point, line, distance,* and *cosine* are applied to certain algebraic entities, and the words *on, between,* and *congruent* express certain relations among our algebraic entities. With the acceptance of these algebraic interpretations of the primitive terms of Hilbert's postulate set we may now convert each postulate of the set into an algebraic statement. It can be shown, by the methods of algebra alone, that each postulate becomes a theorem of algebra. Let us illustrate how this is done.

To verify Postulate I-1 in our interpretation we must show that there is a unique (to within a constant factor) equation in x and y of the form

$$ax + by + c = 0,$$

where a, b, c are real and a and b are not both 0, which is satisfied by two distinct pairs of real values (x_1, y_1) and (x_2, y_2) of the variables x and y. Substitution of (x_1, y_1) and (x_2, y_2) for (x, y) in the equation

$$(y_2 - y_1)x - (x_2 - x_1)y + (x_2 y_1 - x_1 y_2) = 0 \qquad (1)$$

shows that this equation is satisfied by the pairs of values (x_1, y_1) and (x_2, y_2). Also, since (x_1, y_1) and (x_2, y_2) are distinct pairs of numbers, either $y_2 - y_1 \neq 0$ or $x_2 - x_1 \neq 0$; moreover, since x_1, y_1, x_2, y_2 are real, the coefficients in the equation are real. We have thus shown that an equation of the desired form is satisfied by the pairs of values (x_1, y_1) and (x_2, y_2). We must now show that, to within a constant nonzero factor, this is the only equation of the desired form satisfied by the distinct pairs of values (x_1, y_1) and (x_2, y_2). To this end, suppose (x_1, y_1) and (x_2, y_2) satisfy the equation

$$ax + by + c = 0, \qquad (2)$$

where a, b, c are real and a and b are not both 0. Then we have

$$ax_1 + by_1 + c = 0 \quad \text{and} \quad ax_2 + by_2 + c = 0, \qquad (3)$$

or, by subtraction,

$$a(x_2 - x_1) + b(y_2 - y_1) = 0. \qquad (4)$$

Now suppose $a \neq 0$. Then equations (3) become

$$x_1 + By_1 + C = 0 \quad \text{and} \quad x_2 + By_2 + C = 0, \qquad (5)$$

where $B = b/a$ and $C = c/a$. Since $a \neq 0$, we cannot have $y_2 - y_1 = 0$, for otherwise (4) would reduce to $a(x_2 - x_1) = 0$ or $x_2 - x_1 = 0$, a situation which is impossible. Solving equations (5) simultaneously for B and C, we now find

$$B = -\frac{x_2 - x_1}{y_2 - y_1}, \qquad C = \frac{x_2 y_1 - x_1 y_2}{y_2 - y_1},$$

which results in

$$a : b : c = (y_2 - y_1) : -(x_2 - x_1) : (x_2 y_1 - x_1 y_2),$$

and, except for a possible constant nonzero factor, equation (2) becomes our equation (1). A similar argument can be carried out if, instead of supposing $a \neq 0$, we suppose $b \neq 0$. Thus Postulate I-1 becomes, in our interpretation, a theorem of algebra.

Postulate I-2 is easily shown to be, in our interpretation, a theorem of algebra.

To verify Postulate II-1 in our interpretation, we note that no matter what real value t takes on, the point (x, y) given by

$$x = (1 - t)x_1 + tx_2, \qquad y = (1 - t)y_1 + ty_2, \tag{6}$$

satisfies the equation (1) of the line determined by the points (x_1, y_1) and (x_2, y_2). This shows that the points (x, y), (x_1, y_1), (x_2, y_2) are on the same line. Further, by setting $s = 1 - t$, we find

$$x = (1 - s)x_2 + sx_1, \qquad y = (1 - s)y_2 + sy_1. \tag{7}$$

Now suppose t is between 0 and 1. Then s is also between 0 and 1, and equations (7) guarantee that point (x, y) is between points (x_2, y_2) and (x_1, y_1). Solving equations (6) for x_2 and y_2, we obtain

$$x_2 = \left(\frac{1}{t}\right)x + \left(1 - \frac{1}{t}\right)x_1, \qquad y_2 = \left(\frac{1}{t}\right)y + \left(1 - \frac{1}{t}\right)y_1,$$

or

$$x_2 = rx + (1 - r)x_1, \qquad y_2 = ry + (1 - r)y_1,$$

where r is not between 0 and 1. Hence point (x_2, y_2) is not between points (x, y) and (x_1, y_1). We may similarly show that point (x_1, y_1) is not between points (x, y) and (x_2, y_2).

Postulate II-2 is readily verified by taking, in equations (6), $t = 1/2$ and $t = 2$ in turn. In the first case we obtain a point (x, y) between the points (x_1, y_1) and (x_2, y_2). In the second case we obtain a point (x, y) such that point (x_2, y_2) is between points (x_1, y_1) and (x, y).

Perhaps enough has been done to indicate how the interpretations of the various postulates of Hilbert's set can be shown, by algebraic methods alone, to be theorems of algebra. Some of the proofs are very simple, such as that for Postulate III-2, whereas others are more involved, such as that for Postulate II-3. We might point out that if, in the parallel postulate (Postulate IV-1), A is the point (x_1, y_1) and m is the line $ax + by + c = 0$, then the unique line through A that does not intersect m is given by

$$ax + by - (ax_1 + by_1) = 0.$$

Also, it is easy to show that our interpretations of *point*, *line*, and *on* imply a one-to-one correspondence between the points of any line and the set of all real numbers, and this, as we have pointed out, is essentially Postulate V-2. Actually, there are points having coordinates belonging to a suitable subset of the set of all real numbers that satisfy all the rest of Hilbert's postulates, and such a subset of the real numbers would be sufficient to give an algebraic interpretation of these postulates. However, since we desire to deal with *all* real numbers, Postulate V-2 is implied. To carry out the program of verifying the interpretations of all the postulates of the Hilbert set would occupy too much space, and so we drop the matter at this point and invite the interested reader to supply the remaining proofs.[14]

Let us now complete this section on analytic geometry by stating briefly what has been accomplished by setting up our algebraic model, or interpretation, of the Hilbert postulate set for plane Euclidean geometry.

In the first place, we have shown that any theorem of plane Euclidean geometry can be established by the methods of algebra alone, without the use of any propositions of geometry. The explanation is as follows: The assignments 1 through 6 furnish algebraic correspondents for each of the six primitive geometric terms. Since any other geometric term can be defined by means of the six primitive ones, the algebraic correspondent of such another geometric term can be expressed by means of the algebraic correspondents of the primitive terms. It follows that every geometric term has its algebraic correspondent or representation. This enables one to translate every geometric theorem into a corresponding algebraic statement, and step by step to translate the geometric proof of a geometric theorem into an algebraic proof of the corresponding algebraic statement (though, as often happens, a shorter algebraic proof may be found in some other way).

In the second place, we have perhaps clarified the meaning of *analytic geometry*. Plane (Cartesian) analytic geometry is not to be confused with the algebraic model that we have set up of plane Euclidean geometry. It is, rather, the translation process whereby we use the algebraic model in order to solve problems and establish theorems in our geometry. Analytic geometry, then, is not a *branch* of geometry but rather a *method* of geometry. It amounts to thinking through a geometric study by employing a different set of mental images in a different language, and then translating back into the original geometric form the results of this thinking. Such is the powerful idea behind analytic geometry; it permits us to employ a field of study in which we may be more facile in order to obtain information about another quite distinct field of study in which we may be less facile. A person not too familiar with a foreign language resorts to just such a method; being asked a question in a foreign language, he is likely first to translate the question into his native language, think out an answer in his native language, and then translate back into the foreign language the answer that he has constructed. Since so many students are considerably more able as algebraists than as geometers, analytic geometry has been described as the "royal road in geometry" that Euclid thought did not exist.[15]

In the third place, we have shown that plane Euclidean geometry and, therefore, along with it, plane Lobachevskian geometry are consistent if the algebra of real numbers is consistent. The consistency of other branches of mathematics can similarly be shown to depend on the consistency of this algebra. It would seem important, then, for us to take up a study of the real number system. In a later chapter we do this, and we there consider a postulational

---

[14] See, for example, L. P. Eisenhart, Appendix to Chapter I, pp. 279–292; also C. J. Keyser [1], Lecture VI, pp. 86–103.

[15] When asked by King Ptolemy for a short cut to geometric knowledge, Euclid is said to have replied, "There is no royal road in geometry."

treatment of the algebra of real numbers and also take up the question of the consistency of this algebra.

The idea behind analytic geometry—namely, that of using a model of a system in order to establish consequences of the system—can be strikingly illustrated by the Poincaré model of plane Lobachevskian geometry considered in the previous section. Since the Poincaré model is constructed within Euclidean geometry, and since we are so much more familiar with Euclidean rather than Lobachevskian geometry, it is very conceivable that some theorems of plane Lobachevskian geometry may be more readily established by proving the corresponding Euclidean theorems in the Poincaré model than by proving the theorems themselves directly from a postulate set for Lobachevskian geometry.[16]

The device of using a model of a system in order to establish theorems of the system cannot be employed for all systems; the given system must be *categorical*. This concept is not introduced until Section 6.4, but both (plane) Euclidean and (plane) Lobachevskian geometry are categorical and hence susceptible of application of the device of models.

Just as there is an algebraic model of plane Euclidean geometry, and hence the method of *analytic*, or *algebraic, geometry*, so also is there a geometric model of the algebra of real numbers and hence a method of *geometric algebra*. The ancient Greeks established a good deal of elementary algebra, including the solution of quadratic equations, by essentially using a geometric model in this way.[17]

4.5   Projective Geometry and the Principle of Duality

After the discovery of non-Euclidean geometry, a handicap to certain additional developments in the field of geometry was the general belief that mathematics is basically concerned with magnitude. This point of view is somewhat surprising, for even in antiquity there were geometrical studies that considered only relationships of mutual position and order of figures and that did not depend at all on relations involving equalities or inequalities of magnitudes. In the early seventeenth century a small group of French mathematicians actually considerably developed the nonmetrical aspects of geometry. The motivator among these French mathematicians was Gérard Desargues, an engineer and architect, who was born in Lyons in 1593 and who died in the same city about 1662. Perhaps influenced by the growing needs of artists and architects for a satisfactory theory of perspective, Desargues published, in Paris in 1639, a remarkably original treatise on the conic sections that exploited the nonmetrical idea of projection. But this work was so neglected by most other mathematicians of the time that it was soon forgotten and all copies of the publication disappeared. Two centuries later, when the French geometer Michel Chasles (1793–1880) wrote a history of

---

[16] See, for example, H. S. Carslaw, Chapter 8; also H. Eves and V. E. Hoggatt.

[17] See, for example, H. Eves [1], Sections 3.6 and 3.7.

geometry, there was no way to estimate the value of Desargues's work. Six years later, however, in 1845, Chasles happened on a manuscript copy of Desargues's treatise, made by one of Desargues's few followers, and since that time the work has been regarded as one of the classics in the early development of projective geometry.

There are several reasons for the initial neglect of Desargues's little volume. It was overshadowed by the more supple analytic geometry introduced by Descartes two years earlier. Geometers were generally either developing this new powerful tool or trying to apply infinitesimals to geometry. Also, Desargues unfortunately adopted a style of writing that was so eccentric that it beclouded his work and discouraged others from attempting properly to evaluate his accomplishments.

The reintroduction of nonmetrical considerations into geometry did not occur until the nineteenth century, with the work in this field by Jean Victor Poncelet (1788–1867). As a Russian prisoner of war, taken during Napoleon's retreat from Moscow, and with no books at hand, Poncelet planned his great work on projective geometry, which, after his release and return to France, he published in Paris in 1822.[18] This work gave tremendous impetus to the study of the subject and inaugurated the so-called great period in the history of projective geometry. There followed into the field a host of French and German mathematicians, among whom were Gergonne, Brianchon, Chasles, Plücker, Steiner, and Staudt—great names in the history of geometry and in the history of projective geometry in particular. It is interesting to note that at approximately the same time that non-Euclidean geometry was being developed it was first generally appreciated that geometry could no longer be viewed exclusively as a study of quantity.

The work of Desargues and of Poncelet, and their followers, led geometers to classify geometric properties into two categories, the *metric properties*, in which the measure of distances and of angles intervenes, and the *descriptive properties*, in which such measure is unessential. The Pythagorean theorem, that *the square on the hypotenuse of a right triangle is equal to the sum of the squares on the two legs*, is a metric property. As an example of a descriptive property we might mention the remarkable "mystic hexagram" theorem of Blaise Pascal (1623–1662), which was inspired by the work of Desargues: *If a hexagon be inscribed in a conic, then the points of intersection of the three pairs of opposite sides are collinear, and, conversely, if the points of intersection of the three pairs of opposite sides of a hexagon are collinear, then the hexagon is inscribed in a conic* (see Figure 4.5).

The distinction between the two types of geometric properties, at least in the case of plane figures, becomes clearer when viewed from the fact that the descriptive properties are unaltered when the figure is subjected to a central projection, whereas the metric properties may no longer hold when the figure is so projected. Thus, under a central projection from one plane to another, a right triangle does not necessarily remain a right triangle, and so the Pythagorean relation does not necessarily hold for the projected figure; the Pythagorean theorem is a metric theorem. In the case of Pascal's theorem, however, a hexagon

---

[18] J. V. Poncelet.

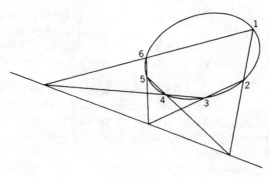

FIGURE 4.5

inscribed in a conic projects into a hexagon inscribed in a conic and collinear points project into collinear points, and hence the theorem itself is preserved; Pascal's theorem is a descriptive theorem.

Many descriptive properties present themselves in the seeming form of metric properties. For example, if we have four points A, B, T, S on a line, the value of the compound ratio, $(AT/BT)/(AS/BS)$, would seem to be a metric relation. The value of this compound ratio, however, can be shown to be unaltered when the line containing the four points A, B, T, S is centrally projected (as in Figure 4.6) into another line and the four points A', B', T', S'. In other words, although the lengths of the various corresponding segments on the two lines are not necessarily equal to one another, the two compound ratios

$$\frac{A'T'/B'T'}{A'S'/B'S'}, \qquad \frac{AT/BT}{AS/BS},$$

are equal in value. The value of the compound ratio $(AT/BT)/(AS/BS)$ is the so-called *cross ratio* (AB, TS) of the four collinear points A, B, T, S (in this order), which we considered in Section 4.3 in connection with our study of the Poincaré model. The cross ratio of four collinear points is a descriptive property of those points.

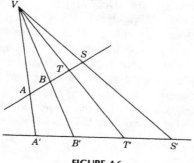

FIGURE 4.6

The study of the descriptive properties of geometric figures is known as *projective geometry*.

In plane projective geometry, a remarkable symmetry exists between points and lines (when ideal elements at infinity are utilized), such that if in a descriptive theorem about *points* and *lines* we should interchange these two words, and perhaps smooth out the language, we obtain another descriptive theorem about *lines* and *points*. As a simple example, consider the following two propositions related in this way:

> *Any two distinct points determine one and only one line on which they both lie.*

> *Any two distinct lines determine one and only one point through which they both pass.*

This symmetry, which results in the pairing of the theorems of plane projective geometry, is a principle of far-reaching consequence known as the *principle of duality* and was perhaps first clearly stated by Joseph-Diez Gergonne (1771–1859). Once the principle of duality is established, then the proof of one theorem of a dual pair carries with it the proof of the other.

Let us dualize Pascal's theorem. We first restate Pascal's theorem in a form that is perhaps more easily dualized.

> *The six points 1, 2, 3, 4, 5, 6 lie on a conic if and only if the points determined by the three pairs of lines (12), (45); (23), (56); (34), (61) lie on a line.*

Dualizing this, we obtain

> *The six lines 1, 2, 3, 4, 5, 6 lie on (that is, are tangent to) a conic if and only if the lines determined by the three pairs of points (12), (45); (23), (56); (34), (61) lie on (that is, intersect in) a point.*

Using less artificial language, Pascal's theorem and its dual may now be stated as

> *A hexagon is inscribed in a conic if and only if the points of intersection of the three pairs of opposite sides are collinear.*

> *A hexagon is circumscribed about a conic if and only if the lines joining the three pairs of opposite vertices are concurrent.*

The dual of Pascal's theorem was first published by C. J. Brianchon (1785–1864) when a student at the Ecole Polytechnique in Paris in 1806, nearly 200 years after Pascal had stated his theorem. Brianchon's theorem is illustrated in Figure 4.7.

Poncelet maintained that the principle of duality is a consequence of the *theory of poles and polars*, which, though known earlier, was first developed by him in a systematic manner. Toward explaining this theory, let Γ (see Figure 4.8) be a given conic. To each point P in the plane of Γ can be made to correspond a line p in the plane of Γ in the following way. If P is on Γ, take the tangent to Γ at P as the associated line p. If P is not on Γ, draw through P two lines cutting Γ in the points A and A', B and B', respectively. Then the line p is the line determined by the points of intersection of AB and $A'B'$, AB' and $A'B$. It can be shown that to a point P there corresponds, in this way, a unique

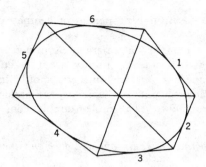

FIGURE 4.7

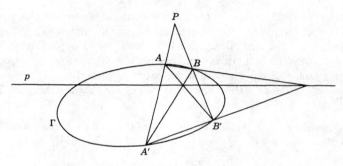

FIGURE 4.8

line p, and conversely, to a line p there corresponds a unique point P. The point P is called the *pole* of the line p, and the line p is called the *polar* of the point P.

This correspondence between the points P and the lines p of the plane of Γ possesses some important properties. First of all, the correspondence is invariant under projection. Further, if p is the polar of a point P, the polar of any point P' on p passes through the point P. It follows that if some points are collinear, then the polars of these points are concurrent.

Now imagine that we have established a projective property about a plane figure F composed of points and lines. Let F' be the figure that one obtains by replacing the points and lines of F by their polars and poles with respect to a given conic Γ lying in the plane of F. Then one will obtain from the projective property of figure F a corresponding projective property of figure F', in which the roles played by the words *point* and *line* have been interchanged. The two properties will be the duals of each other.

The preceding account of plane projective geometry and of its principle of duality is that of the early days of the subject's development. Much is wanting in this early treatment, however. For example, the definition of plane projective geometry, involving, as it does, the distinction between metric properties and descriptive properties, lacks a desired precision and makes projective geometry in some ways subservient to Euclidean geometry. Also, the principle of duality seems to need a more general handling than is given by Poncelet's treatment, and there are objections to the use of ideal elements at infinity. These faults have been remedied in more recent times, and we now have at least two quite acceptable and self-contained approaches to the subject. One of these approaches is based on analytic geometry and is indicated in the problems at the end of the chapter. For the present we consider briefly a very satisfying approach based on the modern postulational method. Our procedure will be to put down a set of postulates and then simply define plane projective geometry as the deductive consequences of the system of postulates. If these postulates, which will be statements about certain primitive entities called *points* and *lines*, can be shown to imply the dual of each postulate, then the principle of duality for plane projective geometry follows, because the dual of any theorem that has been derived from the postulate set may be established by simply dualizing the steps in the proof of the original theorem.

There are many postulate sets for plane projective geometry that we could give, but the following,[19] in which each postulate is self-dual, most neatly carries out our present purpose. Here *point*, *line*, and *on* are primitive terms.

I.   *There is one and only one line on every two distinct points, and one and only one point on every two distinct lines.*

II.   *There exist two points and two lines such that each of the points is on just one of the lines and each of the lines is on just one of the points.*

III.   *There exist two points and two lines, the points not on the lines, such that the point on the two lines is on the line on the two points.*

The fact that each postulate is self-dual saves us the task of deriving as theorems the system of dual statements and automatically vouchsafes us the principle of duality for plane projective geometry. The idea of a principle of duality and of a self-dual postulate set is not peculiar to projective geometry; we shall meet this situation again, in our study of Boolean algebra. The aspect of a principle of duality that is of concern to us is that the proofs of two dual theorems are precisely the same when proper changes are made in those terms that possess duals. In other words, the deductive organizations set up in proving dual theorems are essentially the same if we ignore the superficial difference in language. Thus, if the dual terms are replaced by undefined symbols, it is immediately apparent that the original system of theorems and the dual system are identical. In the concept of duality, therefore, we find the idea reinforced that mathematical studies pertain essentially to the treatment of symbolic structures, and the study of abstract relations between symbols is perhaps more important to mathematics than the consideration of possible meanings that may be associated with the symbols.

It is a significant fact that the postulates for plane projective geometry given above do not require that the number of points in our geometry be infinite. In fact, consider the situation where there are seven *points*—namely, the letters A, B, C, D, E, F, G—and seven *lines*—namely, the letter trios (AFB), (BDC), (CEA), (AGD), (BGE), (CGF), (DEF). Postulate I is easily verified by considering each pair of *points* and each pair of *lines*. Postulate II is verified by the two *points B, C* and the two *lines* (AFB), (AEC). Postulate III is verified by the two *points B, C* and the two *lines* (AGD), (DEF).

A projective geometry containing only a finite number of distinct points is called a *finite projective geometry*. Our model establishes the existence of a finite plane projective geometry. It also shows that our postulates are consistent with one another, for an inconsistency among the postulates would render a concrete model of them impossible. Again, the model furnishes additional evidence for the argument that a mathematical system deduced from a set of postulates is independent of any meanings that may be associated with the primitive terms employed in the postulates; the same system may be capable of a great variety of interpretations. Finally, it shows that if we wish to assure that our plane projective geometry resemble more closely the earlier historical concept of the subject we shall have to introduce further postulates. In fact, by gradually adding appropriate further postulates and introducing certain modifications it is possible to convert the postulate set for plane projective geometry into a postulate set for plane Euclidean geometry, passing through postulate sets for various intermediate geometries on the way. Such a passage from projective geometry to Euclidean geometry is an interesting but lengthy procedure that we must forego.[20] We also forego commenting here on solid projective geometry and *its* principle of duality and pass now to a chapter devoted to the important subject of algebraic structure.

PROBLEMS

4.1.1 **(a)** Consider the following definitions taken from an elementary geometry text: (1) The *diagonals* of a quadrilateral are the two straight line segments joining the two pairs of opposite vertices of the quadrilateral. (2) *Parallel lines* are straight lines that lie in the same plane and that never meet, however far they are extended in either direction. (3) A *parallelogram* is a quadrilateral having its opposite sides parallel.

Now, without using any of the italicized words above, restate the proposition, "The diagonals of a parallelogram bisect each other."

(b) In elementary algebra, if n represents a positive integer and if k represents any real number, we define k^n as a symbol to represent the product $(k)(k) \cdots (k)$, in which k appears as a factor n times. Rewrite, without using exponents, the following expression in which a, b, c represent real numbers:

$$[(a + b)^5 (a - c)^3]^7.$$

(c) By means of appropriate explicit definitions reduce the following sentence to one containing not more than seven words: "The movable seats with four legs and a back were restored to a sound state by the person who takes care of the building."

These exercises illustrate the convenience of explicit definitions.

4.1.2 Give the customary definitions of the following mathematical symbols and illustrate their convenience:

(a) $\displaystyle\sum_{n=1}^{\infty} a_n$,

---

[20] For such a program see, for example, G. de B. Robinson, or B. E. Meserve [2].

(b) $n!$, n a positive integer,

(c) $\binom{m}{n}$, m and n positive integers with $m \geq n$.

4.1.3 Trace the following words through a standard dictionary until a circular chain has been established: **(a)** dead, **(b)** noisy, **(c)** line (in the mathematical sense).

4.1.4 Answer the following questions intuitively, and then check your answers by calculation:

(a) A car travels from P to Q at the rate of 40 miles per hour and then returns from Q to P at the rate of 60 miles per hour. What is the average rate for the round trip?

(b) A can do a job in 4 days, and B can do it in 6 days. How long will it take A and B together to do the job?

(c) A man sells half of his apples at 3 for 68 cents and then sells the other half at 5 for 68 cents. At what rate should he sell all of his apples in order to make the same income?

(d) If a ball of yarn 4 inches in diameter costs 20 cents, how much should you pay for a ball of yarn 6 inches in diameter?

(e) Two jobs have the same starting salary of $18,000 per year and the same maximum salary of $36,000 per year. One job offers an annual raise of $2,400 and the other offers a semiannual raise of $600. Which is the better paying job?

(f) Each bacterium in a certain culture divides into two bacteria once a minute. If there are 20 million bacteria present at the end of one hour, when were there exactly 10 million bacteria present?

(g) Is a salary of 1 cent for the first half month, 2 cents for the second half month, 4 cents for the third half month, 8 cents for the fourth half month, and so on until the year is used up, a good or a poor total salary for the year?

(h) A clock strikes six in 5 seconds. How long will it take to strike twelve?

(i) A bottle and a cork together cost $1.10. If the bottle costs a dollar more than the cork, how much does the cork cost?

(j) Suppose that in one glass there is a certain quantity of a liquid A, and in a second glass an equal quantity of another liquid B. A spoonful of liquid A is taken from the first glass and put into the second glass, then a spoonful of the mixture from the second glass is put back into the first glass. Is there now more or less liquid A in the second glass than there is liquid B in the first glass?

(k) Suppose that a large sheet of paper one one-thousandth of an inch thick is torn in half and the two pieces put together, one on top of the other. These are then torn in half, and the four pieces put together in a pile. If this process of tearing in half and piling is done 50 times, will the final pile of paper be more or less than a mile high?

(l) Is a discount of 15 percent on the selling price of an article the same as a discount of 10 percent on the selling price followed by a discount of 5 percent on the reduced price?

(m) Four-fourths exceeds three-fourths by what fractional part?

(n) A boy wants the arithmetical average of his eight grades. He averages the first four grades, then the last four grades, and then finds the average of these averages. Is this correct?

These exercises illustrate the danger of using intuition in arriving at results.

4.1.5 In each of the following, is the given conclusion a valid deduction from the given pair of premises?

(a) If today is Saturday, then tomorrow will be Sunday.
But tomorrow will be Sunday.
Therefore, today is Saturday.

(b) Germans are heavy drinkers.
Germans are Europeans.
Therefore, Europeans are heavy drinkers.

(c) If *a* is *b*, then *c* is *d*.
But *c* is *d*.
Therefore, *a* is *b*.

(d) All *a*'s are *b*'s.
All *a*'s are *c*'s.
Therefore, all *c*'s are *b*'s.

These exercises illustrate how a person may allow the meanings that are associated with words or expressions to dominate logical analysis. There is a greater tendency to go wrong in (a) and (b) than in (c) and (d), which are symbolic counterparts of (a) and (b).

4.1.6 We repeat here the postulate set of Problem 3.5.8.

P1. *Every abba is a collection of dabbas.*

P2. *There exist at least two dabbas.*

P3. *If* p *and* q *are two dabbas, then there exists one and only one abba containing both* p *and* q.

P4. *If* L *is an abba, then there exists a dabba not in* L.

P5. *If* L *is an abba, and* p *is a dabba not in* L, *then there exists one and only one abba containing* p *and not containing any dabba that is in* L.

(a) What are the primitive terms in this postulate set?

(b) Deduce the following theorems from this postulate set:

1. *Every dabba is contained in at least two abbas.*

2. *Every abba contains at least two dabbas.*

3. *There exist at least four distinct dabbas.*

4. *There exist at least six distinct abbas.*

(c) Define a *kurple* as any three dabbas not contained in the same abba. What is a kurple if we interpret *dabba* as "point" and *abba* as "straight line"?

4.1.7 (a) Establish the following consequences of the first five postulates of Pieri's postulate set for Euclidean geometry (as given in Section 4.1). (1) If *C* and *D* are two distinct points of the straight line *AB*, then *A* and *B* are points of the straight line *CD*. (2) If three points are on a straight line, then the three points corresponding to them in any motion are also on a straight line.

(b) Pieri defines a sphere as follows: "If *A* and *B* are two distinct points, then the aggregate of all points *P* such that for each *P* there exists a motion that leaves *A* fixed but makes *P* correspond to *B* is called *the sphere of center* A *passing through* B." Establish the following consequences of this definition and Pieri's first five postulates: (1) A sphere transforms into a sphere in every motion. (2) A motion that leaves the center of a sphere fixed transforms the sphere into itself. (3) If two spheres with centers *A* and *B* have only one point *C* in common, then the three points *A*, *B*, *C* lie on a line.

(c) Try to formulate, in terms of motion, a suitable definition of *perpendicularity*.

4.2.1 Prove the theorem that follows Postulate III-3 in Hilbert's postulate set for plane Euclidean geometry.

4.2.2 Deduce the following theorems from Hilbert's postulate set for plane Euclidean geometry:

(a) If point B is between points A and D, and point C is between B and D, then C is between A and D.

(b) There is no limit to the number of distinct points between two given distinct points.

(c) If neither of two distinct lines, a and b, intersects a third line c, then a and b do not intersect.

(d) If two sides and the included angle of one triangle are congruent, respectively, to two sides and the included angle of another triangle, then all the parts (angles and sides) of the first triangle are congruent to the corresponding parts of the second triangle.

(e) If two angles and the included side of one triangle are congruent, respectively, to two angles and the included side of another triangle, then all the parts (angles and sides) of the first triangle are congruent to the corresponding parts of the second triangle.

4.2.3 Try to deduce the following proposition from Hilbert's postulate set for plane Euclidean geometry: Given any four points on a line, it is always possible to denote them by letters A, B, C, D in such a way that B is between A and C and also between A and D, and that C is between A and D and also between B and D.

This proposition was included as a postulate in the first edition of Hilbert's work, but was later proved by E. H. Moore to be a consequence of Hilbert's other postulates; that is, in the technical jargon of mathematics, the proposition given above is not independent of the remaining propositions listed by Hilbert as postulates.[21]

4.2.4 (a) Consider the configuration formed by the positive x-axis and a directed circular arc C of radius r radiating from the origin O and such that the arc C is convex when viewed from its right-hand side. We shall call such a configuration a (special kind of) *horn angle* and denote it by h. Let T be the directed tangent to C at O and designate the positive (counterclockwise) angle from the positive x-axis around to T by θ. We shall compare two such horn angles h and h' in the following way. If $\theta = \theta'$ and $r = r'$, then we say $h = h'$; if $\theta > \theta'$, we say $h > h'$; if $\theta = \theta'$ but $r < r'$, then again we say $h > h'$. We further say $h' = nh$, where n is a positive integer, if and only if $\theta' = n\theta$ and $r' = r/n$. Show that our horn angles now form a non-Archimedean system of entities; that is, show that there exist horn angles h and h' such that $nh < h'$ for every positive integer n.

(b) Show that a horn angle for which $\theta = 0$ may be trisected with straightedge and compasses.

(c) Consider pairs of power series of the forms $y = a_1 x + a_2 x^2 + a_3 x^3 + \cdots$ and $y' = a'_1 x + a'_2 x^2 + a'_3 x^3 + \cdots$, where the coefficients are real numbers. We shall compare two such power series as follows. We say $y = y'$ if and only if $a_i = a'_i$ for all i; we say $y > y'$ if and only if there exists some positive integer k such that $a_1 = a'_1$, $a_2 = a'_2, \ldots, a_{k-1} = a'_{k-1}$, $a_k > a'_k$. We further say that $y' = ny$, where n is a positive integer, if and only if $a'_i = na_i$ for all i. Show that the set of all such power series form, under the above definitions, a non-Archimedean system of entities. [This can be interpreted as a generalization of part (a), where the circular arcs C have been replaced by analytic curves passing through O.]

(d) Consider a set of entities of which a typical member M is composed of the

[21] See E. H. Moore [1].

segment $-a \leqq x \leqq 0 (a \geqq 0)$, and the isolated points $x = 1, 2, \ldots, k$. Devise a method of comparing such entities, and define nM, where n is a positive integer, such that the entities form a non-Archimedean system.

(e) Let $z = a + ib$ and $z' = a' + ib'$ $(a, b, a', b'$ real and $i = \sqrt{-1})$ be two complex numbers. Set $z = z'$ if and only if $a = a'$ and $b = b'$. If $a > a'$, set $z > z'$; if $a = a'$ but $b > b'$, again set $z > z'$. Define $nz = na + i(nb)$. Show that the complex numbers, under the above definitions, form a non-Archimedean system of entities.

(f) Show that the set of all coplanar vectors radiating from a point O can be formed into a non-Archimedean system.

4.2.5 It is pointed out in Section 4.2 that Huntington has given a postulate set for Euclidean geometry of space in which *sphere* is taken as a primitive element and *inclusion* as a primitive relation. Try to formulate, in terms of these primitive terms, suitable definitions for *point*, *segment*, *ray*, and *line*.

4.3.1 In Figure 4.1 show that $\lim_{B \to T} \mathbf{AB} = \infty$.

4.3.2 Let Π be a fixed circle of center O and radius ρ, and let P be any point in the plane of Π. Then the point P' on the ray OP such that $OP \cdot OP' = \rho^2$ is called the *inverse* of P with respect to circle Π.[22] Circle Π is called the *circle of inversion*, point O the *center of inversion*, and ρ^2 the *power of inversion*. There is set up a one-to-one correspondence between the points of the plane of Π; to every point there is a corresponding point. The points of a curve C will invert into the points of a curve C', called the *inverse* of C. Establish the following theorems concerning this inversion transformation.

(a) If P' is the inverse of P, then P is the inverse of P'.

(b) A point inside the circle of inversion inverts into a point outside the circle of inversion; a point outside the circle of inversion inverts into a point inside the circle of inversion; a point on the circle of inversion inverts into itself.

(c) A straight line through the center of inversion inverts into itself.

(d) A circle orthogonal to the circle of inversion inverts into itself.

(e) If P, P' and Q, Q' are any two pairs of inverse points that do not lie on the same diameter of the circle of inversion, then these four points lie on a circle and $\angle OPQ = \angle OQ'P'$ and $\angle OQP = \angle OP'Q'$.

(f) A straight line that does not pass through the center of inversion inverts into a circle passing through the center of inversion, and conversely.

(g) A circle that does not pass through the center of inversion inverts into a circle that does not pass through the center of inversion.

(h) Any circle through a pair of inverse points P and P' cuts the circle of inversion orthogonally.

(i) A given circle may be inverted into itself by the use of any given exterior point as center of inversion.

4.3.3 Prove that in an inversion the angle between two intersecting curves is equal to the corresponding angle between the two inverse curves.

A transformation that preserves angles between curves is called a *conformal* transformation. More briefly stated, then, the theorem concerned says that "inversion is a conformal transformation."

4.3.4 (a) Show that the cross ratio of four points on a circle is invariant under inversion.

---

[22] We add to the plane a single ideal point at infinity. If $P \equiv O$, then P' is taken as this ideal point.

(b) Let P, Q be two **points** in the Poincaré model, and let P', Q' be their inverses with respect to some **line**. Show that $\mathbf{P'Q'} = \mathbf{PQ}$.

4.3.5 Let ABC be any triangle in the Poincaré model such that C is not at the center of Σ. Show that there is a suitable **line** m such that inversion with respect to circle m will carry Σ into itself and **triangle** ABC into a **congruent triangle** $A'B'C'$, where C' is at the center O of Σ and $A'C'$, $B'C'$ are straight line segments.

4.3.6 Show that if, in **triangles** ABC, $A'B'C'$ of the Poincaré model, $\mathbf{AB} = \mathbf{A'B'}$, $\mathbf{AC} = \mathbf{A'C'}$, $\measuredangle \mathbf{A} = \measuredangle \mathbf{A'}$, then $\measuredangle \mathbf{B} = \measuredangle \mathbf{B'}$. (This verifies Postulate III-6 in the Poincaré model.)

4.4.1 Prove the following theorems connected with the algebraic model of Section 4.4:
   (a) If a point satisfies an equation of a line, then it satisfies every equation of the line.
   (b) The segment $(x_1, y_1)(x_2, y_2)$ is congruent to the segment $(x_2, y_2)(x_1, y_1)$.
   (c) If t is between 0 and 1, then $1/t$ is not between 0 and 1.
   (d) If (x_1, y_1) and (x_2, y_2) are distinct points, and if
$$x = (1 - t)x_1 + tx_2, \qquad y = (1 - t)y_1 + ty_2,$$
   and also
$$x = (1 - r)x_1 + rx_2, \qquad y = (1 - r)y_1 + ry_2,$$
   then $r = t$.
   (e) If in (4) of Section 4.4 we take $t = 1/2$, then the distance between (x_1, y_1) and (x, y) is equal to the distance between (x, y) and (x_2, y_2). [The point (x, y) is called the *midpoint* between (x_1, y_1) and (x_2, y_2).]
   (f) The lines $ax + by + c_1 = 0$ and $ax + by + c_2 = 0$, where $c_1 \neq c_2$, do not intersect.

4.4.2 There is a very useful inequality, often referred to as *Schwarz's inequality*, which asserts that if a_1, a_2, b_1, b_2 are any four real numbers, then
$$(a_1 b_1 + a_2 b_2)^2 \leqq (a_1^2 + a_2^2)(b_1^2 + b_2^2).$$

Using Schwarz's inequality, establish the following inequalities connected with the algebraic model of Section 4.4:
   (a) The cosine of angle $(x_2, y_2)(x_1, y_1)(x_3, y_3)$ is numerically less than or equal to 1.
   (b) For any three points $P_1 : (x_1, y_1)$, $P_2 : (x_2, y_2)$, $P_3 : (x_3, y_3)$,
$$\text{distance } P_1 P_3 \leqq \text{distance } P_1 P_2 + \text{distance } P_2 P_3.$$
   (This is referred to as the *triangle inequality*.)

4.4.3 Verify Postulate III-1 in the algebraic model of Section 4.4.

4.4.4 Verify Postulate III-2 in the algebraic model of Section 4.4.

4.4.5 Verify Postulate III-3 in the algebraic model of Section 4.4.

4.4.6 Verify Postulate III-5 in the algebraic model of Section 4.4.

4.4.7 Verify Postulate III-6 in the algebraic model of Section 4.4.

4.4.8 Verify Postulate V-1 in the algebraic model of Section 4.4.

4.4.9 Verify Postulates II-3 and III-4 in the algebraic model of Section 4.4. These two postulates are somewhat more difficult to establish than are the other postulates. See, for example, L. P. Eisenhart, *Coordinate Geometry* (Boston: Ginn, 1939), appendix to chap. 1.

4.4.10 Prove, by using the Poincaré model, that in plane Lobachevskian geometry the sum of the angles of a triangle is always less than two right angles.

4.5.1 The consequences of Pascal's "mystic hexagram" theorem are very numerous and attractive, and an almost unbelievable amount of research has been expended on the configuration. There are sixty possible ways of forming a hexagon from six points on a conic, and, by Pascal's theorem, to each hexagon corresponds a *Pascal line*. These sixty Pascal lines pass three by three through twenty points, called *Steiner points*, which in turn lie four by four on fifteen lines, called *Plücker lines*. The Pascal lines also concur three by three in another set of points, called *Kirkman points*, of which there are sixty. Corresponding to each Steiner point there are three Kirkman points such that all four lie on a line, called a *Cayley line*. There are twenty of these Cayley lines, and they pass four by four through fifteen points, called *Salmon points*. There are many further such extensions and properties of the configuration, and they are all descriptive properties.

In this set of problems we shall consider a few of the many corollaries of the "mystic hexagram" theorem that can be obtained by making some of the six points coincide with one another. For simplicity we shall number the points 1, 2, 3, 4, 5, 6. Then Pascal's theorem says that the intersections of the pairs of lines 12, 45; 23, 56; 34, 61 are collinear if and only if the six points lie on a conic.

(a) If a pentagon 12345 is inscribed in a conic, show that the pairs of lines 12, 45; 23, 51; 34 and the tangent at 1, intersect in three collinear points.

(b) Given five points, draw at any one of them the tangent to the conic determined by the five points.

(c) Given four points of a conic and the tangent at any one of them, construct further points on the conic.

(d) Show that the pairs of opposite sides of a quadrangle inscribed in a conic, together with the pairs of tangents at opposite vertices, intersect in four collinear points.

(e) Show that if a triangle is inscribed in a conic, then the tangents at the vertices intersect the opposite sides in three collinear points.

(f) Given three points on a conic and the tangents at two of them, construct the tangent at the third.

4.5.2 Prove the following sequence of theorems:

(a) If the vertex V of a triangle VAB is joined to any point P on the line AB, then

$$\frac{AP}{PB} = \frac{VA \sin AVP}{BV \sin PVB}.$$

(b) If A, B, T, S are four points on a line m, and if V is a fifth point not on m, then

$$\frac{AT/BT}{AS/BS} = \frac{\sin AVT/\sin BVT}{\sin AVS/\sin BVS}.$$

(c) The cross ratio of four collinear points is invariant under projection.

4.5.3 **(a)** Dualize 4.5.1(a).

(b) Given five lines, find on any one of them the point of contact of the conic touching the five lines.

(c) Given four tangents to a conic and the point of contact of any one of them, construct further tangents to the conic.

(d) Dualize 4.5.1(d).

(e) Dualize 4.5.1(e).

(f) Given three tangents to a conic and the points of contact of two of them, construct the point of contact of the third.

(g) Dualize the following theorem, known as *Desargues's two-triangle theorem*: "If two triangles (in the same plane) are so situated that lines joining pairs of corresponding vertices are concurrent, then the points of intersection of pairs of corresponding sides are collinear, and conversely." Actually, the theorem holds whether the two triangles are in the same plane or not.

4.5.4 **(a)** If m is a given line in a given plane π, and O is a given center of projection (not on π), show how to find a plane π' such that the projection of m onto π' will be the line at infinity on π'. (The operation of selecting a suitable center of projection O and plane of projection π' so that a given line on a given plane shall project into the line at infinity on π' is called "projecting the given line to infinity." Many descriptive theorems concerning a given figure may be proved more readily in a specialized figure obtained by projecting a particular line of the given figure to infinity. This device is to be used in the following parts.)

(b) Let UP, UQ, UR be three concurrent coplanar lines, cut by two lines OX and OY in P_1, Q_1, R_1 and P_2, Q_2, R_2, respectively. Prove that the intersections of $Q_1 R_2$ and $Q_2 R_1$, $R_1 P_2$ and $R_2 P_1$, $P_1 Q_2$ and $P_2 Q_1$ are collinear.

(c) Prove that if $A_1 B_1 C_1$ and $A_2 B_2 C_2$ are two coplanar triangles such that $B_1 C_1$ and $B_2 C_2$ meet in L, $C_1 A_1$ and $C_2 A_2$ meet in M, $A_1 B_1$, and $A_2 B_2$ meet in N, where L, M, N are collinear, then $A_1 A_2$, $B_1 B_2$, $C_1 C_2$ are concurrent. [This is the converse part of the statement of Desargues's two-triangle theorem as given in Problem 4.5.3(g).]

(d) Show that by parallel projection (a projection where the center of projection is at infinity) an ellipse may always be projected into a circle.

(e) In 1678 the Italian Giovanni Ceva (*ca.* 1647–1736) published a work containing the following theorem now known by his name: *The three lines that join three points* L, M, N *on the sides* BC, CA, AB *of a triangle* ABC *to the opposite vertices are concurrent if and only if*

$$\left(\frac{AN}{NB}\right)\left(\frac{BL}{LC}\right)\left(\frac{CM}{MA}\right) = +1.$$

Using Ceva's theorem, prove that the lines joining the vertices of a triangle to the opposite points of contact of the inscribed circle are concurrent. Then, by means of part (d), prove that the lines joining the vertices of a triangle to the opposite points of contact of an inscribed ellipse are concurrent.

4.5.5 Using the pole and polar theory as sketched in Section 4.5, show how to draw, with straightedge alone, the tangents to a given ellipse from a given external point.

4.5.6 Consider the following set of postulates:

P1. *There exist a point and a line such that the point is not on the line.*

P2. *Every line contains at least three distinct points.*

P3. *Any two distinct points are on just one line.*

P4. *Any two distinct lines contain at least one common point.*

(a) Show that the postulate set implies the dual of each postulate of the set.

(b) Show that the postulate set is equivalent to the postulate set for plane projective geometry given in Section 4.5.

(c) Since there are plane geometries satisfying the above postulates but in which Desargues's theorem does not hold (such geometries are called *non-Desarguesian*), the following postulate is often added.

P5. *If two triangles have the joins of corresponding vertices concurrent, then the intersections of corresponding sides are collinear.*

Deduce the dual of this postulate.

4.5.7 If, in a trigonometric equation, each trigonometric function that appears is replaced by its cofunction, the new equation obtained is called the *dual* of the original equation. Establish the following *principle of duality* of trigonometry: If a trigonometric equation involving a single angle is an identity, then its dual is also an identity.

4.5.8 Look up, in a text on solid geometry, the definition of *polar triangles* on a sphere. Can this concept be used to establish a *principle of duality* in the geometry of spherical triangles on a given sphere?

4.5.9 An interesting development in coordinate systems was inaugurated by Julius Plücker (1801–1868) in 1829, when he noted that our fundamental element need not be the point but can be any geometric entity. Thus if we choose the straight line as our fundamental element, we might locate any straight line not passing through the origin of a given rectangular Cartesian frame of reference by recording, say, the x and y intercepts of the given line. Plücker actually chose the negative reciprocals of these intercepts as the location numbers of the line and considerably exploited the analytic geometry of these so-called *line coordinates*. A point now, instead of having coordinates, possesses a linear equation—namely, the equation satisfied by the coordinates of all the lines passing through the point. The double interpretation of a pair of coordinates as either point coordinates or line coordinates and of a linear equation as either the equation of a line or the equation of a point furnishes the basis of Plücker's analytical proof of the principle of duality of plane projective geometry.

(a) Find the Plücker coordinates of the lines whose Cartesian equations are $5x + 3y - 6 = 0$ and $ax + by + 1 = 0$. Write the Cartesian equation of the line having Plücker coordinates $(1, 3)$.

(b) Show that the Plücker coordinates u, v of all lines passing through the point with Cartesian coordinates $(2, 3)$ satisfy the linear equation $2u + 3v + 1 = 0$. This equation is taken as the Plücker equation of the point $(2, 3)$. What are the Cartesian coordinates of the points whose Plücker equations are $5u + 3v - 6 = 0$ and $au + bv + 1 = 0$? Write the Plücker equation of the point having Cartesian coordinates $(1, 3)$.

ALGEBRAIC STRUCTURE

As remarked at the beginning of the last chapter, another factor, besides the discovery of non-Euclidean geometry, greatly influenced the development of the axiomatic method and hence the character of much of modern mathematical research. This other factor was the recognition, first by British mathematicians in the earlier half of the nineteenth century, of the existence of structure in algebra.

Let us clarify what we mean by the phrase, *algebraic structure*. In studying the ordinary arithmetic of the positive integers, one encounters two operations called *addition* and *multiplication*. The notion of an operation, which is fundamental in mathematics, stems from the more general idea of function. A formal definition may be phrased as follows:

Definition An *operation* on a set S of elements is a *rule* that assigns to each ordered subset of n elements of S a uniquely defined element of the same set S; according to whether $n = 1, 2, 3, \ldots, n, \ldots$, the operation is said to be *unary, binary, ternary, ..., n-ary, ...*

We see, then, three criteria for a rule to be called an n-ary operation on a set S. First, the rule must define a result for *every* ordered subset of n elements of S. Thus division is not a binary operation on the set of all real numbers, since $\frac{3}{0}$, for example, is not defined. Second, the result must always be *unique*. Finding a solution, then, to the equation $x^2 = a^2$ is not a unary operation on the set of real numbers a, since we can have either $x = a$ or $x = -a$. Third, the result must always be an element of the *given* set S. Thus finding $\sqrt{a}$ is not a unary operation on the set of all real numbers a, since $\sqrt{a}$ might not be real as in the case of $\sqrt{-1}$.

An example of a unary operation performed on the set of positive integers is that of squaring, whereby to each positive integer p is assigned the positive integer p^2. As other examples of unary operations performed on the set of positive integers we might mention that of taking the successor and that of taking the factorial, whereby to each positive integer p are assigned, respectively, the positive integers $p + 1$ and $(1)(2)(3) \cdots (p)$.

Ordinary addition and multiplication performed on the set of positive integers are binary operations; to each ordered pair of positive integers a and b are assigned unique positive integers c and d, called, respectively, the *sum* of a and b and the *product* of a and b, and denoted by the symbols

$$c = a + b, \qquad d = a \times b.$$

These two binary operations of addition and multiplication performed on the set of positive integers possess certain basic properties. For example, if a, b, c denote arbitrary positive integers, we have

1. $a + b = b + a$, the so-called *commutative law for addition.*

2. $a \times b = b \times a$, the *commutative law for multiplication.*

3. $(a + b) + c = a + (b + c)$, the *associative law for addition.*

4. $(a \times b) \times c = a \times (b \times c)$, the *associative law for multiplication.*

5. $a \times (b + c) = (a \times b) + (a \times c)$, the *(left) distributive law for multiplication over addition.*

In the early nineteenth century, algebra was considered simply symbolized arithmetic.[1] In other words, instead of working with specific numbers, as in arithmetic, in algebra letters are used that represent these numbers. The above five properties, then, are statements that always hold in the algebra of positive integers. But since the statements are symbolic, it is conceivable that they might be applicable to some set of elements other than the positive integers, provided we supply appropriate definitions for the two binary operations involved. This is indeed the case, as the following examples amply illustrate. If in each case we denote the set of elements by S, it is an easy matter to verify that the elements, under the two given binary operations of $+$ and $\times$, satisfy all five of the above basic properties. In each example, equality is employed in the sense of identity.

EXAMPLES

(a) Let S be the set of all *even* positive integers, and let $+$ and $\times$ denote the usual addition and multiplication.

(b) Let S be the set of all rational numbers (integers and fractions, positive, negative, and zero), and let $+$ and $\times$ denote the usual addition and multiplication.

(c) Let S be the set of all real numbers, and let $+$ and $\times$ denote the usual addition and multiplication.

(d) Let S be the set of all real numbers of the form $m + n\sqrt{2}$, where m and n are integers, and let $+$ and $\times$ denote the usual addition and multiplication.

---

[1] This is still the view of algebra as taught in many high schools and frequently in the freshman year at college.

(e)   Let S be the set of *Gaussian integers* (complex numbers $m + in$, where m and n are ordinary integers and $i = \sqrt{-1}$), and let $+$ and $\times$ denote the usual addition and multiplication of complex numbers.

(f)   Let S be the set of all ordered pairs (m, n) of integers, and let (a, b) $+ (c, d) = (a + c, b + d)$ and $(a, b) \times (c, d) = (ac, bd)$.

(g)   Let S be the set of all ordered pairs (m, n) of integers, and let (a, b) $+ (c, d) = (a + c, b + d)$ and $(a, b) \times (c, d) = (ac - bd, ad + bc)$.

(h)   Let S be the set of all real polynomials in the real variable x, and let $+$ and $\times$ denote the ordinary addition and multiplication of polynomials.

(i)   Let S be the set of all real-valued continuous functions of the variable x defined on the closed interval $0 \leq x \leq 1$, and let $+$ and $\times$ denote ordinary addition and multiplication.

(j)   Let S be the set consisting of just two distinct elements m and n, where we define

$$m + m = m, \qquad\qquad m \times m = m,$$

$$m + n = n + m = n, \qquad m \times n = n \times m = m,$$

$$n + n = m, \qquad\qquad n \times n = n.$$

(k)   Let S be the set of all sets of points in a plane. If a and b are two members of S, define $a + b$ to be the set of points belonging to either set a or set b (or to both a and b), and define $a \times b$ to be the set of points belonging to both set a and set b. As a special member of S we introduce an ideal set, the *null set*, which has no points in it.

In view of these examples, which can easily be extended in number, it is apparent that the five basic properties of positive integers listed above may also be regarded as properties of many other entirely different systems of elements. The consequences of the above five properties constitute an algebra applicable to the positive integers, but it is also evident that the consequences of the five properties also constitute an algebra applicable to many other systems. That is to say, there is a common *algebraic structure* (the five basic properties and their consequences) attached to many different systems. The five basic properties may be regarded as postulates for a particular type of algebraic structure, and any theorem formally implied by these postulates would be applicable to each of the examples given above or to any other interpretation satisfying the five basic properties. Considered from this view, then, algebra is severed from its tie to arithmetic, and an algebra becomes a purely formal hypothetico-deductive study.

The earliest glimmerings of the above modern view of algebra appeared about 1830 in England, with the work of George Peacock (1791–1858). Peacock was one of the first to study seriously the fundamental principles of algebra.[2] He made a distinction between what he called "arithmetical algebra" and "symbolical algebra." The former was regarded by Peacock as the study that results from the use of symbols to denote ordinary positive decimal numbers, together with

---

[2] G. Peacock [1] and [2].

signs for the operations, like addition and subtraction, to which these numbers may be subjected. Now, in "arithmetical algebra," certain operations are limited in their applicability. For example, in a subtraction, $a - b$, we must have a greater than b. Peacock's "symbolical algebra," on the other hand, adopts the operations of "arithmetical algebra" but ignores their restrictions. Thus subtraction in "symbolical algebra" differs from the same operation in "arithmetical algebra" in that it is to be regarded as always applicable. The justification of this extension of the rules of "arithmetical algebra" to "symbolical algebra" was called, by Peacock, the *principle of the permanence of equivalent forms*. Peacock's "symbolical algebra" is a universal "arithmetical algebra" whose operations are determined by those of "arithmetical algebra," so far as the two algebras proceed in common, and by the principle of the permanence of equivalent forms in all other cases. The principle of permanence of equivalent forms was regarded as a powerful concept in mathematics, and it played a historical role in such matters as the early development of the arithmetic of the complex number system and the extension of the laws of exponents from positive integral exponents to exponents of a more general kind. In the theory of exponents, for example, if a is a positive rational number and n is a positive integer, then a^n is, by definition, the product of n a's. From this definition it readily follows that, for any two positive integers m and n, $a^m a^n = a^{m+n}$. By the principle of permanence of equivalent forms, Peacock affirmed that in "symbolical algebra," $a^m a^n = a^{m+n}$, no matter what might be the nature of the base a or of the exponents m and n. The hazy principle of permanence of equivalent forms has today been scrapped, but we are still often guided, when attempting to extend a definition, to formulate the more general definition in such a way that most properties of the old definition will be preserved.

British contemporaries of Peacock advanced his studies and pushed the notion of algebra closer to the modern concept of the subject. Thus Duncan Farquharson Gregory (1813–1844) published a paper in 1840[3] in which the commutative and distributive laws in algebra were clearly brought out. Further advances in an understanding of the foundations of algebra were made by Augustus De Morgan (1806–1871),[4] another member of the British school of algebraists. In the somewhat groping work of the British school one can trace the emergence of the idea of algebraic structure and the preparation for the postulational program in the development of algebra. Soon the ideas of the British school spread to continental Europe where, in 1867 they were considered with great thoroughness by the German historian of mathematics, Hermann Hankel (1839–1873).[5] But even before Hankel's treatment appeared, the Irish mathematician William Rowan Hamilton (1805–1865) and the German mathematician Hermann Günther Grassmann (1809–1877) had published results which were of a far-reaching character, results that led to the liberation of algebra in much the same way that the discoveries of Lobachevsky and Bolyai led to the liberation of geometry and that opened the flood gates of modern abstract

---

[3] D. F. Gregory.

[4] A. De Morgan.

[5] H. Hankel.

algebra. This remarkable work of Hamilton and Grassmann will be considered in the next section.

Before passing to the liberation of algebra by Hamilton and Grassmann, it might be well to consider a modern set of postulates for common algebra. This structure is today technically called an *ordered field*.

A *field* is a set S of elements, along with two binary operations on S, here denoted by $\oplus$ and $\otimes$ so that we will not necessarily think of them as ordinary addition and multiplication, satisfying the following postulates.[6] For the present, equality is used in the sense of identity; thus $a = b$ means a and b are the same element.

P1. *If* a *and* b *are in* S, *then* $a \oplus b = b \oplus a$.

P2. *If* a *and* b *are in* S, *then* $a \otimes b = b \otimes a$.

P3. *If* a, b, c *are in* S, *then* $(a \oplus b) \oplus c = a \oplus (b \oplus c)$.

P4. *If* a, b, c *are in* S, *then* $(a \otimes b) \otimes c = a \otimes (b \otimes c)$.

P5. *If* a, b, c *are in* S, *then* $a \otimes (b \oplus c) = (a \otimes b) \oplus (a \otimes c)$ *and*
$(b \oplus c) \otimes a = (b \otimes a) \oplus (c \otimes a)$.

P6. S *contains an element* z (zero) *such that for any element* a *of* S,
$a \oplus z = a$.

P7. S *contains an element* u (unity), *different from* z, *such that for any element* a *of* S, $a \otimes u = a$.

P8. *For each element* a *in* S *there exists an element* ā *in* S *such that*
$a \oplus \bar{a} = z$.

P9. *If* a, b, c *are in* S, $c \neq z$, *and* $c \otimes a = c \otimes b$ *or* $a \otimes c = b \otimes c$, *then*
$a = b$. (These are the *cancellation laws* for the operation $\otimes$.)

P10. *For each element* $a \neq z$ *in* S, *there exists an element* a^{-1} *in* S *such that*
$a \otimes a^{-1} = u$.

If, in addition to the above ten postulates, the following two postulates hold, the field is called an *ordered field*.

P11. *There exists a subset* P, *not containing* z, *of the set* S *such that if* $a \neq z$
then one and only one of a *and* ā *is in* P.

P12. *If* a *and* b *are in* P, *then* $a \oplus b$ *and* $a \otimes b$ *are in* P.

Definition 1　The elements of P are known as the *positive* elements of S; all other nonzero elements of S are known as the *negative* elements of S.

Definition 2　If a and b are elements of S and if $a \oplus \bar{b}$ is positive, then we write $a \ominus b$ and $b \oslash a$.

The postulate set for a field has been made somewhat redundant in order that it may serve a future purpose. For example, in view of Postulate P2, only one of the distributive laws in Postulate P5 needed to be given. Also it can be shown (see Appendix, Section A.3) that the entire Postulate P1 and the entire Postulate

---

[6] It is customary to place the symbol designating a binary operation between the two elements on which it operates.

P9 are redundant. Note that our original five basic properties of the positive integers appear as Postulates P1, P2, P3, P4, P5, above.

The twelve postulates for an ordered field constitute a postulate set for elementary algebra. The development of elementary algebra from these postulates would be an edifying, though tedious, undertaking. Such a development would be too abstract to employ at the high school level, where elementary algebra is first taught, but its understanding is essential to a comprehension of trends in modern mathematics.

It is an easy matter for the reader to show that any set of at least two complex numbers, in which the sum, difference, product, and quotient (with denominator different from 0) of any two numbers in the set are again in the set, constitutes a field under the binary operations $+$ and $\times$. Such a field is known as a *number field*. This explains why ordinary addition, subtraction, multiplication, and division, applied to numbers, are sometimes referred to as the *four field operations*. In the Appendix, Section A.2, we employ some elementary theory of number fields to prove the impossibility of trisecting an arbitrary angle, of duplicating a cube, and of squaring a circle with only the Euclidean straightedge and compasses.

Two very important examples of ordered fields are the set of all rational numbers and the set of all real numbers, together with the familiar operations of addition and multiplication performed on these numbers. We leave it to the reader to verify this by checking the twelve postulates listed above.

5.2   The Liberation of Algebra

Geometry, we have seen, remained shackled to Euclid's version of the subject until Lobachevsky and Bolyai, in 1829 and 1832, liberated it from its bonds by creating an equally consistent geometry in which one of Euclid's postulates fails to hold. With this accomplishment, a deep-rooted and centuries-old conviction that there could be only the one possible geometry was shattered, and the way was opened for the creation of many different systems of geometry. The same sort of story can be told of algebra. It seemed inconceivable, in the early nineteenth century, that an algebra could exist that was different from the common algebra of arithmetic. To attempt, for example, the construction of a consistent algebra in which the commutative law of multiplication fails to hold probably did not occur to anyone of the time, but had it occurred it surely would have been dropped as a purely ridiculous idea; after all, how could one possibly have a logical algebra in which $a \times b$ is not equal to $b \times a$? Such was the feeling about algebra when, in 1843, William Rowan Hamilton was forced, in efforts to work out an algebra suitable to the physicist for space analysis, to invent an algebra in which the commutative law of multiplication does not hold.[7] The radical step of abandoning the commutative law did not come easily to Hamilton; it dawned on him only after years of cogitation on a particular problem.

It would take us too far afield to go into the physical motivation that lay behind Hamilton's creation. Perhaps the best approach, for our purposes, is

---

[7] Hamilton later considerably expanded these ideas. See W. R. Hamilton [1] and [2].

through Hamilton's elegant treatment of complex numbers as real number pairs.[8] To the mathematicians of his time, a complex number was considered a number of the form $a + bi$, where a and b are real numbers and $i^2 = -1$. Addition and multiplication of complex numbers was accomplished by treating $a + bi$ as a linear polynomial in i, and by replacing i^2, wherever it might occur, by -1. One finds in this way, for addition,

$$(a + bi) + (c + di) = (a + c) + (b + d)i,$$

and, for multiplication,

$$(a + bi)(c + di) = ac + adi + bci + bdi^2 = (ac - bd) + (ad + bc)i.$$

If these results should be taken as definitions for the addition and multiplication of pairs of complex numbers, it is not difficult to show that addition and multiplication are commutative and associative, and that multiplication is distributive over addition.

Now, since a complex number $a + bi$ is completely determined by the two real numbers a and b, it occurred to Hamilton to represent the complex number by the ordered real number pair (a, b). He defined two such number pairs (a, b) and (c, d) to be equal if and only if $a = c$ and $b = d$. Addition and multiplication of such number pairs he defined (to agree with the above results) to be $(a, b) + (c, d) = (a + c, b + d)$ and $(a, b)(c, d) = (ac - bd, ad + bc)$. With these definitions it is easy to show that addition and multiplication of the ordered real number pairs are commutative and associative, and that multiplication is distributive over addition, if one assumes, of course, that these laws hold for the ordinary addition and multiplication of real numbers. In fact, assuming that the real numbers, under ordinary addition and multiplication, satisfy all the postulates of a field (as given in the last section), one can show that Hamilton's number pairs, under his definitions of addition and multiplication, also satisfy all the postulates of a field.

It is to be noted that the real number system is *embedded* in the complex number system. By this statement is meant that if each real number r is identified with the corresponding number pair $(r, 0)$, then the correspondence is preserved under addition and multiplication of complex numbers, for we have

$$(a, 0) + (b, 0) = (a + b, 0) \qquad \text{and} \qquad (a, 0)(b, 0) = (ab, 0).$$

In practice, a complex number of the form $(r, 0)$ can be replaced by its corresponding real number r.

To obtain the older form of a complex number from Hamilton's form, we note that any complex number (a, b) can be written as

$$(a, b) = (a, 0) + (0, b) = (a, 0) + (b, 0)(0, 1) = a + bi,$$

where $(0, 1)$ is represented by the symbol i, and $(a, 0)$ and $(b, 0)$ are identified with the real numbers a and b. Finally, we see that

$$i^2 = (0, 1)(0, 1) = (-1, 0) = -1.$$

---

[8] This treatment was communicated by Hamilton in 1833 and 1835 to the Royal Irish Academy and published in 1837.

The complex number system is a very convenient number system for the study of vectors and rotations in the plane.[9] Hamilton was attempting to devise an analogous system of numbers for the study of vectors and rotations in three-dimensional space. In his researches he was led to the consideration, not of ordered real number pairs (a, b) having the real numbers embedded within them, but of ordered real number quadruples (a, b, c, d) having both the real and the complex numbers embedded within them. In other words, defining two such quadruples (a, b, c, d) and (e, f, g, h) to be equal if and only if $a = e, b = f, c = g, d = h$, Hamilton found it necessary to define an addition and multiplication of ordered real number quadruples in such a way that, in particular, he would have

$$(a, 0, 0, 0) + (b, 0, 0, 0) = (a + b, 0, 0, 0),$$

$$(a, 0, 0, 0)(b, 0, 0, 0) = (ab, 0, 0, 0),$$

$$(a, b, 0, 0) + (c, d, 0, 0) = (a + c, b + d, 0, 0),$$

$$(a, b, 0, 0)(c, d, 0, 0) = (ac - bd, ad + bc, 0, 0).$$

Calling such ordered real number quadruples, (real) *quaternions*, Hamilton found he had to formulate the following definitions for addition and multiplication of his quaternions:

$$(a, b, c, d) + (e, f, g, h) = (a + e, b + f, c + g, d + h),$$

$$(a, b, c, d)(e, f, g, h) = (ae - bf - cg - dh, af + be + ch - dg,$$

$$ag + ce + df - bh, ah + bg + de - cf).$$

It can be shown, with these definitions, that the real numbers and the complex numbers are embedded among the quaternions, and that if we identify the quaternion $(m, 0, 0, 0)$ with the real number m, then

$$m(a, b, c, d) = (a, b, c, d)m = (ma, mb, mc, md).$$

It also can be shown that addition of quaternions is commutative and associative, and that multiplication of quaternions is associative and distributive over addition. But the commutative law for multiplication fails to hold. To see this, consider, in particular, the two quaternions $(0, 1, 0, 0)$ and $(0, 0, 1, 0)$. One finds that

$$(0, 1, 0, 0)(0, 0, 1, 0) = (0, 0, 0, 1),$$

while

$$(0, 0, 1, 0)(0, 1, 0, 0) = (0, 0, 0, -1) = -(0, 0, 0, 1);$$

that is, the commutative law for multiplication is broken. In fact, if we represent by the symbols $1, i, j, k$, respectively, the *quaternionic units* $(1, 0, 0, 0), (0, 1, 0, 0), (0, 0, 1, 0), (0, 0, 0, 1)$, we can verify that the following multiplication table

---

[9]This convenience results from the fact that when a complex number $z = a + bi$ is considered as representing the point Z having rectangular Cartesian coordinates (a, b), then the complex number z may also be regarded as representing the vector OZ, where O is the origin of coordinates.

prevails; that is, the desired product is found in the box common to the row headed by the first factor and the column headed by the second factor:

×	1	i	j	k
1	1	i	j	k
i	i	-1	k	$-j$
j	j	$-k$	-1	i
k	k	j	$-i$	-1

Hamilton has told the story that the idea of abandoning the commutative law of multiplication came to him in a flash, after fifteen years of fruitless meditation, while standing by a bridge near Dublin. He was so struck by the unorthodoxy of the idea that it is said he scratched the gist of the above multiplication table into the stone of the bridge.

We can write the quaternion (a, b, c, d) in the form $a + bi + cj + dk$. When two quaternions are written in this form they may be multiplied like polynomials in i, j, k, and then the resulting product put into the same form by means of the above multiplication table.

In the year 1844, Hermann Grassmann published the first edition of his remarkable *Ausdehnungslehre*,[10] in which were developed classes of algebras of much greater generality than Hamilton's quaternion algebra. Instead of considering just ordered sets of quadruples of real numbers, Grassmann considered ordered sets of n real numbers. To each such set $(x_1, x_2, \ldots, x_n)$ Grassmann associated a hypercomplex number of the form $x_1 e_1 + x_2 e_2 + \cdots + x_n e_n$, where $e_1, e_2, \ldots, e_n$ are the fundamental units of his algebra. Two such hypercomplex numbers are added and multiplied like polynomials in $e_1, e_2, \ldots, e_n$. The addition of two such numbers yields, then, a number of the same kind. To make the product of two such numbers a number of the same kind requires the construction of a multiplication table for the units $e_1, \ldots, e_n$ similar to Hamilton's multiplication table for his units $1, i, j, k$. Here one has considerable freedom, and different algebras can be created by making different multiplication tables. The multiplication table is governed by the desired application of the algebra and by the laws of common algebra one wishes to preserve.

We shall not go deeper here into either Hamilton's or Grassmann's work. By developing algebras satisfying laws different from those obeyed by common algebra, these men opened the way for the study of innumerable algebraic structures. By weakening or deleting various postulates of common algebra, or by replacing one or more of the postulates by others, which are consistent with

---

[10] H. G. Grassmann.

the remaining postulates, an enormous variety of systems can be studied. For example,[11] the deletion of Postulate P2 from the postulates for a field, as listed in the previous section, leads to an algebraic structure technically known as a *division ring*, or *sfield*.[12] The deletion of Postulates P2, P7, P10 leads to an *integral domain*; the deletion of Postulates P2, P9, P10 to a *ring with unity*; the deletion of Postulates P7, P9, P10 to a *commutative ring*; and the deletion of Postulates P2, P7, P9, P10 to a *ring* (with no qualifying phrase or adjective). In the next section, we shall consider a particularly basic and important algebraic structure, obtained in a similar way, known as a *group*. It is probably correct to say that mathematicians have studied well over 200 such algebraic structures. As the American algebraists Garrett Birkhoff and Saunders MacLane have written, "Modern algebra has exposed for the first time the full variety and richness of possible mathematical systems."[13]

Before closing this section, let us consider one more noncommutative algebra—the matric algebra devised by the English mathematician Arthur Cayley (1821–1895) in 1857.[14] Matrices arose with Cayley in connection with linear transformations of the type

$$x' = ax + by,$$

$$y' = cx + dy,$$

where a, b, c, d are real numbers, and which may be thought of as mapping the point (x, y) of a Cartesian plane into the point (x', y'). Clearly, the above transformation is completely determined by the four coefficients a, b, c, d, and so the transformation can be symbolized by the square array

$$\begin{bmatrix} a & b \\ c & d \end{bmatrix},$$

which we shall call a *(square) matrix (of order* 2). Since two transformations of the kind under consideration are identical if and only if they possess the same coefficients, we define two matrices

$$\begin{bmatrix} a & b \\ c & d \end{bmatrix} \quad \text{and} \quad \begin{bmatrix} e & f \\ g & h \end{bmatrix}$$

to be equal if and only if $a = e$, $b = f$, $c = g$, $d = h$. If the transformation given above is followed by the transformation

$$x'' = ex' + fy',$$

$$y'' = gx' + hy',$$

---

[11] The reader should be warned that there are slight variations in these definitions as given by different writers.

[12] It can be shown that quaternions constitute a division ring.

[13] Quoted by permission from G. Birkhoff and S. MacLane, p. 1.

[14] A. Cayley.

the result can be shown, by elementary algebra, to be the transformation

$$x'' = (ea + fc)x + (eb + fd)y,$$
$$y'' = (ga + hc)x + (gb + hd)y.$$

This motivates the following definition for the product of two matrices:

$$\begin{bmatrix} e & f \\ g & h \end{bmatrix} \begin{bmatrix} a & b \\ c & d \end{bmatrix} = \begin{bmatrix} ea + fc & eb + fd \\ ga + hc & gb + hd \end{bmatrix}.$$

Addition of matrices is defined by

$$\begin{bmatrix} a & b \\ c & d \end{bmatrix} + \begin{bmatrix} e & f \\ g & h \end{bmatrix} = \begin{bmatrix} a + e & b + f \\ c + g & d + h \end{bmatrix},$$

and, if m is any real number, we define

$$m \begin{bmatrix} a & b \\ c & d \end{bmatrix} = \begin{bmatrix} a & b \\ c & d \end{bmatrix} m = \begin{bmatrix} ma & mb \\ mc & md \end{bmatrix}.$$

In the resulting algebra of matrices, it may be shown that addition is both commutative and associative and that multiplication is associative and distributive over addition. But multiplication is not commutative, as is shown by the simple example:

$$\begin{bmatrix} 1 & 0 \\ 0 & 0 \end{bmatrix} \begin{bmatrix} 0 & 1 \\ 0 & 1 \end{bmatrix} = \begin{bmatrix} 0 & 1 \\ 0 & 0 \end{bmatrix}, \quad \begin{bmatrix} 0 & 1 \\ 0 & 1 \end{bmatrix} \begin{bmatrix} 1 & 0 \\ 0 & 0 \end{bmatrix} = \begin{bmatrix} 0 & 0 \\ 0 & 0 \end{bmatrix}.$$

There is another law of common algebra, besides the commutative law of multiplication, that is broken in Cayley's matric algebra, and this is the cancellation law for multiplication. Thus, although[15]

$$\begin{bmatrix} 1 & 0 \\ 0 & 0 \end{bmatrix} \begin{bmatrix} 0 & 1 \\ 0 & 1 \end{bmatrix} = \begin{bmatrix} 1 & 0 \\ 0 & 0 \end{bmatrix} \begin{bmatrix} 0 & 1 \\ 1 & 0 \end{bmatrix} = \begin{bmatrix} 0 & 1 \\ 0 & 0 \end{bmatrix},$$

we do not have

$$\begin{bmatrix} 0 & 1 \\ 0 & 1 \end{bmatrix} \quad \text{and} \quad \begin{bmatrix} 0 & 1 \\ 1 & 0 \end{bmatrix}$$

equal to each other.

Although a number of algebras in which multiplication is noncommutative were devised in the middle nineteenth century, algebras in which multiplication is nonassociative are, for the most part, of rather recent origin. As examples of such algebras we have *Jordan algebras* and *Lie algebras*. A (special) Jordan algebra, which is used in quantum mechanics, has matrices for elements, with equality and addition defined as in Cayley's matric algebra, but with the product

---

[15] In Cayley's matric algebra, the matrix $\begin{bmatrix} 0 & 0 \\ 0 & 0 \end{bmatrix}$ plays the role of zero.

of two matrices A and B defined as $(AB + BA)/2$, where AB stands for Cayley's product of the two matrices A and B. Although multiplication in this algebra can be shown to be nonassociative, it is obviously commutative. A Lie algebra differs from the above Jordan algebra in that the product of the two matrices A and B is defined by $AB - BA$, where again AB denotes the Cayley product of the matrices A and B. In this algebra, multiplication is neither associative nor commutative.

5.3   Groups

A *group*, which is one of the simplest algebraic structures of consequence, is a set G of elements in which a binary operation $*$ is defined satisfying the following three postulates:

G1. *For all* a, b, c *in* G, $(a * b) * c = a * (b * c)$.

G2. *There exists an element* i *of* G *such that, for all* a *in* G, $a * i = a$.

(The element i is called an *identity element* of the group. Later we shall prove that a group possesses only one identity element.)

G3. *For each element* a *of* G *there exists an element* a^{-1} *of* G *such that* $a * a^{-1} = i$.

(The element a^{-1} is called an *inverse element* of a. Later we shall prove that an element a of a group possesses only one inverse element.)

If, in addition to the above three postulates, the following postulate is satisfied, the group is called a *commutative*, or an *Abelian*, *group*.

G4. *For all* a, b *in* G, $a * b = b * a$.

A group for which Postulate G4 does *not* hold is called a *non-Abelian group*. If the set G of a group contains only a finite number of distinct elements the group is called a *finite group*; otherwise it is called an *infinite group*. For some purposes the much simpler concept of a *semigroup* is important; it is a nonempty set G of elements in which a binary operation $*$ is defined satisfying the single Postulate G1. If, in addition, Postulate G4 is satisfied, the semigroup is called an *Abelian semigroup*. Illustrations of groups are numerous and diverse, as the following examples testify.

EXAMPLES

(*a*)   Let G be the set of all integers, and let $*$ denote ordinary addition. Here the integer 0 is the identity element, and the inverse element of a given integer a is its negative. This is an example of an infinite Abelian group.

(*b*)   Let G be the set of all rational numbers with 0 omitted, and let $*$ denote ordinary multiplication. Here the rational number 1 is the identity element, and the inverse element of a given rational number a is its reciprocal $1/a$. This is another example of an infinite Abelian group.

(c)   Let G be the set of all translations

$$T: \begin{matrix} x' = x + h, \\ y' = y + k, \end{matrix}$$

where h and k are real numbers, of the coordinate plane, and let $T_2 * T_1$ denote the result of performing first translation T_1 and then translation T_2. If T_1 and T_2 are the translations

$$T_1: \begin{matrix} x' = x + h_1, \\ y' = y + k_1, \end{matrix} \qquad T_2: \begin{matrix} x' = x + h_2, \\ y' = y + k_2, \end{matrix}$$

then it is easy to show that $T_2 * T_1$ is the transformation

$$x' = x + (h_1 + h_2),$$

$$y' = y + (k_1 + k_2),$$

which is again a translation. One can easily show that the operation $*$ is associative. The identity element is the translation in which $h = k = 0$, and the inverse of translation T is the translation

$$T^{-1}: \begin{matrix} x' = x - h, \\ y' = y - k. \end{matrix}$$

This, too, is an example of an infinite Abelian group.

(d)   Let G be the set of four numbers $1, -1, i, -i$, where $i^2 = -1$, and let $*$ denote ordinary multiplication. Here 1 is the identity element, and the inverse elements of $1, -1, i, -i$ are $1, -1, -i, i$, respectively. This is an example of a finite Abelian group.

(e)   Let G be the set of four integers $1, 2, 3, 4$, and let $a * b$ denote the remainder obtained by dividing the ordinary product of a and b by 5. We may represent all possible values of $a * b$ by means of the operation table:

$*$	1	2	3	4
1	1	2	3	4
2	2	4	1	3
3	3	1	4	2
4	4	3	2	1

Here the value of $a * b$ is found in the box common to the row headed by a and the column headed by b. Note that 1 is the identity element, and that the inverse elements of $1, 2, 3, 4$ are $1, 3, 2, 4$, respectively. This is another example of a finite Abelian group. An operation table, like the above, can be made for any finite group. If the table is symmetrical in its principal diagonal, as in the present example, the group is Abelian; otherwise it is non-Abelian. The identity element is the element that heads the column that exactly repeats the column of row headings. To find the inverse of an element a, we merely travel across the row headed by a until we come to the identity element; the element heading the column we are now in is a^{-1}. There is no simple way of checking the associativity of the operation $*$ from the table.

(f) Let G be the set of all rotations of a wheel about its axis through angles which are non-negative integral multiples of $60°$, where a rotation of $(6n + k) 60°$, where n and k are non-negative integers with $k < 6$, is considered the same as the rotation $k \cdot 60°$. Let $a * b$ denote the rotation b followed by the rotation a. This is an example of a finite Abelian group.

(g) Let G be the set of all 2 by 2 matrices

$$\begin{bmatrix} a & b \\ c & d \end{bmatrix},$$

where a, b, c, d are rational numbers such that $ad - bc \neq 0$, and let $*$ denote ordinary (Cayley) matric multiplication (see the last section). Here the matrix

$$\begin{bmatrix} 1 & 0 \\ 0 & 1 \end{bmatrix}$$

is the identity element, and the inverse element of matrix

$$\begin{bmatrix} a & b \\ c & d \end{bmatrix}$$

is the matrix

$$\begin{bmatrix} d/(ad - bc) & -b/(ad - bc) \\ -c/(ad - bc) & a/(ad - bc) \end{bmatrix}.$$

This is an example of an infinite non-Abelian group.

(h) Let G be the set of six expressions

$$r, \quad \frac{1}{r}, \quad 1 - r, \quad \frac{1}{1 - r}, \quad \frac{r - 1}{r}, \quad \frac{r}{r - 1},$$

and let $a * b$ denote the result of substituting the expression b in place of r in the expression a. For example,

$$[1 - r] * \frac{r}{r - 1} = 1 - \frac{r}{r - 1} = \frac{1}{1 - r}.$$

This is an example of a finite non-Abelian group.

Several things are accomplished by the above exhibition of examples of groups. First of all, the examples establish the existence of both finite and infinite Abelian groups and of both finite and infinite non-Abelian groups. Next, the existence of non-Abelian groups establishes the independence of Postulate G4 from the other three postulates. Finally, the various representations of Abelian groups establish the consistency of Postulates G1, G2, G3, G4. Since some of these representations (see Examples (d), (e), (f)) contain only a finite number of objects, which can be explicitly exhibited one by one and denoted by symbols, it follows that we have established *absolute* consistency of the postulate set rather than just *relative* consistency. This is the case for any postulate set for which a model containing only a finite number of objects can be exhibited, as, for instance, in the case of the postulates for projective geometry in Section 4.5, where a "finite" model was exhibited. In all other cases the method of models can do no more than reduce the consistency of one system to that of another. The important matter of consistancy of postulate sets will be considered more fully in the next chapter.

Mathematicians have devised many postulate sets for a group, all of which, of course, are equivalent to one another. Frequently more is assumed in the postulates than need be. For example, it is quite common to assume, in G2 and G3, the commutativity of i with every element of G and the commutativity of each element a of G with its inverse element a^{-1}. Since it seems to be aesthetically more satisfying to assume as little as possible, we have adopted the weaker set of postulates and shall prove as theorems the additional assumptions about commutativity mentioned above.[16] We now proceed to do this; some of the proofs we shall leave as exercises for the reader.

Some Fundamental Theorems of Groups

Theorem 1   *If* a, b, c, *are in* G *and* a $*$ c = b $*$ c, *then* a = b.

By G3 there exists c^{-1}. From $a*c=b*c$, we then have $(a*c)*c^{-1} = (b*c)*c^{-1}$, or, by G1, $a*(c*c^{-1}) = b*(c*c^{-1})$. Employing G3, we now have $a*i = b*i$, whence finally, by G2, $a = b$.

Theorem 2   *For all* a *in* G, i $*$ a = a $*$ i.

By G3 there exists a^{-1}. Hence, applying G1, G3, G2, G3 in turn, we have $(i*a)*a^{-1} = i*(a*a^{-1}) = i*i = i = a*a^{-1}$. By Theorem 1 we then have $i*a = a$. But, by G2, we have $a*i = a$. It now follows that $i*a = a*i$.

Theorem 3   *A group has a unique identity element.*

Let i and j be two identity elements for the group. Then, by G2 applied to the identity element j, $i*j = i$. Also, by Theorem 2, $i*j = j*i$. But, by G2 applied to the identity element i, $j*i = j$. It now follows that $i = j$.

Theorem 4   *For each element* a *of* G, $a^{-1} *$ a = a $* a^{-1}$.

By G1, G3, and Theorem 2, applied in turn, $(a^{-1}*a)*a^{-1} = a^{-1}*(a*a^{-1}) = a^{-1}*i = i*a^{-1}$. Therefore, by Theorem 1, $a^{-1}*a = i$. But, by G3, $a*a^{-1} = i$. It now follows that $a^{-1}*a = a*a^{-1}$.

Theorem 5   *If* a, b, c *are in* G *and* c $*$ a = c $*$ b, *then* a = b.
Theorem 6   *Each element of a group has a unique inverse element.*
Theorem 7   *If* a *is in* G, *then* $(a^{-1})^{-1}$ = a.
Theorem 8   *If* a *and* b *are in* G, *then there exist unique elements* x *and* y *of* G *such that* a $*$ x = b *and* y $*$ a = b.

---

[16] Actually, however, our postulate set for a group can itself be further weakened. See, for example, E. V. Huntington [1].

5.4   The Significance of Groups in Algebra and Geometry

Groups and semigroups have considerable significance in the foundations of mathematics. We shall devote the present section to a brief treatment of the fundamental role played by these structures in algebra and geometry.

The importance of groups and semigroups in algebra lies largely in the fact that many algebraic systems are actually groups or semigroups with respect to one or more of the binary operations of the systems. In other words, many algebraic structures contain the group structure or the semigroup structure within them as a substructure. Groups and semigroups are like algebraic atoms from which many algebraic systems can be constructed. These ideas are illustrated by the following alternative definitions[17] of *ring, commutative ring, ring with identity, integral domain, division ring,* and *field* (see Section 5.2).

A *ring* is a set S of elements for which two binary operations, here called addition and multiplication, are defined such that (1) S is an Abelian group under addition, with identity element called *zero*, (2) S is a semigroup under multiplication, (3) the two distributive laws of multiplication over addition hold.

A *commutative ring* is a ring in which the multiplicative semigroup is Abelian.

A *ring with unity* is a ring in which the multiplicative semigroup has an identity element.

An *integral domain* is a ring in which the nonzero elements constitute a subsemigroup of the multiplicative semigroup.

A *division ring* (or *sfield*) is a ring of more than one element in which the nonzero elements constitute a subgroup of the multiplicative semigroup.

A *field* is a division ring in which the multiplicative semigroup is Abelian. Otherwise stated, a *field* is a set S of at least two elements for which two binary operations, here called addition and multiplication, are defined such that (1) the elements of S constitute an Abelian group under addition, (2) the nonzero elements of S constitute an Abelian group under multiplication, (3) multiplication is both right and left distributive over addition.[18]

In 1872 Felix Klein made a stunning application of groups to geometry, which introduced a beautiful order into the then existing chaos of geometrical information.

Klein's application of groups to geometry depends on the concept of a *nonsingular transformation* of a set S of elements onto itself, by which we simply mean a correspondence under which each element of S corresponds to a unique element of S, and each element of S is the correspondent of a unique element

---

[17] These definitions may be found in N. Jacobson, chap. 2.
[18] See A. H. Lightstone.

of S. Such a correspondence is said to be *one-to-one*[19] and may be described as a permutation of the elements of S among themselves; it is a special kind of unary operation defined on the set S. If, in a nonsingular transformation T of a set S onto itself, element a of S corresponds to element b of S, we say that, under the transformation T, a *is carried into* b.

By the *product*, $T_2 T_1$, of two nonsingular transformations T_1 and T_2 of a set S of elements onto itself, we mean the resultant transformation obtained by first performing transformation T_1 and then transformation T_2. The product of two nonsingular transformations is not necessarily commutative, as is instanced by taking T_1 to be a translation of a distance of one unit in the direction of the positive x-axis applied to the set S of all points in the (x, y)-plane, and T_2 a counterclockwise rotation of the set S through $90°$ about the origin of co-ordinates. Under $T_2 T_1$ the point $(1, 0)$ is carried into the point $(0, 2)$, whereas under $T_1 T_2$ it is carried into the point $(1, 1)$. But a product of nonsingular transformations is associative, for if T_1, T_2, T_3 are any three nonsingular transformations of a set S onto itself, $T_3(T_2 T_1)$ and $(T_3 T_2)T_1$ both denote the resultant transformation obtained by first performing T_1, then T_2, then T_3. This can be seen by following the fate under these transformations of some arbitrary element a of S. Thus, suppose T_1 carries element a into element b, T_2 carries element b into element c, and T_3 carries element c into element d. Then $T_2 T_1$ carries element a directly into element c and T_3 carries element c into element d, whence $T_3(T_2 T_1)$ carries element a directly into element d. On the other hand, T_1 carries element a into element b and $T_3 T_2$ carries element b directly into element d, whence $(T_3 T_2)T_1$ also carries element a directly into element d.

Let T be any nonsingular transformation of a set S onto itself that carries each element a of S into its corresponding element b of S. The transformation that undoes transformation T, by carrying each element b of S back into its original element a of S, is called the *inverse transformation* of transformation T, and is denoted by T^{-1}. The product of T and T^{-1} is a transformation that clearly leaves all elements of S unchanged; such a transformation is called an *identical transformation* and will be denoted by I. We note that $TI = T$ for all T.

We may now prove the following important theorem:

Theorem   *A nonempty set Γ of nonsingular transformations of a set* S *onto itself constitutes a group under multiplication of transformations if* (1) *the product of any two transformations of the set* Γ *is in the set* Γ, (2) *the inverse of any transformation of the set* Γ *is in the set* Γ.

To establish this theorem we first note that, since the product of any two transformations in the set Γ is in the set Γ, multiplication of transformations is a binary operation defined in the set Γ. This binary operation is associative, as we have shown above. Also, if T is a transformation in Γ, then T^{-1} is in Γ. But $TT^{-1} = I$ and $TI = T$ for all T of Γ. All three postulates for a group are thus

---

[19] More generally, if S and T are two sets, then by a *one-to-one correspondence between the elements of* S *and the elements of* T we mean a collection of ordered pairs (s, t) such that each element s of S occurs as a first element exactly once, and each element t of T occurs as a second element exactly once.

satisfied, and the theorem is established. Such a group of nonsingular transformations is briefly referred to as a *transformation group*.

We are now ready to give Felix Klein's famous definition of a geometry: *A geometry is the study of those properties of a set* S *that remain invariant when the elements of set* S *are subjected to the transformations of some transformation group* Γ. This very general concept of a geometry was developed by Felix Klein in his celebrated Erlanger Programm,[20] distributed by him when he accepted a chair at the University of Erlangen in 1872.

To illustrate Klein's definition of a geometry, let S be the set of all points of an ordinary plane, and consider the set Γ of all transformations of S compounded from translations, rotations, and reflections in lines. Since the product of any two such transformations and the inverse of any such transformation are also such transformations, it follows that Γ is a transformation group. The resulting geometry is ordinary *plane Euclidean metric geometry*. Since such properties as length, area, congruence, parallelism, perpendicularity, similarity of figures, collinearity of points, and concurrence of lines are invariant under the group Γ, these properties are studied in plane Euclidean metric geometry. If Γ is enlarged by including, together with the translations, rotations, and reflections in lines, the homothety transformations (in which each point P is carried into a point P' such that $AP = k \cdot AP'$, where A is some fixed point, k is some fixed positive constant, and A, P, P' are collinear), we obtain *plane similarity*, or *plane equiform*, *geometry*. Under this enlarged group such properties as length, area, and congruence no longer remain invariant, and hence are no longer subjects of study, but parallelism, perpendicularity, similarity of figures, collinearity of points, and concurrence of lines are still invariant properties, and hence do constitute subject matter for study in this geometry. Considered from Klein's point of view, plane projective geometry is the study of those properties of the points of a plane that remain invariant when the points are subjected to the group of so-called projective transformations. Of the previously mentioned properties, only collinearity of points and concurrence of lines still remain invariant. An important invariant under this group of transformations is the cross ratio of four collinear points; this invariant plays an important role in the study of projective geometry. The plane non-Euclidean metric geometries, considered in earlier chapters, can be thought of as the study of those properties of the points of a *non-Euclidean* plane which remain invariant under the group of transformations compounded from translations, rotations, and reflections in lines.

In all of the above geometries, the fundamental elements on which the transformations of some transformation group are made to act are points; hence the above geometries are all examples of so-called point geometries. There are, as one might expect, geometries in which entities other than points are chosen for fundamental elements. Thus geometers have studied line geometries, circle geometries, sphere geometries, and various other geometries. In building up a geometry one is at liberty to choose, first of all, the fundamental element of the geometry (point, line, circle, etc.); next, the manifold or space of these elements (plane of points, ordinary space of points, spherical surface of points, plane of

---

lines, pencil of circles, etc.); and finally, the group of transformations to which the fundamental elements are to be subjected. The construction of a new geometry becomes, in this way, a rather simple matter.

Another interesting feature is the way in which some geometries embrace others. Thus, since the transformation group of plane Euclidean metric geometry is a subgroup of the transformation group of plane equiform geometry, it follows that any theorem holding in the latter geometry must hold in the former. From this point of view it can be shown that projective geometry lies within each of the former, and we have a sort of sequence of nesting geometries. Until recent times, the transformation group of projective geometry contained as subgroups the transformation groups of practically all other geometries that had been studied. This is essentially what Cayley meant when he remarked that "projective geometry contains all geometry." Actually, as far as the theorems of the geometries are concerned, it is the other way about—the theorems of projective geometry are contained among the theorems of each of the other geometries.

The various geometries when considered from Klein's point of view can perhaps best be studied by the methods of analytic geometry, where the transformations of each underlying transformation group Γ are given by formulas connecting the new and the old coordinates of the fundamental elements of the space under consideration. For example, the transformation group of plane projective point geometry is the totality of all transformations of the form

$$x' = \frac{ax + by + c}{gx + hy + i},$$

$$y' = \frac{dx + ey + f}{gx + hy + i},$$

where $a, b, c, \ldots$ are any real numbers satisfying the condition

$$\begin{vmatrix} a & b & c \\ d & e & f \\ g & h & i \end{vmatrix} \neq 0.$$

If Γ is taken as the totality of all transformations of the form

$$x' = f(x, y), \qquad y' = g(x, y),$$

where $f(x, y)$ and $g(x, y)$ are continuous and single-valued, and where the inverse transformation exists and is also continuous and single-valued, we obtain the rather recently developed branch of geometry known as *topology*, or *analysis situs, of the Euclidean plane*. From this definition one can see why topology of the plane is frequently referred to as "rubber-sheet" geometry, for in stretching or contracting a rubber sheet, the points of the sheet undergo just such a bicontinuous single-valued transformation.

Topology is perhaps the deepest geometry in the sense considered above; its transformation group embraces all previous transformation groups mentioned, and any theorem of topology thus holds within these other geometries. Since the topological transformations are so broad, it is fair to wonder what properties of the plane of points, say, can possibly remain invariant under these transformations. To give a few simple examples, we might mention, first of all, that a simple closed curve (a closed curve that does not cut itself) remains a simple closed curve under all topological transformations. Again, the fact that the deletion of only one point from a simple closed curve does not disconnect the curve is a topological property, and the fact that the deletion of two points from a simple closed curve separates the curve into two pieces is also a topological property. Frequent use is made of such topological properties in modern treatments of function theory.

A deeper and more rigorous consideration of Klein's remarkable synthesis of geometries must be reserved for a course in the foundations of geometry. The concept is very useful when it applies, and many geometries can be considered neatly in this way. But shortly after the turn of the century, bodies of mathematical propositions, which mathematicians felt should be called geometries, came to light that cannot be fitted into this codification, and a new point of view on the matter was developed based on the idea of abstract space with a superimposed structure that may or may not be definable in terms of some transformation group. We return to this point of view in a later chapter and close our present discussion by noting that these new geometries have found application in the modern theory of physical space that is incorporated in Einstein's general theory of relativity.

5.5 Relations

We have seen that the postulates of a postulational discourse are primary, or assumed, statements about a set of primitive terms, the term *primary* being employed to indicate that the statements are not derived logically from other statements within the discourse. The primitive terms may consist of unspecified relations connecting unspecified elements, as is the situation in Hilbert's postulational treatment of plane Euclidean geometry, where the primitive elements are *points* and *lines*, and the primitive relations connecting the elements are those denoted by *on*, *between*, and *congruent*; or the primitive terms may consist of unspecified operations on unspecified elements, as is the case, for example, in the postulate set for a field; or, again, the primitive terms may consist of all three kinds—elements, relations connecting the elements, and operations on the elements. Much of the present chapter has been devoted to a consideration of operations. Thus we have classified operations into unary, binary, ternary, and so on, and we have described those properties of binary operations known as commutativity, associativity, and distributivity of one operation over another. So far, however, we have said nothing of a similar nature about relations. This is an important matter, and it will now receive our attention.

Everyone, of course, is familiar with the idea of a relation as a form of connectivity between two or more things, for it is a concept that is not peculiar to mathematics but permeates everyday life and conversation. Thus it is common to hear of such relations as "is the father of," "is a cousin of," "is married to," "admires," "is to the left of," "is above," "is smaller than," "is taller than," "is heavier than," "is the same color as," "implies," "is between," and so on.

Elementary geometry is interested in such relations as "is similar to," "coincides with," "is congruent to," "is parallel to," "is perpendicular to," and "is collinear with," and the arithmetic of numbers is concerned with such relations as "is equal to," "is greater than," "is less than," "is the sum of," and "is the product of."

Relations, like operations, can be classified according to the number of elements involved. Here, however, it is customary to use Greek derivatives instead of Latin derivatives, and we call a relation connecting two elements a *dyadic* relation, and one connecting three elements a *triadic* relation. Similarly, there are *tetradic*, *pentadic*, and higher *polyadic* relations. The most commonly occurring relations, in both general and mathematical use, are dyadic relations, as the above list exemplifies. The relation "is between" is, however, an example of a triadic relation, for it is always used in the form "*a* is between *b* and *c*." Other triadic relations appearing in the above list are "is collinear with," "is the sum of," and "is the product of," when used in the forms "*a* is collinear with *b* and *c*," "*a* is the sum of *b* and *c*," and "*a* is the product of *b* and *c*."

Although relations are always used in connection with terms, we can consider the general properties of relations independently of the particular terms related. We limit ourselves to only the commonest properties of dyadic relations, namely, *reflexiveness*, *irreflexiveness*, *nonreflexiveness*, *symmetry*, *asymmetry*, *nonsymmetry*, *transitivity*, *intransitivity*, and *nontransitivity*. Since in our consideration we are not interested in particular dyadic relations but in their possible properties, we symbolize any dyadic relation whatsoever by R and the elements that it connects by a, b, c, and so on. If a is related to b by the R relation, we shall write $a R b$; if a is *not* related to b by the R relation, we shall write $a \not R b$.

A dyadic relation R is said to be *reflexive* in a set S of elements if, for each element a of S, it must be true that $a R a$. For example, the relation "is equal to," when applied to numbers, is reflexive, since $a = a$ for every number a. But the relation "is the father of," applied to people, is not reflexive, since no person is the father of himself. A dyadic relation R is said to be *irreflexive* in a set S of elements if, for each element a of S, it is true that $a \not R a$. The relation "is greater than," when applied to numbers, is obviously irreflexive. A dyadic relation R is said to be *nonreflexive* in a set S if for some a in S we have $a R a$ and for some b in S we have $b \not R b$. The relation "admires," applied to people, is nonreflexive, for if a is a person, a may or may not admire himself. Likewise, "is the square of," applied to real numbers, is nonreflexive, for 0 is the square of 0, but 2 is not the square of 2.

A dyadic relation R is said to be *symmetric* in a set S if whenever $a R b$ for a and b in S, we also have $b R a$. For example, the dyadic relations "is equal to," applied to numbers, and "is married to," applied to people, are symmetric relations. For if a and b are two numbers and if $a = b$, then $b = a$; also if a and b are two people and a is married to b, then b is married to a. A dyadic relation R is said to be *asymmetric* in a set S if whenever $a R b$ for a and b in S, we have $b \not R a$; that is, when $a R b$ is incompatible with $b R a$. The relations "is greater than," applied to numbers, and "is the father of," applied to people, are examples of asymmetric relations. For if a and b are two numbers and if $a > b$, then we cannot have $b > a$; also, if a and b are two people and if a is the father of b, then we cannot have that b is the father of a. A dyadic relation R is said to be *nonsymmetric* in a set S if for some a, b in S we have both $a R b$ and $b R a$, and for some c, d in S we have both $c R d$ and $d \not\!R c$. Thus the dyadic relation R is nonsymmetric if $a R b$ is compatible with, but does not imply, $b R a$. As examples of nonsymmetric dyadic relations we have "is a brother of" and "admires," when applied to people, and "implies," when applied to propositions. If a is a brother of b, b may or may not be a brother of a; if John is a brother of Jane, then Jane is not a brother of John, but if John is a brother of James, then James is a brother of John. If a admires b, b may or may not admire a. Finally, if proposition a implies proposition b, proposition b may or may not imply proposition a. An arithmetical example of a nonsymmetric relation is the relation "is not greater than," applied to numbers, for if a and b are numbers and if a is not greater than b, then b may or may not be not greater than a.

A dyadic relation R is said to be *transitive* in a set S if whenever $a R b$ and $b R c$, for a, b, c in S, we also have $a R c$. Examples of transitive relations are "is equal to" and "is greater than," applied to numbers, "is included in," applied to classes, and "implies," applied to propositions. If a, b, c are numbers and if $a = b$ and $b = c$, then $a = c$, and if $a > b$ and $b > c$, then $a > c$. If a, b, c are classes of things, and if class a is included in class b and class b is included in class c, then class a is included in class c. Finally, if a, b, c are propositions, and if a implies b and b implies c, then a implies c. The relationship of class inclusion is fundamental in the construction of syllogisms, and the relationship of propositional implication is fundamental in the establishment of a deductive chain of argument. Since both of these latter relations are transitive, it is no great exaggeration to say that the structure of transitivity is the mainspring of deductive reasoning. Also, this property of transitivity makes elimination possible; if $a R b$ and $b R c$, we can eliminate b and assert $a R c$. A dyadic relation R is said to be *intransitive* in a set S if whenever $a R b$ and $b R c$, for a, b, c in S, we have $a \not\!R c$. For example, the relation "is the father of," applied to people, is intransitive, for if a is the father of b and b is the father of c, then a cannot be the father of c (a must be a grandfather of c). A dyadic relation R is said to be *nontransitive* in a set S if for some a, b, c in S we have $a R b$, $b R c$, $a R c$, and for some c, d, e in S we have $c R d$, $d R e$, $c \not\!R e$. An example is the dyadic relation "admires," applied to people. If a admires b and b admires c, then a may or may not admire c.

Note that the properties above depend not only on the particular dyadic relation involved but also on the set of elements in which the relation is defined. Thus "is the square of" is reflexive, symmetric, and transitive in the set consisting of the numbers 0 and 1 only; it is nonreflexive, symmetric, and transitive in the set consisting of 0, 1, and 2 only; it is nonreflexive, nonsymmetric, and nontransitive in the set of all real numbers; it is irreflexive, asymmetric, and intransitive in the set consisting of all the real numbers except 0 and 1.

It should be clear that the three sets of properties—the reflexive set, the symmetric set, and the transitive set—are largely independent of one another. That is to say, a dyadic relation may at the same time possess almost any one of the three reflexive properties, any one of the three symmetric properties, and any one of the three transitive properties. For example, the relation "is contemporaneous with," when applied to people, is reflexive, symmetric, and transitive; the relation "is an ancestor of," applied to people, is irreflexive, asymmetric, and transitive; the relation "is married to," applied to people, is irreflexive, symmetric, and intransitive; the relation "admires," applied to people, is nonreflexive, nonsymmetric, and nontransitive; the relation "is the father of," applied to people, is irreflexive, asymmetric, and intransitive. Since there are three reflexive properties, three symmetric properties, and three transitive properties, there are twenty-seven triples of these properties that dyadic relations might possibly possess. It is an interesting exercise to try to construct examples illustrating as many of these possibilities as one can.

Of the twenty-seven triples of dyadic relations, one is of paramount importance in mathematics. This is the so-called *equivalence relation*, which is the type that is reflexive, symmetric, and transitive. In other words, a dyadic relation, which we here denote by the symbol $\approx$, is called an equivalence relation in a set S of elements if and only if the following postulates hold:

E1. *If* a *is an element of* S, *then* a $\approx$ a.

E2. *If* a *and* b *are elements of* S *and if* a $\approx$ b, *then* b $\approx$ a.

E3. *If* a, b, c *are elements of* S *and if* a $\approx$ b *and* b $\approx$ c, *then* a $\approx$ c.

Examples of equivalence relations are abundant. Thus we have "is contemporaneous with" and "is the same age as," applied to people, "is identical with," applied to any set of elements, "is equal to," applied to numbers, "is similar to," and "is congruent to," applied to triangles in geometry, and so on. All of these diverse examples have the common algebraic structure implied by the above postulate set. The examples, incidentally, since they are interpretations of the set of postulates, establish the consistency of the postulate set. It is evident that the concept of an equivalence relation is a generalization of the equals relation, considered in the sense of identity, which we have used frequently in the preceding sections of this chapter.

An equivalence relation $\approx$ in a set S separates the elements of S into classes. Thus all elements equivalent to some particular element a of S belong in one class; all elements equivalent to some other element b of S, not already in the first class, belong to a second class; and so on. It can be shown that this process leads to a unique decomposition of S into nonoverlapping classes. This separation of S into classes is called *the class decomposition of* S *corresponding to the equivalence relation* $\approx$.

As a simple example of the class decomposition of a set S corresponding to an equivalence relation in S, consider the following. Let S be the set of positive integers and let "$a \approx b$" mean "a differs from b by a multiple of 3." This is easily seen to be an equivalence relation if we keep in mind that 0 is a multiple of 3, namely, the 0th multiple. Decomposing the positive integers into classes, by putting into the same class all pairs of positive integers that differ by a multiple of 3, we obtain the three unique and nonoverlapping classes: $\{1, 4, 7, \ldots\}$, $\{2, 5, 8, \ldots\}$, and $\{3, 6, 9, \ldots\}$.

A significant application of the decomposition of a set into classes by means of an equivalence relation will be considered in the next chapter with the important concept of *isomorphism*.

In this chapter the recognition of algebraic structure in the early half of the nineteenth century was shown to have led to the liberation of algebra from common arithmetic and thence to the postulational treatment of many different algebras, so that an algebra, like a geometry, became a purely hypothetico-deductive study. At the same time, some of the flavor and terminology of modern abstract algebra was introduced. Use will be made in later chapters of many of the algebraic concepts considered in this chapter.

PROBLEMS

5.1.1 Sometimes an n-*ary operation* on a set S of elements is defined, less restrictively than in Section 5.1, as a rule that assigns to each ordered subset of n elements of S a uniquely defined element of some set T; if T is S, or a subset of S, then S is said to be *closed* under the operation. Thus the set of all even integers is closed under the binary operation of addition, for the sum of two even integers is always an even integer, but the set of all odd integers is not closed under the binary operation of addition, for the sum of two odd integers is not always (in fact, is never) an odd integer. Which of the following sets are closed under the given operations?

(a) The set of integers from 1 to 20, inclusive, under the binary operation of addition.

(b) The set of all odd integers under the binary operation of multiplication.

(c) The set of all positive integers under the binary operation of division.

(d) The set of all positive rational numbers under the binary operation of division.

(e) The set consisting of the single number 1 under the binary operation of multiplication.

(f) The set of all positive rational numbers under the unary operation of taking the positive square root.

(g) The set of all positive integers under the binary operation of subtraction.

(h) The set of all prime numbers p under the unary operation $p^2 - p + 41$.

(i) The set of all integral multiples $a, b, c, \ldots$ of 3 under the ternary operation $ab + c$.

(j) The set of all odd integers $a, b, c, \ldots$ under the ternary operation $ab + c + 1$.

(k) The set of all gases at 70° Fahrenheit under the binary operation of chemical combination of pairs of gases.

(l) The set of all colors under the binary operation of mixing pairs of colors.

5.1.2 **(a)** Is the operation + distributive over the operation × in the set of all integers?

(b) Is the operation + distributive over the operation × in Example (k) of Section 5.1?

5.1.3 Two unary operations (defined for the same set) are said to be *commutative with respect to each other* if the same final result is achieved regardless of the order of performance of the two operations. Which of the following pairs of unary operations are commutative with respect to one another?

(a) (1) Adding 3, (2) adding 5, applied to the set of positive integers.

(b) (1) Squaring, (2) cubing, applied to the set of positive integers.

(c) (1) Adding 3, (2) multiplying by 5, applied to the set of positive integers.

(d) (1) Taking the cosine, (2) taking the sine, applied to the set of real numbers.

(e) (*for students who have studied calculus*) (1) $dy/dx + ay$, (2) $dy/dx - ay$, applied to the set of analytic functions y of x.

(f) (1) Putting on a pair of socks, (2) putting on a pair of shoes, applied to the set of a person's appearances.

(g) (1) Insuring an automobile, (2) colliding the automobile with that of a struggling lawyer, applied to the set of legal demands on one's pocketbook.[21]

(h) (1) Shelling an egg, (2) scrambling the egg, applied to the set of appetizing foods.

(i) (1) Disrobing, (2) taking a bath, applied to the set of the appearances of one's clothing.

5.1.4 Reduce the left member of each of the following equalities to the right member by using successively an associative, commutative, or distributive law. Following custom, multiplication is here sometimes indicated by a dot ($\cdot$) and sometimes by mere juxtaposition of the factors.

(a) $5(6 + 3) = 3 \cdot 5 + 5 \cdot 6$.

(b) $5(6 \cdot 3) = (3 \cdot 5)6$.

(c) $4 \cdot 6 + 5 \cdot 4 = 4(5 + 6)$.

(d) $a[b + (c + d)] = (ab + ac) + ad$.

(e) $a[b(cd)] = (bc)(ad)$.

(f) $a[b(cd)] = (cd)(ab)$.

(g) $(ad + ca) + ab = a[(b + c) + d]$.

(h) $a + [b + (c + d)] = [(a + b) + c] + d$.

5.1.5 **(a)** through **(k)** Actually show that the examples of Section 5.1 satisfy the five basic properties of that section. Which of the examples satisfy the postulates for a field? for an ordered field?

5.1.6 Establish the following theorems for any field S:

(a) If $a \oplus z = a$ and $a \oplus z' = a$ for all elements a of S, then $z = z'$. (This proves that the zero element is unique.)

(b) If a, b, c are elements of S and if $a \oplus b = a \oplus c$, then $b = c$. (This is the (left) *cancellation law for the operation* $\oplus$.)

(c) Given two elements a and b of S, then there exists a unique element x of S such that $a \oplus x = b$. (This proves that "subtraction" is always possible in a field.)

(d) If a is any element of S, then $a \otimes z = z$.

(e) If a and b are any two elements of S, then $\bar{a} \otimes \bar{b} = a \otimes b$.

(f) If a and b are elements of S and if $a \otimes b = z$, then either $a = z$ or $b = z$.

(g) Given two elements a and b of S, $a \neq z$, then there exists a unique element x of S such that $a \otimes x = b$. (This proves that "division," except by zero, is always possible in a field.)

(h) Show that $\bar{a} \otimes b = a \otimes \bar{b} = \overline{a \otimes b}$.

5.1.7 Establish the following theorems for any *ordered*-field S:

(a) If a is a positive element of S, then $\bar{a}$ is negative.

(b) Element a of S is positive if and only if $a \ominus z$.

(c) Element a of S is negative if and only if $a \oslash z$.

(d) If a and b are distinct elements of S, then either $a \ominus b$ or $a \oslash b$, but not both.

(e) If a, b, c are elements of S and if $a \ominus b$ and $b \ominus c$, then $a \ominus c$.

(f) If $a \ominus b$ and c is any element of S, then $(a \oplus c) \ominus (b \oplus c)$.

(g) If $a \ominus b$ and c is positive, then $(a \otimes c) \ominus (b \otimes c)$.

(h) If a and b are positive elements of S, then (1) $a \otimes \bar{b} = \bar{a} \otimes b =$ the negative element, $\overline{a \otimes b}$, (2) $\bar{a} \otimes \bar{b} =$ the positive element, $a \otimes b$.

(i) If $a \neq z$ is an element of S, then $a \otimes a$ is positive.

(j) The element u is positive.

5.1.8 Show that any set of at least two complex numbers, in which the sum, difference,

---

[21] R. P. Agnew, p. 98.

product, and quotient (with denominator different from 0) of any two numbers in the set are again in the set, constitutes a field under addition and multiplication.

5.2.1 Let S be the set of all ordered real number pairs and define equality, addition, and multiplication of the number pairs by (1) $(a, b) = (c, d)$ if and only if $a = c$ and $b = d$, (2) $(a, b) + (c, d) = (a + c, b + d)$, (3) $(a, b)(c, d) = (ac - bd, ad + bc)$.

   (a) Show that addition is commutative and associative.

   (b) Show that multiplication is commutative and associative.

   (c) Show that multiplication is distributive over addition.

   (d) Show, in fact, that the number pairs constitute a field under addition and multiplication.

5.2.2 In the set S of Problem 5.2.1, with the same definitions of equality, addition, and multiplication, show that

   (a) $(a, 0) + (b, 0) = (a + b, 0)$,

   (b) $(a, 0)(b, 0) = (ab, 0)$,

   (c) $(0, b) = (b, 0)(0, 1)$,

   (d) $(0, 1)(0, 1) = (-1, 0)$.

5.2.3 **(a)** Add the two quaternions $(1, 0, -2, 3)$ and $(1, 1, 2, -2)$.

   (b) Multiply, in both orders, the two quaternions $(1, 0, -2, 3)$ and $(1, 1, 2, -2)$.

5.2.4 **(a)** Show that addition of quaternions is commutative and associative.

   (b) Show that multiplication of quaternions is associative and distributive over addition.

   (c) Show that the real numbers are embedded within the quaternions.

   (d) Show that the complex numbers are embedded within the quaternions.

5.2.5 **(a)** Verify the multiplication table for quaternionic units given in Section 5.2.

   (b) Multiply the two quaternions $a + bi + cj + dk$ and $e + fi + gj + hk$ like polynomials in i, j, k, and, by means of the multiplication table for the quaternionic units, check into the defined product of the two quaternions.

5.2.6 Show that any set of at least two complex numbers, in which the difference and product of any two numbers in the set are again in the set, constitutes a commutative ring under addition and multiplication.

5.2.7 Show that the set of even integers, under ordinary addition and multiplication, constitutes an integral domain without a unity.

5.2.8 Show that the set of all integers, under ordinary addition and multiplication, constitutes an integral domain but does not constitute a field.

5.2.9 Show that Example (i) of Section 5.1 is not an example of an integral domain.

5.2.10 If

$$x' = ax + by, \qquad x'' = ex' + fy',$$

$$y' = cx + dy, \qquad y'' = gx' + hy',$$

show that

$$x'' = (ea + fc)x + (eb + fd)y,$$

$$y'' = (ga + hc)x + (gb + hd)y.$$

5.2.11 Given the matrices

$$A = \begin{bmatrix} 2 & -3 \\ 4 & 1 \end{bmatrix}, \qquad B = \begin{bmatrix} -2 & 2 \\ 0 & 3 \end{bmatrix},$$

calculate $A + B$, AB, BA, and A^2.

5.2.12 Show that the matrix $\begin{bmatrix} 0 & 1 \\ 0 & 0 \end{bmatrix}$ has no square root.

5.2.13 Show that the set of all 2 by 2 matrices with real elements constitutes a noncommutative ring with unity under the usual addition and multiplication of matrices.

5.2.14 Show that we may define complex numbers as matrices of the form

$$\begin{bmatrix} a & b \\ -b & a \end{bmatrix},$$

where a and b are real, subject to the usual definitions of addition and multiplication of matrices.

5.2.15 Show that we may define real quaternions as matrices of the form

$$\begin{bmatrix} a + bi & c + di \\ -c + di & a - bi \end{bmatrix},$$

where a, b, c, d are real and $i^2 = -1$, subject to the usual definitions of addition and multiplication of matrices.

5.2.16 Taking $A = \begin{bmatrix} 1 & 0 \\ -1 & 0 \end{bmatrix}$, $B = \begin{bmatrix} 1 & 1 \\ -1 & 1 \end{bmatrix}$, $C = \begin{bmatrix} 1 & 1 \\ 0 & 1 \end{bmatrix}$ as elements of a Jordan algebra, calculate $A + B$, AB, BA, $A(BC)$, and $(AB)C$.

5.2.17 Taking A, B, C of Problem 5.2.16 as elements of a Lie algebra, calculate $A + B$, AB, BA, $A(BC)$, and $(AB)C$.

5.2.18 Determine whether the following binary operations $*$ and $|$, defined for positive integers, obey the commutative and associative laws, and whether the operation $|$ is distributive over the operation $*$:

(a) $a * b = a + 2b$, $a|b = 2ab$.

(b) $a * b = a + b^2$, $a|b = ab^2$.

(c) $a * b = a^2 + b^2$, $a|b = a^2b^2$.

(d) $a * b = a^b$, $a|b = b$.

5.2.19 Let S be the set of all ordered real number pairs and define: (1) $(a, b) = (c, d)$ if and only if $a = c$ and $b = d$; (2) $(a, b) + (c, d) = (a + c, b + d)$; (3) $(a, b)(c, d) = (0, ac)$; (4) $k(a, b) = (ka, kb)$.

(a) Show that multiplication is commutative, associative, and distributive over addition.

(b) Show that the product of three or more factors is always equal to $(0, 0)$.

(c) Construct a multiplication table for the units $u = (1, 0)$ and $v = (0, 1)$.

5.2.20 Hamilton's quaternions and, to some extent, Grassmann's calculus of extension were devised by their creators as mathematical tools for the exploration of physical space. These tools proved to be too complicated for quick mastery and easy application, but from them emerged the much more easily learned and more easily applied subject of vector analysis. This work was due principally to the American physicist Josiah Willard Gibbs (1839–1903) and is encountered by every student of elementary physics. In elementary physics, a vector is graphically regarded as a directed line segment, or arrow, and the following definitions of equality, addition, and multiplication of these vectors are made:

(1) Two vectors a and b are *equal* if and only if they have the same length and the same direction.

(2) Let a and b be any two vectors. Through a point in space draw vectors a' and b' equal, respectively, to vectors a and b, and complete the parallelogram determined by a' and b'. Then the *sum*, $a + b$, of vectors a and b is a vector whose length and direction are those of the diagonal running from the common origin of a' and b' to the fourth vertex of the parallelogram.

(3) Let a and b be any two vectors. By the *vector product, $a \times b$*, of these two vectors is meant a vector whose length is numerically equal to the area of the parallelogram in definition (2), and whose direction is that of the progress of an ordinary screw when placed perpendicular to both a' and b' and twisted through the angle of not more than $180°$ that will carry vector a' into vector b'.

(a) Show that vector addition is commutative and associative.

(b) Show that vector multiplication is noncommutative and nonassociative.

(c) Show that vector multiplication is distributive over vector addition.

5.2.21 Show that, in the definition of an integral domain, Postulate P9 may be replaced by

P9′. *If* a *and* b *are in* S *and if* a $\neq$ z *and* b $\neq$ z, *then* a $\otimes$ b $\neq$ z.

5.2.22 Let capital letters $P, Q, R, \ldots$ denote points of the plane. Define *addition* of points P and Q by $P + Q = R$, where triangle PQR is a counterclockwise equilateral triangle.

(a) Show that the addition of points of the plane is noncommutative and nonassociative.

(b) Show that if $P + Q = R$, then $Q + R = P$.

(c) Establish the following identities:

(1) $(P + (P + (P + (P + (P + (P + Q)))))) = Q$,

(2) $P + (P + (P + Q)) = (Q + P) + (P + Q)$,

(3) $(P + Q) + R = (P + (Q + R)) + Q$.

5.3.1 Show that the set of two numbers, $1, -1$, under ordinary multiplication, constitutes a group which is a subgroup of that of Example (d) of Section 5.3.

5.3.2 **(a)** Do the even integers form a group with respect to addition?

(b) Do the odd integers form a group with respect to addition?

(c) Do all the rational numbers form a group with respect to multiplication?

(d) Let $a * b = a - b$, where a and b are integers. Do the integers form a group with respect to this operation?

(e) Do all the integral multiples of 3 form a group with respect to addition?

5.3.3 Let G be the set of all rotations

$$R: \begin{aligned} x' &= x \cos \theta - y \sin \theta, \\ y' &= x \sin \theta + y \cos \theta, \end{aligned}$$

of the plane about the origin, and let $R_2 * R_1$ denote the result of performing first rotation R_1 and then rotation R_2. Show that G, under the operation $*$, constitutes an infinite Abelian group.

5.3.4 Construct the operation table for Example (d) of Section 5.3.

5.3.5 Let G be the set of five integers 0, 1, 2, 3, 4, and let $a * b$ denote the remainder obtained by dividing the ordinary product of a and b by 5. Does G, under the operation $*$, constitute a group?

5.3.6 **(a)** Actually show that Example (g) of Section 5.3 constitutes an infinite non-Abelian group.

(b) Calculate the inverse A^{-1} of matrix $A = \begin{bmatrix} 2 & 1 \\ 3 & -1 \end{bmatrix}$, and show that the product AA^{-1} is the identity matrix.

5.3.7 **(a)** Form the operation table for the group of Example (h) of Section 5.3.

(b) We have defined the cross ratio of four collinear points A, B, C, D, taken in this order, to be

$$r = (AB, CD) = \left(\frac{AC}{BC} \right) \left(\frac{BD}{AD} \right).$$

The value of the cross ratio of four points evidently depends upon the order in which we consider the points. There are, then, twenty-four cross ratios corresponding to the twenty-four permutations (or orders) of the four points. Show that the number of *distinct* cross ratios is at most only six and that they are given by the six expressions of Example (h). For this reason the group of Example (h) is known as the *cross ratio group*.

5.3.8 **(a)** Consider the ordered triple (a, b, c). The substitution of c for a, a for b, and b for c, can be represented by the array

$$S = \begin{pmatrix} a & b & c \\ c & a & b \end{pmatrix}.$$

There are, in all, six possible substitutions (counting the identity substitution) that can be made. If S_1 and S_2 denote any two of these six substitutions, let $S_2 * S_1$ denote the result of substitution S_1 followed by substitution S_2. For example,

$$\begin{pmatrix} a & b & c \\ b & a & c \end{pmatrix} * \begin{pmatrix} a & b & c \\ a & c & b \end{pmatrix} = \begin{pmatrix} a & b & c \\ b & c & a \end{pmatrix}.$$

Show that the six substitutions of the ordered triple (a, b, c), under the operation $*$, constitute a finite non-Abelian group. This group is known as the *symmetric group of degree 3*; there exists such a substitution group for each positive integral degree n.

(b) Can you find some subgroups of the symmetric group of degree 3?

5.3.9 **(a)** Prove Theorem 5 of Section 5.3.

(b) Prove Theorem 6 of Section 5.3.

(c) Prove Theorem 7 of Section 5.3.

(d) Prove Theorem 8 of Section 5.3.

5.3.10 Show that Postulates G2 and G3 for a group may be replaced by

G2′. *If* a *and* b *are any elements of* G, *then there exist elements* x *and* y *of* G *such that* a $*$ x $=$ b *and* y $*$ a $=$ b.

Show that the simpler assumption of just the existence of y such that $y * a = b$ is not sufficient.

5.3.11 Do we still necessarily have a group if Postulate G3 for a group is replaced by the following?

G3′. *For each element* a *of* G *there exists an element* a^{-1} *of* G *such that* $a^{-1} * a = i$.

5.3.12 Let G be a nonempty set of elements $a, b, c, \ldots$, along with a binary operation denoted by juxtaposition and a unary operation denoted by prime such that

G1′. *For all* a, b, c, d, f *of* G, (ab)c $=$ (ad)f *implies* b $=$ d(fc′).

Show that G is a group for which the binary operation is multiplication and the unary operation is inversion.

5.3.13 Of increasing importance in mathematics are nonassociative systems, such as loops. A *loop* is a set L of elements $a, b, c, \ldots$ along with a binary operation $*$ satisfying the following two postulates:

L1. *For each* a *and* b *in* L, *there are unique elements* x *and* y *in* L *such that* a $*$ x $=$ b *and* y $*$ a $=$ b.

L2. *There exists an element* e *in* L *such that* a $*$ e $=$ e $*$ a $=$ a *for every* a *in* L.

(a) Show that the identity element e is unique.

(b) Show that $a * x = a * y$ or $x * a = y * a$ implies $x = y$.

Each of the following questions refers to a loop having the so-called *left inverse property*; that is, for each x in L there is an element x^{-1} in L such that $x^{-1} * (x * y) = y$ for every y in L:

(c) Show that $x^{-1} * x = e$.

(d) Show that x^{-1} is unique.

(e) Show that $x * x^{-1} = e$.

(f) Show that $(x^{-1})^{-1} = x$.

A loop has the *right inverse property* if, for each x in L, there is an element x^r in L such that $(y * x) * x^r = y$ for all y in L.

(g) Show that if both inverse properties hold, then $x^r = x^{-1}$.

(h) Show that if both inverse properties hold, then $(x * y)^{-1} = y^{-1} * x^{-1}$.

5.4.1 Show that the definitions of *ring, commutative ring, ring with unity, integral domain, division ring,* and *field,* as given in Section 5.4 in terms of groups and semigroups, are equivalent to the definitions of these algebraic structures as given in Section 5.2 in terms of certain postulate sets.

5.4.2 Find which of the following pairs of transformations of the points of a plane are commutative with respect to multiplication:

(a) two translations;

(b) two rotations about the same fixed point O;

(c) a rotation about a fixed point A and another rotation about a different fixed point B;

(d) a rotation about a fixed point A and a reflection in a line m passing through A;

(e) a rotation about a fixed point A and a reflection in a line m not passing through A;

(f) a translation and a reflection in a line m parallel to the direction of the translation;

(g) a translation and a reflection in a line m not parallel to the direction of the translation.

5.4.3 Prove that the inverse of the product of two nonsingular transformations is the product of the inverses of the transformations taken in reverse order; that is, show that $(TS)^{-1} = S^{-1}T^{-1}$. Extend this to the product of any number of nonsingular transformations.

5.4.4 A transformation T such that $TT = I$, the identical transformation, is called an *involutoric transformation.*

(a) If T is involutoric, show that $T^{-1} = T$.

(b) Give at least three examples of involutoric transformations of the points of the plane.

5.4.5 If S and T are two nonsingular transformations, then the transformation $S' = TST^{-1}$ is called the *transform of* S *by* T.

(a) Show that if R' and S' are the transforms of R and S, respectively, by T, then $R'S'$ is the transform of RS by T.

(b) Show that the transform by T of the inverse of S is the inverse of the transform by T of S.

(c) Show that if the product of two nonsingular transformations is commutative then each is its own transform by the other.

(d) If S is a rotation of the plane about a fixed point A and if T is a translation of the plane, find the transform of S by T and the transform of T by S.

(e) Let G_1 be a subgroup of a transformation group G. If every transformation of G_1 is replaced by its transform by T, where T is a fixed transformation of G, show that the transformations thus found form a subgroup of G.

5.4.6 Although the set of all transformations of the form

$$x' = x \cos \theta - y \sin \theta,$$

$$y' = x \sin \theta + y \cos \theta,$$

constitutes a group of transformations (see Problem 5.3.3), show that the set of all transformations of the form

$$x' = x \cos \theta + y \sin \theta,$$

$$y' = x \sin \theta - y \cos \theta,$$

does not constitute a group of transformations.

5.4.7 Show that the set of transformations of the form

$$x' = \frac{ax + b}{cx + d},$$

where a, b, c, d are real and $ad - bc = 1$, constitutes a group of transformations.

5.4.8 Show that the set of transformations of the form

$$x' = kx + c,$$

$$y' = \frac{y}{k} + d,$$

where c and d are real numbers and k is a positive real number, constitutes a group of transformations. The study of the properties of the points of a plane which remain invariant under this group of transformations is known as a (special) *Lorentz geometry*. Lorentz geometries are of importance in the theory of relativity.

5.4.9 All rigid motions of a plane (rotations, translations, and their compounds) constitute a group of transformations of the set of points of a plane into itself. It can be shown that the analytical representation of a general rigid motion is

$$x' = x \cos \theta - y \sin \theta + a,$$

$$y' = x \sin \theta + y \cos \theta + b,$$

where θ, a, b are arbitrary real numbers. Establish, analytically, the following invariants under the group of rigid motions of the plane:

(a) $(x_1 - x_2)^2 + (y_1 - y_2)^2$, as an invariant of the two points (x_1, y_1), (x_2, y_2). What is the geometrical significance of this invariant?

(b) $(a_1 b_2 - a_2 b_1)/(a_1 b_1 + a_2 b_2)$, as an invariant of the two lines $a_1 x + a_2 y + a_3 = 0$, $b_1 x + b_2 y + b_3 = 0$. What is the geometrical significance of this invariant?

(c) $(a_1 x_0 + a_2 y_0 + a_3)/\sqrt{a_1^2 + a_2^2}$, as an invariant of the point (x_0, y_0) and the line $a_1 x + a_2 y + a_3 = 0$. What is the geometrical significance of this invariant?

(d) $x_0^2 + y_0^2 + a_1 x_0 + a_2 y_0 + a_3$, as an invariant of the point (x_0, y_0) and the circle $x^2 + y^2 + a_1 x + a_2 y + a_3 = 0$. What is the geometrical significance of this invariant?

(e) Express a rigid motion analytically as the product of a rotation about the origin and a translation.

5.4.10 Consider the following nonsingular transformations of the plane:

  R.   a clockwise rotation of 90° about the origin,

  R'. a clockwise rotation of 180° about the origin,

  R''. a clockwise rotation of 270° about the origin,

  H.   a reflection in the x-axis,

  V.   a reflection in the y-axis,

  D.   a reflection in the line $y = x$,

  D'. a reflection in the line $y = -x$,

  I.   the identity motion, in which all points are left unmoved.

(a) Show that the eight transformations constitute a finite non-Abelian transformation group.

(b) Give the inverse of each of the eight transformations.

(c) Imagine a material square having its center at the origin and its sides parallel to the coordinate axes. Show that each of the above eight transformations carries this square into itself. The group of transformations is known as the *group of symmetries of a square*.

5.4.11 Show that any nonempty set Γ of nonsingular transformations of a set S into itself for which $T_2 T_1^{-1}$ is an element of Γ whenever T_1 and T_2 are elements of Γ constitutes a transformation group under transformation multiplication.

5.5.1 Give an example of a tetradic relation and an example of a pentadic relation.

5.5.2 Classify the following dyadic relations as to types of reflexiveness, symmetry, and transitivity:

  (a) "Was born in the same town as," applied to people.

  (b) "Is the spouse of," applied to people.

  (c) "Is the husband of," applied to people.

  (d) "Is a grandfather of," applied to people.

  (e) "Lives within a mile of," applied to people.

  (f) "Disagrees with," applied to people.

  (g) "Is east of," applied to places in America.

  (h) "Has the same longitude as," applied to places anywhere on Earth except the poles.

  (i) "Is less than," applied to the positive integers.

  (j) "Is a multiple of," applied to the positive integers.

  (k) "Is a factor of," applied to the positive integers.

  (l) "Is relatively prime to," applied to the positive integers.

  (m) "Is not equal to," applied to the positive integers.

  (n) "Is perpendicular to," applied to lines in a plane.

  (o) "Is parallel to," applied to lines in a plane.

  (p) "Has the same length as," applied to line segments.

  (q) "Is skew to," applied to lines in space.

  (r) "Is perpendicular to," applied to planes in space.

  (s) "Is tangent to," applied to spheres in space.

  (t) "Is the complement of," applied to angles.

  (u) "Is consistent with," applied to propositions.

  (v) "Contradicts," applied to propositions.

  (w) "Is implied by," applied to propositions.

5.5.3 Construct examples illustrating as many as you can of the twenty-seven types of dyadic relations as far as reflexiveness, symmetry, and transitivity are concerned. Which types cannot exist?

5.5.4 Which of the relations listed in Problem 5.5.2 are equivalence relations?

5.5.5 What is wrong with the following argument showing that a relation which is both symmetric and transitive is necessarily reflexive?

By symmetry, aRb implies bRa; and by transitivity, aRb and bRa imply aRa.

5.5.6 Let S be the set of all ordered pairs of positive integers and define $(a, b) = (c, d)$ if and only if $a + d = b + c$. Show that this definition of equality is an equivalence relation.

5.5.7 Let S be the set of all ordered pairs of integers and define $(a, b) = (c, d)$ if and only if $ad = bc$. Show that this definition of equality is an equivalence relation.

5.5.8 Let S be a set of elements $a, b, c \ldots$ in which an equivalence relation $\approx$ and a binary operation $*$ are defined. We say that the operation $*$ is *well defined* relative to the equivalence relation $\approx$ if and only if for $a' \approx a$ and $b' \approx b$ we always have $a' * b' \approx a * b$.

(a) Let S be the set of all ordered pairs of positive integers, and define $(a, b) + (c, d)$ to be $(a + c, b + d)$ and $(a, b) \cdot (c, d)$ to be $(ac + bd, ad + bc)$. Show that the operations $+$ and $\cdot$ are well defined relative to the equivalence relation of Problem 5.5.6.

(b) Let S be the set of all ordered pairs of integers, and define $(a, b) + (c, d)$ to be $(ad + bc, bd)$ and $(a, b) \cdot (c, d)$ to be (ac, bd). Show that the operations $+$ and $\cdot$ are well defined relative to the equivalence relation of Problem 5.5.7.

(c) Let S be the set of all ordered pairs of integers and define $(a, b) * (c, d)$ to be (a, c). Show that this operation is not well defined relative to the equivalence relation of Problem 5.5.7. It follows that one cannot assume that a given binary operation is well defined relative to a given equivalence relation; proof of such is necessary.

The $=$ sign in the postulates for a field (see Section 5.1) may be replaced by any equivalence relation provided the operations $\oplus$ and $\otimes$ are well defined relative to it.

5.5.9 Let S be the set of all nonnegative integers, and consider the dyadic relation "a has the same remainder as b when divided by 4."

(a) Show that the dyadic relation is an equivalence relation.

(b) Find the class decomposition of S corresponding to this equivalence relation.

5.5.10 Establish the uniqueness and the nonoverlapping of the class decomposition of a set S corresponding to an equivalence relation $\approx$ defined in S by proving the following sequence of theorems:

Definition If x is an element of S, let $S(x)$ denote the class of all elements y of S such that $x \approx y$.

Theorem 1 If x is an element of S, then x is an element of $S(x)$.

Theorem 2 Each element x of S is in some $S(x)$.

Theorem 3 If $x \approx y$, then the classes $S(x)$ and $S(y)$ are identical.

Theorem 4 A class $S(x)$ is completely determined by any one of its elements.

Theorem 5 If two classes $S(x)$ and $S(y)$ have a common element then the two classes are identical.

Theorem 6 The classes $S(x)$ constitute a unique decomposition of S into nonoverlapping classes.

5.5.11 Consider the following definitions:

Definition 1 A dyadic relation R is said to be *determinate* (or *connected*) in a set S of elements if and only if for any two distinct elements a and b of S, either aRb or bRa.

Definition 2   A dyadic relation R is said to be *antisymmetric* in a set S of elements if and only if whenever $a\,R\,b$ and $b\,R\,a$, then a is identical with b.

Definition 3   A set of elements S is said to be *simply*, or *linearly*, *ordered* with respect to a dyadic relation R if and only if the following postulates hold:

    O1. R is determinate in S.

    O2. R is irreflexive in S.

    O3. R is transitive in S.

Definition 4   A set of elements S is said to be *partially ordered* with respect to a dyadic relation R if and only if the following postulates hold:

    Q1. R is reflexive in S.

    Q2. R is antisymmetric in S.

    Q3. R is transitive in S.

Definition 5   If a and b are elements of S and if $a\,R\,b$, then we write $b\,R'\,a$.

(a) Which of the relations listed in Problem 5.5.2 are determinate?
(b) Which are antisymmetric?
(c) Which induce a simple ordering?
(d) Which induce a partial ordering?
(e) If R is the relation "is less than," applied to the positive integers, what is the relation R'? If R is the relation "is not greater than," applied to the positive integers, what is the relation R'?
(f) Show that if S is partially ordered with respect to relation R, then it is also partially ordered with respect to relation R'. (This is known as the *principle of duality* for partially ordered sets; for any theorem about the relation R there is a corresponding theorem about the relation R'.)

5.5.12 A dyadic relation R is said to be *distinctly transitive* in a set S of elements if for any three *distinct* elements x, y, z of S, $x\,R\,y$ and $y\,R\,z$ together imply $x\,R\,z$. Give an example of a set S and a dyadic relation R such that R is distinctly transitive in S but is not transitive in S.

6

FORMAL AXIOMATICS

6.1 Statement of the Modern Axiomatic Method[1]

The discovery of a non-Euclidean geometry and, not long after, of a non-commutative algebra led to a deeper study and refinement of axiomatic procedure; thus, from the material axiomatics of the ancient Greeks evolved the formal axiomatics of the twentieth century. To help clarify the difference between the two forms of axiomatics, we shall first introduce the modern concept of *propositional function*, the fundamental importance of which was first brought to notice by the English mathematician and philosopher Bertrand Russell (1872–1970).

Consider the three statements: (1) Spring is a season. (2) 8 is a prime number. (3) x is a y. Each of these statements has form—the same form; statements (1) and (2) have content as well as form; statement (3) has form only. Clearly, statements (1) and (2) are propositions, one true and the other false. Equally clearly, statement (3) is *not* a proposition, for, since it asserts nothing definite, it is neither true nor false, and a proposition, by definition, is a statement which is true or false. Statement (3), however, though not a proposition, does have the form of a proposition. It has been called a *propositional function*, for if in the form

$$x \text{ is a } y,$$

we substitute terms of definite meaning for the variables x and y, we may obtain propositions, true propositions if the substituted terms verify the propositional function, false propositions if the substituted terms falsify the propositional function. It is apparent that some substitutions for x and y convert the propositional function into so much nonsense; such meanings for the variables

---

[1]This section is based largely on the opening essay in C. J. Keyser, "The Meaning of Mathematics," *Math and the Question of Cosmic Mind, with Other Essays*, Scripta Mathematica, 1935, No. 2, Yeshiva University Press.

are considered inadmissible. The form considered above is a propositional function in two variables, and it has infinitely many verifiers.

A propositional function may contain any number of variables. An example having but one is this: x is a volume in the Library of Congress. Here x evidently has as many verifying values as there are volumes in the Library of Congress. Evidently, too, the variable x has many falsifying values.

There is no need for the variables in a propositional function to be denoted by symbols, such as $x, y, \ldots$; they may be ordinary words. Thus, should a statement whose terms are ordinary words appear in a discourse with no indication as to the senses in which the words are to be understood, then in that discourse the statement is really a propositional function, rather than a proposition, and in the interests of clarity the ambiguous or undefined terms might better be replaced by such symbols as $x, y, \ldots$. Written and spoken discourse often contains such statements, and though asserted by their authors to be propositions, true or false, are in reality propositional functions, devoid of all true or false quality. As Keyser has observed,[2] this fact perhaps accounts for much of the argument and misunderstanding among people.

With the idea of a propositional function firmly in mind, let us return to a discussion of axiomatic procedure. We recall that any logical discourse, in an endeavor to be clear, tries to define explicitly all elements of the discourse, the relations among these elements, and the operations to be performed on them. Such definitions, however, must employ other elements, relations, and operations, and these, too, are subject to explicit definition. If these are defined, it must again be by reference to further elements, relations, and operations. There are two roads open to us; either the chain of definitions must be cut short at some point, or else it must be circular. Since circularity is not to be tolerated in a logical discourse, the definitions must be brought to a close at some point; thus it is necessary that one or more elements, relations, and operations receive no explicit definition. These are known as the *primitive terms* of the discourse. There is likewise an effort logically to deduce the statements of the discourse, and, again, in order to get started and also to avoid the vicious circle, one or more of the statements must remain entirely *unproved*. These are known as the *postulates* (or axioms, or primary statements) of the discourse. Clearly, then, any logical discourse such as we are considering must conform to the following pattern:

Pattern of Formal Axiomatics

1. The discourse contains a set of technical terms (elements, relations among the elements, operations to be performed on the elements) that are deliberately chosen as undefined terms. These are the *primitive terms* of the discourse.

2. The discourse contains a set of statements about the primitive terms that are deliberately chosen as unproved statements. These are called the *postulates* (or *axioms*), P, of the discourse.

---

[2] Ibid.

3. All other technical terms of the discourse are defined by means of previously introduced terms.

4. All other statements of the discourse are logically deduced from previously accepted or established statements. These derived statements are called the *theorems*, *T*, of the discourse.

5. For each theorem T_i of the discourse a corresponding statement (which may or may not be formally expressed) exists that asserts that theorem T_i is logically implied by the postulates *P*. (Often a corresponding statement appears at the end of the proof of the theorem in some such words as, "Hence the theorem," or "This completes the proof of the theorem." In some elementary geometry textbooks the statement appears at the end of the proof of the theorem as *Q.E.D.* (*Quod erat demonstrandum*). The modern symbol □, suggested by Paul R. Halmos, or some variant of it, is frequently used to signal the end of a proof.)

Note in the above pattern that the primitive terms, being undefined terms, might just as well (if such is not already the case) be replaced by symbols like $x, y, \ldots$. Let us suppose this substitution is made. Then the primitive terms are clearly variables. Second, note that the postulates, *P*, since they are statements about the primitive terms, are nothing less than propositional functions. Third, note that the theorems, *T*, since they are but logical implications of the postulates *P*, also are propositional functions. We are thus brought to a fact of cardinal importance—namely, that once the primitive terms are realized to be variables, both the postulates and the theorems of a logical discourse are not propositions but propositional functions.

Since the postulates and the theorems of a logical discourse are propositional functions—that is, are statements of form only and without content—it would seem that the whole discourse is somewhat vacuous and entirely devoid of truth or falseness. Such, however, is not the case, for by (5) of the postulational pattern we have the all-important statement,

6. The postulates *P* imply the theorems *T*.

Now (6) asserts something definite; it is true or false, and so is a proposition—a true one if the theorems *T* are in fact implied by the postulates *P*, and a false one if they are not. Statement (6) is precisely what the discourse is designed for; it is the discourse's sole aim and reason for being.

A discourse conducted according to the above pattern has been called, by some mathematicians, a *branch of pure mathematics*, and the grand total of all such existing branches of pure mathematics, the *pure mathematics up to date*.

If, for the variables (the primitive terms) in a branch of pure mathematics, we should substitute terms of definite meaning that convert all the postulates of the branch into true propositions, then the set of substituted terms is called an *interpretation* of the branch of pure mathematics. The interpretation will also, provided all deductions have been correctly performed, convert the theorems of the discourse into true propositions. The result of such an interpretation is called a *model* of the branch of pure mathematics.

A model of a branch of pure mathematics has been called a *branch of applied mathematics*, and the grand total of all existing branches of applied mathematics, the *applied mathematics up to date*. Thus, according to this definition, the difference between applied and pure mathematics is not one of applicability and inapplicability, but rather of concreteness and abstractness. Behind every branch of applied mathematics lies a branch of pure mathematics, the latter being an abstract development of what formerly was a concrete development. It is conceivable (and indeed such is often the case) that a single branch of pure mathematics may have several models, or associated branches of applied mathematics. This is the "economy" feature of pure mathematics, for the establishment of a branch of pure mathematics automatically ensures the simultaneous establishment of all of its branches of applied mathematics.

The abstract development of some branch of pure mathematics is an instance of *formal axiomatics*, whereas the concrete development of a given branch of applied mathematics is an instance of *material axiomatics*. In the former case we think of the postulates as prior to any specification of the primitive terms, and in the latter we think of the objects and concepts that interpret the primitive terms as being prior to the postulates. In the former case a postulate is simply a basic assumption about some undefined primitive terms; in the latter case a postulate expresses some property of the basic objects and concepts that is taken as initially evident. This latter is the older view of a postulate and was the view held by some of the ancient Greeks. Thus, to these Greeks, geometry was thought of as a study dealing with a unique structure of physical space, in which the elements *points* and *lines* are regarded as idealizations of certain actual physical entities and in which the postulates are readily accepted statements about these idealizations. From the modern point of view, geometry is a purely abstract study devoid of any physical meaning or imagery.

The notion of mathematics as an assemblage of abstract postulational discourses gives considerable sense to Bertrand Russell's facetious statement that "mathematics may be defined as the subject in which we never know what we are talking about, nor whether what we are saying is true."[3] It also accords with Henri Poincaré's saying that mathematics is "the giving of the same name to different things,"[4] and with Benjamin Peirce's (1809–1880) remark that "mathematics is the science which draws necessary conclusions."[5]

═══════════ **6.2 A Simple Example of a Branch** ═══════════
of Pure Mathematics

We propose to give in this section a simple example of a branch of pure mathematics followed by three applications of that branch. In other words, we shall develop a formal axiomatic discourse and then, by appropriate interpretations, obtain three models of it. To this end consider a set K of undefined elements $a, b, c, \ldots$, and an undefined dyadic relation R connecting certain pairs

---

[3]B. Russell [1].

[4]Quoted in C. J. Keyser [1], p. 134.

[5]B. Peirce.

of elements of K. If element a is related to element b by the R relation, we shall write $a\,R\,b$ and read "a is R-related to b"; if element a is not related to element b by the R relation, we shall write $a\,\not{R}\,b$ and read "a is not R-related to b." If elements a and b are identical (the same element), we shall write $a = b$; if elements a and b are distinct, we shall write $a \neq b$. We now list four unproved statements, or postulates, concerning the elements of set K and the dyadic relation R.

P1. *If* $a \neq b$, *then either* $a\,R\,b$ *or* $b\,R\,a$.

P2. *If* $a\,R\,b$, *then* $a \neq b$.

P3. *If* $a\,R\,b$ *and* $b\,R\,c$, *then* $a\,R\,c$.

P4. K *consists of exactly four distinct elements.*

From these postulates we shall now logically deduce further statements, or *theorems*. Theorems will be designated by T1, T2, . . . and definitions by D1, D2, ...

T1. *If* $a\,R\,b$, *then* $b\,\not{R}\,a$.

Suppose both $a\,R\,b$ and $b\,R\,a$. Then, by P3, $a\,R\,a$. But this statement is impossible by P2. Hence the theorem by *reductio ad absurdum*.

T2. *If* $a\,R\,b$, *and* c *is in* K, *then either* $a\,R\,c$ *or* $c\,R\,b$.

If $c = a$, then $c\,R\,b$, and we are done. If $c \neq a$, we have, by P1, either $a\,R\,c$ or $c\,R\,a$. If $c\,R\,a$, since also $a\,R\,b$, we have, by P3, $c\,R\,b$. Hence the theorem.

T3. *There is at least one element of* K *not R-related to any element of* K.

Suppose the contrary case and let a be any element of K. Then, by our supposition, there exists an element b of K such that $a\,R\,b$. By P2, a and b are distinct elements of K.

By our supposition there exists an element c of K such that $b\,R\,c$. By P2, $b \neq c$. By P3, we also have $a\,R\,c$. By P2, $a \neq c$. Thus a, b, c are distinct elements of K.

By our assumption there exists an element d of K such that $c\,R\,d$. By P2, $c \neq d$. By P3, we also have $b\,R\,d$ and $a\,R\,d$. By P2, $b \neq d$, $a \neq d$. Thus a, b, c, d are distinct elements of K.

By our supposition there exists an element e of K such that $d\,R\,e$. By P2, $d \neq e$. By P3, we also have $c\,R\,e$, $b\,R\,e$, $a\,R\,e$. By P2, $c \neq e$, $b \neq e$, $a \neq e$. Thus a, b, c, d, e are distinct elements of K.

We now have a contradiction of P4. Hence the theorem by *reductio ad absurdum*.

T4. *There is only one element of* K *not R-related to any element of* K.

By T3, there is at least one such element, say a. Let $b \neq a$ be any other element of K. By P1, either $a\,R\,b$ or $b\,R\,a$. But by hypothesis, we do not have $a\,R\,b$. Therefore we must have $b\,R\,a$, and the theorem is proved.

D1. If $b\,R\,a$, then we say $a\,D\,b$.

T5. *If* $a\,D\,b$ *and* $b\,D\,c$, *then* $a\,D\,c$.

By D1, bRa and cRb. By P3, we then have cRa, or, by D1, aDc.

D2. If aRb and there is no element c of K such that aRc and cRb, then we say aFb.

T6. *If* a F c *and* b F c, *then* a = b.

Suppose $a \neq b$. Then, by P1, either aRb or bRa.

Case 1 Suppose aRb. Since bFc, by D2, bRc. But this is impossible, since aFc.

Case 2 Suppose bRa. Since aFc, by D2, aRc. But this is impossible, since bFc.

Thus in either case we are led to a contradiction of our hypothesis. Hence the theorem by *reductio ad absurdum*.

T7. *If* a F b *and* b F c, *then* a $\not{F}$ c.

By D2, aRb and bRc. Hence, again by D2, $a\not{F}c$.

D3. If aFb and bFc, then we say aGc.

We shall cut short our abstract postulational discourse at this point. Many other theorems could be established in the system, but these illustrate the notion of a branch of pure mathematics. A number of the postulates and theorems just considered can be more rhetorically and more tersely worded by using some of the theory of relations developed in Section 5.5 and its associated problems. Thus P1, P2, P3, T1, T5, and T7 can be restated as

P1. R *is determinate in* K.

P2. R *is irreflexive in* K.

P3. R *is transitive in* K.

T1. R *is asymmetric in* K.

T5. D *is transitive in* K.

T7. F *is intransitive in* K.

Theorem T3 is an example of an *existence theorem* and Theorem T4 is an example of a *uniqueness theorem*. Such theorems are common and important in many bodies of mathematics.

The following three interpretations of our branch of pure mathematics result in three derived branches of applied mathematics.

Application 1 (Genealogical)

Let the elements of K be four men—some man, his father, his father's father, and his father's father's father—and let R mean "is an ancestor of."

We readily see that this interpretation of the elements of K and of the relation R converts the postulates into true propositions. We are thus led to a concrete model of our abstract postulational discourse—that is, to a branch of applied mathematics derived from our branch of pure mathematics. The

theorems, which now must all become true propositions, and the definitions read
as follows:

T1(1). If a is an ancestor of b, then b is not an ancestor of a.

T2(1). If a is an ancestor of b, and if c is one of the four men, then either a is
an ancestor of c or c is an ancestor of b.

T3(1). At least one man in K is not an ancestor of anyone in K.

T4(1). Only one man in K is not an ancestor of anyone in K.

D1(1). If b is an ancestor of a, we say that a is a *descendant* of b.

T5(1). If a is a descendant of b and b is a descendant of c, then a is a
descendant of c.

D2(1). If a is an ancestor of b and there is no individual c such that a is an
ancestor of c and c is an ancestor of b, we say that a is a *father* of b.

T6(1). A man has at most one father.

T7(1). If a is the father of b and b is the father of c, then a is not the father
of c.

D3(1). If a is the father of b and b is the father of c, we say that a is a
grandfather of c.

Application 2 (Geometrical)

Let the elements of K be four distinct points on a horizontal straight line, and let
R mean "is to the left of."

Again our postulates are satisfied and we have a second branch of applied
mathematics derived from our branch of pure mathematics. The relation D
means "is to the right of," the relation F means "is the first point of K to the left
of," and relation G means "is the second point of K to the left of."

Application 3 (Arithmetical)

Let the elements of K be the four integers 1, 2, 3, 4, and let R mean "is less
than."

Once again our postulates are satisfied and we have a third branch of applied
mathematics derived from our branch of pure mathematics. Here relation D
means "is greater than," relation F means "is 1 less than," and relation G means
"is 2 less than."

Our example of a branch of pure mathematics, with its derived branches of
applied mathematics, illustrates the "economy" feature of the modern axiomatic
method. Any theorem of the branch of pure mathematics yields a corresponding
theorem in each of the applications, and these latter require no proof as long as
the theorem in the abstract system has been proved. The abstract postulate set
studied above is a postulate set for *simple order* among four elements. Any
relation that is an interpretation of the first three postulates[6] is called a *simple
order relation.* "Is less than," "is greater than," "is to the left of," "is to the right
of," "is before," "is after," are all simple order relations.

[6] These postulates were first studied by E. V. Huntington in 1905. See E. V. Huntington [9].

6.3 Properties of Postulate Sets—
Equivalence and Consistency

It must not be thought, in building up an abstract postulational discourse, that we may set down a collection of symbols for undefined terms and then list for postulates an arbitrary system of assumed statements about these terms. This system of assumed statements—our postulates—should possess certain required and certain desired properties. This section and the following section accordingly are devoted to a brief examination of some of the properties of postulate sets Such a study is technically known as *metamathematics* and was first brought into prominence by Hilbert's *Grundlagen der Geometrie*. Of the properties of postulate sets, we consider the four known as *equivalence, consistency, independence,* and *completeness*. The first property applies to pairs of postulate sets, and the remaining three apply to individual postulate sets.

Equivalence

Two postulate systems $P^{(1)}$ and $P^{(2)}$ are said to be *equivalent* if each system implies the other—that is, if the primitive terms in each are definable by means of the primitive terms of the other, and if the postulates of each are deducible from the postulates of the other. If two postulate systems are equivalent, then the two abstract studies implied by them are, of course, the same, and it is merely a matter of "saying the same thing in different ways." The idea of equivalent postulate systems arose in ancient times when geometers, dissatisfied with Euclid's parallel postulate, tried to substitute for it a more acceptable equivalent. The modern studies of Euclidean geometry, with their various and quite different postulational bases, clearly illustrate that a postulate system is by no means uniquely determined by the study in question but depends on which technical terms of the study are chosen as primitive and which statements of the study are taken as unproved.

Since a given study may be built on more than one possible postulate system, the question naturally arises regarding how to determine which of two equivalent postulate systems is the better. Perhaps no simple criterion can be used to compare a pair of equivalent postulate sets with the aim of determining which set is the better; it seems to be largely a matter of personal preference. On grounds of economy of assumption, the set that contains fewer primitive terms and fewer postulates might seem to be preferred. There is certainly something satisfying in reducing the number of primitive terms to the barest minimum, but the idea of reducing the number of postulates to a minimum is somewhat illusory, for one could, if one wished, lump all the postulates of a given set into one big, but complicated, postulate by using a proper number of conjunctions. On the other hand, there might be reason to think a postulate set better, because it would seem to be in a sense simpler, if each postulate were to be broken down into as many separate statements as possible. But this, too, has little significance, as Olaf Helmer amusingly pointed out by the following example:[7] The simple state-

---

[7] O. Helmer.

ment, "There is one and only one x that satisfies $g(x)$," can be replaced by the five statements:

1. The number of x's that will satisfy $g(x)$ is odd.

2. The number of x's that will satisfy $g(x)$ is less than 8.

3. The number of x's that will satisfy $g(x)$ is not equal to 7.

4. The number of x's that will satisfy $g(x)$ is not equal to 5.

5. The number of x's that will satisfy $g(x)$ is not equal to 3.

Such a procedure may be extended indefinitely. Finally, one postulate set may lead more rapidly than another to the key theorems of a study and may for this reason be preferred to the other. But if the concern is to reach certain key results as expeditiously as possible, these key results can be included in the postulate set to avoid delay in reaching them.

As a postulate set equivalent to that of the example developed in Section 6.2 we might use P1, T1, P3, P4. Since T1 has already been derived from P1, P2, P3, P4, it suffices to show that P2 can be derived from P1, T1, P3, P4. This may be accomplished as follows:

P2. *If* $a \, R \, b$, *then* $a \neq b$.

Suppose, on the contrary, that we have $a \, R \, b$ and $a = b$. Then we also have $b \, R \, a$. But this is impossible by T1. Hence P2 follows by *reductio ad absurdum*.

Consistency

A postulate set is said to be *consistent* if contradictory statements are not implied by the set. This is the most important and most fundamental property of a postulate set; without this property the postulate set is worthless, and it is useless to consider any further properties of the set.

The most successful method so far invented for establishing consistency of a postulate set is the method of models. A model of a postulate set, it will be recalled, is obtained if we can assign meanings to the primitive terms of the set that convert the postulates into true statements about some concept. There are two types of models—concrete models and ideal models. A model is said to be *concrete* if the meanings assigned to the primitive terms are objects and relations adapted from the real world, whereas a model is said to be *ideal* if the meanings assigned to the primitive terms are objects and relations adapted from some other postulate system.

Where a concrete model has been exhibited we feel that we have established the *absolute* consistency of our postulate system, for if contradictory theorems are implied by our postulates, then corresponding contradictory statements would hold in our concrete model. But contradictions in the real world we accept as being impossible. As an illustration, consider the postulate set for the miniature branch of mathematics developed in the preceding section. In the genealogical application made there, we interpreted the four primitive elements of K to be a man, his father, his father's father, and his father's father's father, and we interpreted the primitive R-relation to be "is an ancestor of." We noted that these interpretations convert the postulates into true statements about a simple genealogical concept. But the meanings assigned to the four objects and the connecting relation have been adapted from the real world, and we therefore have a resulting concrete model of our postulate system. Any inconsistency in our branch of mathematics would yield a corresponding inconsistency concerning the genealogical relationship of our four men. But since the real world, we believe, will not permit contradictions to exist within it, we also believe that we have established the absolute consistency of our postulate set. Many other concrete models of the same postulate set can easily be given. For example, we might let K consist of four books standing side by side on a bookshelf, and let the R-relation mean "precedes," or we might let K consist of four people walking Indian file, and let the R-relation mean "is in front of."

As another example of a proof of absolute consistency, consider the postulate system for plane projective geometry as given in Section 4.5. A concrete model of the finite geometry consisting of the seven symbolic points A, B, C, D, E, F, G and the seven symbolic lines a, b, c, d, e, f, g is set up by the following table, in which an asterisk at, say, the intersection of row C and column f indicates that point C lies on line f.

	a	b	c	d	e	f	g
A	⋆		⋆	⋆			
B	⋆	⋆			⋆		
C		⋆	⋆			⋆	
D		⋆		⋆			⋆
E			⋆		⋆		⋆
F	⋆					⋆	⋆
G				⋆	⋆	⋆	

Since the postulates for plane projective geometry can be verified for our interpretation by means of this table, it follows that the table furnishes a concrete model of the geometry and that the postulate set is therefore absolutely consistent.

It is not always feasible to try to set up a concrete model of a given postulate set. Thus, if the postulate set contains an infinite number of primitive elements, a concrete model would certainly be impossible, for the real world does not contain an infinite number of objects. In such instances we attempt to set up an ideal model, by assigning to the primitive terms of postulate system A, say, concepts of some other postulate system B, in such a manner that the interpretations of the postulates of system A are logical consequences of the postulates of system B. But now our test of consistency of the postulate set A can no longer claim to be an absolute test, but only a *relative* test. All we can say is that postulate set A is consistent if postulate set B is consistent, and we have reduced the consistency of system A to that of another system B.

Relative consistency is the best we can hope for when we apply the method of models to many of the branches of mathematics, for many of the branches of mathematics contain an infinite number of primitive elements. Consider, as an illustration, the postulate set of Section 6.2, with P4 replaced by P′4: K *consists of an infinite number of distinct elements*. As an interpretation of this postulate set we might let the collection K be the set of all positive integers and let the R-relation mean "is less than." But the resulting model is an ideal one, adapted from the arithmetic of the positive integers, and all we can say is that our postulate set is consistent if the arithmetic of the positive integers is consistent.

The idea of relative consistency was encountered when, in Chapter 4, we endeavored to show the consistency of Lobachevskian plane geometry. By employing certain concepts from Euclidean plane geometry, we were able to set up an ideal model known as the Poincaré model, and we succeeded in showing that Lobachevskian plane geometry is consistent if Euclidean plane geometry is consistent. The idea of analytic geometry, on the other hand, sets up an ideal arithmetical model of Euclidean plane geometry, showing that Euclidean plane geometry is consistent if the real number system is consistent. It also follows, then, that the Lobachevskian plane geometry is consistent if the real number system is consistent. In this way, the consistency of many branches of mathematics can be reduced to the consistency of some basic branch, which, for the larger part of mathematics and the whole of physics, is the real number system. This important idea will be a main topic of interest in the next chapter.

Proof of consistency by the method of models is an indirect process. It is conceivable that absolute consistency may be established by a direct procedure that endeavors to show that by following the rules of deductive inference no two theorems can be arrived at from a given postulate set that will contradict each other. In such a procedure a complete enumeration of the permissible rules of logic is, of course, necessary. In recent years, Hilbert attacked the problem of securing the consistency of the real number system in such a direct manner, but with only partial success. Since this method depends on the rules of logical inference, any change in those rules could upset a previously established consistency proof of this sort. An advantage of the method of models is that it is independent of the "rules of the game."

Although the direct method of establishing consistency is too complicated to illustrate here, we can clarify the idea of the method by an analogue in chess.[8] Suppose we wish to show that in a game of chess, no matter how many moves might be made, one can never, if one plays in accordance with the rules, arrive at the situation in which ten queens of the same color are on the board. Here the direct method is applicable, for we can prove from the rules of the game that no move can increase the sum of the number of queens and pawns of the same color. Since this sum is initially 9, it must remain ≤ 9.

---

[8] See H. Weyl [4], pp. 23, 24.

6.4   Properties of Postulate Sets—Independence, Completeness, and Categoricalness

In this section we consider the two properties of a postulate set known as independence and completeness.[9] These two properties differ from that of consistency, considered in the previous section, in that they are not required properties but rather are properties that are sometimes desired and sometimes not desired. We henceforth assume that our postulate set is consistent, for there is little point in studying independence and completeness of an inconsistent postulate set.

Independence

A postulate of a postulate set is said to be *independent* in the set if it is not a logical consequence of the other postulates of the set, and the entire postulate set is said to be *independent* if each of its postulates is independent.

The most famous consideration in the history of mathematics of the independence of a postulate is that associated with the study of Euclid's parallel postulate. For centuries mathematicians had difficulty in regarding the parallel postulate as independent of Euclid's other postulates (and axioms) and accordingly made repeated attempts to show that it was a consequence of these other assumptions. The discovery of, and the ultimate proof of the relative consistency of, Lobachevskian non-Euclidean geometry finally established the independence of Euclid's parallel postulate. In fact, it is no exaggeration to say that the historical consideration of the independence of Euclid's parallel postulate is responsible for initiating the entire study of properties of postulate sets and hence for the shaping of much of the modern axiomatic method.

The manner in which Euclid's parallel postulate was finally shown to be independent furnishes us with a general test for the independence of a postulate in a postulate set. The test consists in finding meanings for the primitive terms of the postulate set that falsify the concerned postulate but that verify each of the remaining postulates. If we are successful in finding such meanings, then the concerned postulate cannot be a logical consequence of the remaining postulates, for if it were a logical consequence of the remaining postulates, then the meanings that convert all the other postulates into true propositions would have to convert it also into a true proposition. A test, along these lines, of the independence of an entire set of postulates can apparently be a lengthy business, for if there are n postulates in the set, n separate tests (one for each postulate) will have to be formulated.

Let us show that the postulate set studied in Section 6.2 is an independent set.

To show the independence of Postulate P1, let us interpret K as consisting of two brothers, their father, and their father's father, and interpret the R-relation as "is an ancestor of." This interpretation verifies P2, P3, and P4 but falsifies P1.

---

[9] These properties are given somewhat varying definitions by different writers.

To show the independence of Postulate P2, we may interpret K as the set of integers 1, 2, 3, 4, and the R-relation as "is not greater than." This interpretation verifies each postulate except P2.

To show the independence of Postulate P3, let us interpret K as a set of any four distinct elements and the R-relation as "is not identical with." Now all the postulates except P3 are verified.

Finally, to show the independence of Postulate P4, one may interpret K as the set of five integers 1, 2, 3, 4, 5 and the R-relation as "is less than." All the postulates except P4 are verified by this interpretation.

Independence of a postulate set is by no means necessary, and a postulate set clearly is not invalidated just because it lacks independence. Generally speaking, a mathematician prefers a postulate set to be independent, for he wants to build his theory on a minimum amount of assumption. A postulate set that is not independent is merely redundant in that it contains one or more statements that ought perhaps to appear as theorems instead of as postulates. On the other hand, on occasion it is a decided advantage to have a postulate set not independent. For example, in the teaching of mathematics, one may wish, for pedagogical reasons, to develop a subject from a postulational basis. However, as sometimes happens, an early theorem in the development may be very difficult to prove. This theorem can then be stated as one of the postulates; later, when the students have gained the requisite mathematical maturity and familiarity with the subject, it can be pointed out that the postulate is really not independent, and a demonstration of it from the other postulates can be given.

Some well-known postulate sets, when first published, unknowingly contained postulates that were not independent. Such was the situation with Hilbert's original set of postulates for Euclidean geometry. This set was later shown to possess two postulates that are implied by the others. The finding of these two dependent postulates in no way invalidated Hilbert's system; in a subsequent amendment these postulates were merely changed to theorems, and their proofs supplied.

Similarly, R. L. Wilder was able to show that R. L. Moore's famous set of eight postulates, which virtually inaugurated modern set-theoretic topology, could be reduced to seven by the elimination of Moore's sixth postulate.[10] The suspicion that the sixth postulate was not independent arose from the fact that the independence proof for this postulate was found to be at fault, and a subsequent search for a satisfactory proof turned out to be fruitless. Of course, Moore's mathematical theory remained intact in spite of Wilder's discovery, but the reduction of an eight-postulate system to an equally effective seven-postulate system has an aesthetic appeal to the mathematician.

Again, however, as in the above discussion of equivalence of postulate systems, the actual number of postulates used as a basis is of little significance in connection with the property of independence. Any postulate set, independent or not, can easily be converted into an independent set having but one postulate by the simple device of lumping all the separate postulates of the set into one big postulate by means of conjunctions.

---

[10] R. L. Moore [2] and R. L. Wilder [1].

Completeness

The property of completeness is more recondite than the three properties already described, and so it might be well to precede the definition of completeness with some motivation for it. Consider for the moment, then, the hypothetical task of building up a postulate set for plane Euclidean geometry. The first thing we might do is to select our primitive terms. These must constitute a collection of technical terms of the geometry such that all other technical terms of the study can be defined by means of them. The next thing we might do is to begin formulating a growing list of mutually compatible statements about the primitive terms. These will be our postulates. Here a problem arises. When can we cut short our growing list of postulates? We want our postulate set to be ample enough to imply the "truth" or the "falseness" of any possible statement in plane Euclidean geometry. In other words, we want to have a sufficient number of postulates so that if S is any statement whatever concerning the primitive terms, then either S or its contradictory statement, not-S, is implied by our postulates. If the postulate set is not sufficiently ample, certainly some statements of the geometry will not be able to be reached from our postulates. This condition would exist, for example, if we had chosen Hilbert's collection of primitive terms and all of Hilbert's postulates except, say, the parallel postulate. This slightly truncated postulate set could never decide for us whether or not the sum of the angles of a triangle is always equal to 180°, for our truncated postulate set is common to the postulate sets of both Euclidean and Lobachevskian plane geometry. Apparently our list of postulates will be sufficiently ample if a point is reached when it becomes impossible to add to the list any further statement that will be both independent of, and consistent with, the postulates so far formulated. If such a point can be reached, then we have what may be called a complete postulate set for the geometry.

We may now give the following formal definition of completeness. A consistent postulate set is said to be *complete* if it is impossible to add to the set, without extending the collection of primitive terms, another postulate which is both independent of, and consistent with, the given postulates.

Categoricalness

The definition of completeness, like that of consistency, is not easily adapted to testing a postulate set for the property concerned. A test for consistency was devised that depends on the concept of an interpretation of a postulate set. A test for completeness can be devised that depends on a concept called *categoricalness* of a postulate set, explained here.

Among the primitive terms of a postulate set P we have a collection of E's, say, that denote elements, perhaps some relations R_1, R_2, ... among the elements, and perhaps some operations O_1, O_2, ... on the elements. Accordingly, an interpretation of the postulate set is composed at least in part of element constants (the meanings assigned to the E's), perhaps in part of relation constants (the meanings assigned to the R's), and perhaps in part of operation constants (the meanings assigned to the O's). In any given interpretation I of P, let a collection of e's be the element constants (representing the E's), $r_1, r_2, \ldots$

the relation constants (representing the R's), and $o_1, o_2, \ldots$ the operation constants (representing the O's); and in any other interpretation I' of P let the element constants be a collection of e''s, the relation constants $r'_1, r'_2, \ldots$, and the operation constants $o'_1, o'_2, \ldots$. If it is possible to set up a one-to-one correspondence between the elements e of I and the elements e' of I' in such a way that, if two or more of the e's are related by some r, the corresponding e''s are related by the corresponding r', and if an o operating on one or more of the e's yields an e, the corresponding o' operating on the corresponding e''s yields the corresponding e', then we say that the two interpretations I and I' of P are *isomorphic*. This definition is often more briefly stated by saying that two interpretations I and I' of a postulate set P are isomorphic if one can set up a one-to-one correspondence between the elements of I and those of I' in such a way as "to be preserved by the relations and the operations of P." It follows that if two interpretations I and I' of a postulate set P are isomorphic, then any true (false) proposition p in interpretation I becomes a true (false) proposition p' in interpretation I' when we replace the e's, r's, and o's in p by their corresponding e''s, r''s, and o''s. Two isomorphic interpretations of a postulate set P are, except for superficial differences in terminology and notation, identical; they differ from each other no more than does the multiplication table up to 10×10 when correctly written first in English and then in French.

With the notion of isomorphic interpretations of a postulate set established, we are prepared to define categoricalness of a postulate set. A postulate set P is said to be *categorical* if every two interpretations of P are isomorphic.[11]

We shall now show that categoricalness of a consistent postulate set P implies completeness of P. To this end suppose that P is categorical but incomplete. Since P is incomplete, a statement S exists about the primitive terms of P such that both S and not-S are consistent with P. Let I be an interpretation of the consistent set of statements $P + S$, and let I' be an interpretation of the consistent set of statements $P +$ not-S.[12] Since P is categorical, there is a one-to-one correspondence between the elements of I and the elements of I' such that corresponding propositions in the two resulting models are either both true or both false. But this is impossible, since S is a true proposition for interpretation I but a false proposition for interpretation I'.

In practice it is difficult to establish completeness of a set of postulates. To show incompleteness one has merely to produce two nonisomorphic interpretations of the postulate set. Sometimes completeness can be established by showing that any interpretation of the postulate set is isomorphic to some given interpretation. This is the process that has been applied to Hilbert's postulate set for plane Euclidean geometry; it can be shown that any interpretation of Hilbert's postulates is isomorphic to the algebraic interpretation provided by Descartes's analytic geometry. In this way it can also be shown that the simple postulate set of Section 6.2 is complete. If, on the other hand, we delete Postulate

---

[11] The adjective *categorical* seems to have been introduced into the literature by Veblen, who received the suggestion from John Dewey. Huntington had earlier employed the adjective *sufficient*, and Hilbert used the adjective *complete*.

[12] It should be noted that the proof rests on the (doubtful) assumption that a consistent set of statements always possesses at least one interpretation.

P4 from the postulate set of Section 6.2, then the remaining postulate set is incomplete. This is readily seen by interpreting R as "is less than" and by taking K to consist in turn of the three integers 1, 2, 3 and of the four integers 1, 2, 3, 4. Each of these interpretations verifies Postulates P1, P2, and P3, but the interpretations are not isomorphic, since it is impossible to set up a one-to-one correspondence between the three elements of the one interpretation and the four elements of the other interpretation.

There are advantages and disadvantages in having a postulate set complete. Perhaps the most desirable feature of an incomplete postulate set is its wide range of applicability. This feature is strikingly seen in the postulate set of Section 6.2, where the deletion of Postulate P4 yields an incomplete postulate set that may be interpreted for any n sequentially related elements—indeed, we may even take n infinite. When we wish to study a structure that is common to a number of more complicated structures, an incomplete postulate system is, of course, necessary. For example, many structures contain the group as a substructure, and it is therefore convenient and economical to have the group properties worked out once and for all from an appropriate incomplete postulate set. A similar remark can be made about the postulate sets governing such important structures as fields, rings, simple order, equivalence, and so on. Frequently, in building up a given theory by the postulational method, it is desirable to introduce the postulates one by one or in related sets, so that it becomes clear just how the theorems of the study depend on the underlying postulates.[13] For example, the theorems common to Euclidean and Lobachevskian plane geometry may be found from Hilbert's postulate set with the parallel postulate deleted. Such partial postulate sets are, of course, incomplete, and may be verified by nonisomorphic interpretations. On the other hand, one certainly hopes to have complete postulate sets for certain significant branches of mathematics, like Euclidean geometry and the real number system (which lies at the basis of analysis). If p is any statement about the primitive terms of such a postulate set, then of p and its contradictory statement, not-p, one and only one is implied by the postulate set.

6.5 Miscellaneous Comments

We shall conclude the present chapter with a few additional miscellaneous comments about postulate sets and the modern axiomatic method.

Mathematicians have considered properties of postulate sets other than those we have already described. Thus E. H. Moore introduced, in 1910, a property that he called "complete independence" of a postulate set.[14] Moore defined a postulate set P to be *completely independent* if, for every proper subsystem $P_i, P_j, \ldots, P_k$ of the postulates, both P and P with each of the postulates $P_i, P_j, \ldots, P_k$ replaced by its contradictory statement are consistent. This definition is a generalization of the definition of independence of a postulate

---

[13] See, for example, D. Hilbert [1], H. G. Forder [1], and W. Sierpinsky.

[14] E. H. Moore [2], p. 82.

set as given in Section 6.4. It can be shown that an independent postulate set is not necessarily completely independent.[15]

Closely related to the concepts of independence of a postulate and of a postulate set are the concepts of relative "weakness" of a postulate and of a postulate set. Given two postulates P_i and P'_i, we say that P'_i is *weaker* than P_i (and P_i is *stronger* than P'_i) if P'_i is implied by P_i but P_i is not implied by P'_i; and a postulate set containing P'_i is then said to be *weaker* than the same postulate set with P'_i replaced by P_i. Thus, of two equivalent postulate sets it is possible for one to be weaker than the other. For example, we gave, toward the end of Section 5.1, a postulate set for a field. At the time, we pointed out that the postulate set given there is somewhat redundant, and we pointed out some of the redundancies. Nevertheless, the redundant postulate set and the allied set with the redundancies removed are equivalent; we merely listed a stronger postulate set than was needed. The idea of relative weakness of equivalent postulate sets was also encountered in Section 5.3 in connection with our postulate set for a group. This idea of relative weakness of pairs of equivalent postulate sets is often useful when we consider which of two equivalent postulate sets may be thought the simpler. Forder[16] has discussed a number of parallel postulates, of varying strengths, for Euclidean plane geometry.

It is interesting to note that the concept of independence, and the method of testing for independence, are applicable in the formulation of definitions.[17] Suppose, for example, a square is defined as "a quadrilateral that is both equilateral and equiangular." The three properties in the definition—that of being a quadrilateral, of being equilateral, and of being equiangular—are consistent in Euclidean geometry, for a figure possessing all three properties can be constructed. Furthermore, none of the three properties is redundant, for figures exist in Euclidean geometry that possess any two of the three properties but lack the remaining one, as is instanced by an equilateral triangle (which possesses the second and third properties, but lacks the first), an oblong (which possesses the first and third properties, but lacks the second), and a non-rectangular rhombus (which possesses the first and second properties, but lacks the third). In contrast to the above definition of a square, consider the definition of a parallelogram as "a quadrilateral whose pairs of opposite sides are parallel and equal." This definition also involves three properties, and the three properties are certainly consistent in Euclidean geometry. But the three properties are not independent, inasmuch as the property of equal opposite sides, for example, can be proved from the property of parallel opposite sides. The definition, then, although not incorrect, is undesirable because it contains redundancies and is therefore unnecessarily cumbersome.

Many interesting comments can be made in connection with the postulate property of completeness; we shall here content ourselves with but one.[18] It should be noticed that the concept of isomorphism of interpretations of a

---

[15] For a simple example, see O. Helmer.

[16] H. G. Forder [1].

[17] See E. R. Stabler [4], p. 158.

[18] See R. L. Wilder [3], p. 49.

postulate set is an equivalence relation. Thus, if P is a postulate system and if I_1 and I_2 represent two interpretations of P, we may let "$I_1 \approx I_2$" mean "I_1 and I_2 are isomorphic interpretations." On checking the equivalence postulates, as given in Section 5.5, we find that they are satisfied by this meaning of $\approx$. It follows that we can decompose the set of all interpretations of P into classes, putting two interpretations into the same class if and only if they are isomorphic. Categoricalness of P corresponds to the condition where there is only one class of interpretations. In the case of noncategoricalness, we may have any number of classes, and the number of such classes may be able to serve as a quantitative measure of the incompleteness of a postulate set.

The tests that we have discussed for consistency, independence, and completeness of a postulate set raise some deep questions that have only recently been at least partly answered. For example, might not a postulate set be inconsistent without our being able to discover the fact? And, on the other hand, might not a postulate set be consistent, though we should never know it? Can a consistent postulate set containing only a finite number of primitive elements exist for which there is no interpretation? If such a consistent postulate set should have no known interpretation, it nevertheless can be thought of as describing the structure of some *possible* system of entities, though in this case it would have little interest in applied mathematics. In connection with the property of completeness we can ask the following very significant question. For a given deductive theory, can one find a procedure for deciding, in a finite number of explicitly described steps, whether a given proposition formulated in terms of the technical terms of the theory is or is not within the theory? In other words, can one find a procedure to decide of any particular proposition whether it, its denial, or neither belongs to the theory? Questions like this have led in recent times to the development of an intricate but important *decision theory*, into which we cannot enter here. We content ourselves by merely stating that the decision problem has been solved for some deductive theories and not for others; indeed, an argument has been developed that demonstrates that there is a large class of deductive theories for which a decision procedure cannot exist. As still another important question in the theory of postulate sets, and one that is receiving increased attention, we might mention the following. In interpreting a postulate set, what are the allowable ranges of the various variables involved in the propositional functions? Studies related to this question have been made by A. N. Whitehead and Bertrand Russell in their so-called theory of logical types.[19] Different approaches have been employed by others, notably W. V. Quine.[20] Without some limitation imposed on these ranges, it can be shown that paradoxes may creep into one's work. At the present time this subject is in a state of uncertainty, and it definitely is not a matter for elementary consideration.

Some of the above matters cause little or no trouble in connection with most of the postulate sets that are employed by mathematicians and others because most postulate sets originate with a model of the abstract system. That is, in most cases, the individual researcher has in mind some specific model and then

---

[19]A. N. Whitehead, and B. Russell.

[20]W. V. Quine.

proceeds to build a postulate system that will cover his model, so that, in practice, the model or interpretation generally comes first and the postulate set later. This is particularly true in such fields as biology, chemistry, economics, ethics, law, mechanics, philosophy, physics, zoology, and so on, where someone familiar with the field might choose to set down some postulates for the field, or for some part of the field, and then see what theorems can be logically deduced from the postulates. Theoretically, of course, the model does not need to precede the abstract development, though this is the usual source of a postulate set. If one were to put down some arbitrary symbols for primitive terms and then to try to formulate postulates about these terms, it would be difficult, without some model in mind, to think of anything to say, and, moreover, one would have to take great care to ensure consistency of the postulates. Sometimes a new postulate system is derived from a given postulate system by altering one or more postulates of the given system. In this way postulate systems for Lobachevskian and Riemannian non-Euclidean geometries were derived from the postulates for Euclidean geometry.

Note that there are, in general, three distinct levels in axiomatic study: the more or less informal theories of specific fields of knowledge, the formal abstract postulational developments having these specific fields as models, and finally a theory that studies the properties possessed by formal abstract postulational developments. It is this third and highest of the three levels that Hilbert christened *metamathematics*.[21]

One very important feature of the postulational method has as yet scarcely been touched on. In part (4) of the pattern of formal axiomatics, as outlined in Section 6.1, we note that the theorems of the discourse must be *logically deduced* from the postulates of the discourse. That is, a formal mathematical organization starts with a set of postulates, which are then expanded into a more or less extensive mathematical theory through the use of a logic or set of rules of inference accepted as an adjunct to the theory. In a sense, the use of a logic does not add to the statements of the postulates but merely extracts from the postulates what is already in a way implied by them and is perhaps hidden by them in a subtle manner. In any event, however, the mathematical theory results from the interplay of two factors—a set of postulates and a logic.

We have given considerable attention to the first of the two factors of a logical discourse (namely, the postulates of the discourse), but as yet we have paid scant attention to the second factor—namely, the logic that is adopted for deducing the theorems of the discourse from the postulates of the discourse. The logic employed by mathematicians is usually the classical logic of Aristotle, developed by the Greek philosopher in rudimentary form in the fourth century B.C. In fact, Aristotle is said to have received suggestions for his science of deduction from a study of mathematical reasoning then in vogue. Most people believe that logical thinking, according to the Aristotelian tradition, is virtually instinctive in a person, whereas serious students of the subject realize that the

---

[21] Perhaps the chief problems of metamathematics center around the permissible proof paraphernalia of the axiomatic method so that consistency will be ensured. For an elementary discussion of some of the problems of metamathematics see, for example, A. Tarski [2], chap. 6. Also see R. Carnap [1] and S. C. Kleene [2].

ability to construct valid arguments must largely be developed through diligent effort. In truth, instructors of subjects requiring a knowledge of deductive reasoning recognize that many students have very little ability in obtaining logical consequences from given hypotheses.

Until quite recent times the classical logic of Aristotle was believed to be the only possible system of deductive logic. If this were the case, then the logical constituent of a logical discourse would be a constant one; that is, it would be the same for all logical discourses. One of the achievements of recent times is the realization that there is not just one system of deductive logic but an unlimited number of possible systems. The significance of this achievement to mathematics may in the future be very great, for whereas it has been appreciated for some decades that the axiomatic constituent in formal mathematical organization is a variable one, now the task of the mathematician may be considerably complicated by the fact that the logical constituent may also be variable. The *Principia mathematica* of Whitehead and Russell employs a logic that is not precisely Aristotelian in nature, yet the deductions made there are acceptable. Certainly mathematicians will do well to stay abreast of the work that is being done in formal logic. It is clear that our own neglect of the logical element of mathematics must be remedied. We shall accordingly return to this matter in Chapter 9 after further preliminary groundwork has been laid. Even a casual reader of these pages must have observed the dependence of the properties of postulate sets on such laws of classical logic as the law of contradiction and the law of the excluded middle. Should a logic be adopted in which these laws are in some sense questioned, the whole treatment of consistency, independence, and completeness of a postulate set may need recasting.

PROBLEMS

6.1.1   (a) Construct an example of a propositional function containing three variables.

(b) Obtain, by appropriate substitutions, three true propositions from the propositional function.

(c) Obtain, by appropriate substitutions, three false propositions from the propositional function.

(d) Discuss the statement "God is love" in terms of the concepts *proposition* and *propositional function*.

6.1.2   In elementary mathematics, an equation in one variable x, say, can be regarded as a propositional function in which the only admissible meanings for x are complex numbers. A fundamental problem in elementary mathematics is the determination of those complex numbers, if there are such, for which the equation becomes a true proposition. What are the verifying values of x for each of the following equations?[22]

(a) $3150 + 15x = 3456 - 33/2x$.

(b) $\sin x = \cos x$.

(c) $2x^2 - 5x - 3 = 0$.

(d) $x^2 = 3 - x$.

---

[22] See C. V. Newsom and H. Eves, chap. 12.

(e) $x^2 + 3x + 6 = 0$.

(f) $\dfrac{2x}{x^2 - 9} + \dfrac{2}{x + 3} = \dfrac{1}{x - 3}$.

(g) $\sqrt{x + 2} = -\sqrt{x^2 - 4}$.

6.1.3 Discuss the following definitions (or descriptions) of mathematics in relation to the notion of the subject as an assemblage of discourses developed by formal axiomatics:[23]

(a) "The purely formal sciences, logic and mathematics, deal with those relations which are, or can be, independent of the particular content or the substance of objects." (Hermann Hankel, 114)

(b) "Perhaps the least inadequate description of the general scope of modern Pure Mathematics—I will not call it a definition—would be to say that it deals with *form*, in a very general sense of the term." (E. W. Hobson, 118)

(c) "Mathematics in its widest signification is the development of all types of formal, necessary, deductive reasoning." (A. N. Whitehead, 122)

(d) "A mathematical science is any body of propositions which is capable of an abstract formulation and arrangement in such a way that every proposition of the set after a certain one is a formal logical consequence of some or all the preceding propositions. Mathematics consists of all such mathematical sciences." (J. W. Young, 124)

(e) "Pure mathematics is a collection of hypothetical, deductive theories, each consisting of a definite system of primitive, *undefined*, concepts or symbols and primitive, *unproved*, but self-consistent assumptions (commonly called axioms) together with their logically deducible consequences following by rigidly deductive processes without appeal to intuition." (G. D. Fitch, 125)

(f) "Pure mathematics consists entirely of such asservations as that, if such and such a proposition is true of *anything*, then such and such another proposition is true of that thing. It is essential not to discuss whether the first proposition is really true, and not to mention what the anything is of which it is supposed to be true. . . . If our hypothesis is about *anything* and not about some one or more particular things, then our deductions constitute mathematics. Thus mathematics may be defined as the subject in which we never know what we are talking about, nor whether what we are saying is true." (Bertrand Russell, 127)

(g) "The critical mathematician has abandoned the search for truth. He no longer flatters himself that his propositions are or can be known to him or to any other human being to be true; and he contents himself with aiming at the correct, or the consistent. The distinction is not annulled nor even blurred by the reflection that consistency contains immanently a kind of truth. He is not absolutely certain, but he believes profoundly that it is possible to find various sets of a few propositions each such that the propositions of each set are compatible, that the propositions of each such set imply other propositions, and that the latter can be deduced from the former with certainty. That is to say, he believes that there are systems of coherent or consistent propositions, and he regards it his business to discover such systems. Any such system is a branch of mathematics." (C. J. Keyser, 132)

(h) "[Mathematics is] the study of ideal constructions (often applicable to real

---

[23] All of these quotations, except the last, are taken with the permission of Mrs. R. E. Moritz from R. E. Moritz. The number following each quotation is that assigned to it in Moritz's collection.

problems), and the discovery thereby of relations between the parts of these constructions, before unknown." (C. S. Peirce, 133)

(i) "Mathematics is preferably free in its development and is subject only to the obvious consideration, that its concepts must be free from contradictions in themselves, as well as definitely and orderly related by means of definitions to the previously existing and established concepts." (Georg Cantor, 207)

(j) "Mathematicians assume the right to choose, within the limits of logical contradiction, what path they please in reaching their results." (Henry Adams, 208)

(k) "[W]e cannot get more out of the mathematical mill than we put into it, though we may get it in a form infinitely more useful for our purpose." (John Hopkinson, 239)

(l) "In Pure Mathematics, where all the various truths are necessarily connected with each other (being all necessarily connected with those *hypotheses* which are the principles of the science), an arrangement is beautiful in proportion as the principles are few; and what we admire perhaps chiefly in the science, is the astonishing variety of consequences which may be demonstrably deduced from so small a number of premises." (Dugald Stewart, 242)

(m) "The mathematician may be compared to a designer of garments, who is utterly oblivious of the creatures whom his garments may fit. To be sure, his art originated in the necessity for clothing such creatures, but this was long ago; to this day a shape will occasionally appear which will fit into the garment as if the garment has been made for it. Then there is no end of surprise and delight.[24]"

6.2.1 **(a)** Establish the following theorem of the simple branch of pure mathematics of Section 6.2: *If* a G c *and* b G c, *then* a = b.

(b) Define the triadic relation $B(abc)$ to mean either ($a\,R\,b$ and $b\,R\,c$) or ($c\,R\,b$ and $b\,R\,a$). Now prove: *If* B(abc) *holds, then* B(acb) *does not hold.*

(c) What is the meaning of $B(abc)$ in the three applications of the branch of pure mathematics given in Section 6.2?

6.2.2 Write out the statements of the theorems and definitions of the geometrical application of the branch of pure mathematics of Section 6.2.

6.2.3 Write out the statements of the theorems and definitions of the arithmetical application of the branch of pure mathematics of Section 6.2.

6.2.4 In the branch of pure mathematics of Section 6.2, let K be a set of four concentric circles of different radii, and let R mean "is contained within." Show that these meanings yield an interpretation of the postulate set. Write out the statements of the theorems and definitions in the resulting model.

6.3.1 Show that equivalence of postulate sets is an equivalence relation as defined in Section 5.5.

6.3.2 Show that the ten postulates for a field, as given in Section 5.1, are equivalent to the same set of postulates with Postulates P5 and P9 replaced by

P′5. *If* a, b, c *are in* S, *then* $a \otimes (b \oplus c) = (a \otimes b) \oplus (a \otimes c)$.

P′9. *If* a, b, c *are in* S, $c \neq z$, *and* $c \otimes a = c \otimes b$, *then* a = b.

6.3.3 Establish the absolute consistency of the postulate set for a field (as given in Section 5.1).

---

[24] Quoted by permission from T. Dantzig, pp. 231–2.

6.3.4 Which of the following proposed postulate sets are consistent?

    (a) The ten field postulates of Section 5.1 plus the statement, "For every element a in S, $a \oplus a = z$."

    (b) The ten field postulates of Section 5.1 plus the statement, "For every element a in S, $(a \oplus a) \oplus a = z$."

    (c) The ten field postulates of Section 5.1 plus the statement, "For every element a in S, $a \otimes a = a$."

    (d) The ten field postulates of Section 5.1 plus the statement, "S contains infinitely many elements."

    (e) The ten field postulates of Section 5.1 plus the statement, "S contains only a finite number of elements."

    (f) The ten field postulates of Section 5.1 plus the statement, "If a and b are elements of S, then $a \oplus b = a \oplus \bar{b}$."

    (g) The ten field postulates of Section 5.1 plus the statement, "There exist elements a and b of S such that $a \neq z$, $b \neq z$, and $a \otimes b = z$."

    (h) The ten field postulates of Section 5.1 with Postulate P10 replaced by P'10: *For each element* a *in* S *there exists an element* a^{-1} *in* S *such that* $a \otimes a^{-1} = u$.

6.3.5 If p, q, r represent propositions show that the following set of four statements is inconsistent: (1) If q is true, then r is false. (2) If q is false, then p is true. (3) r is true. (4) p is false.

6.3.6 Compare the concept of consistency and inconsistency of a set of simultaneous equations with the concept of consistency and inconsistency of a postulate set.

6.3.7 Show that the table given in Section 6.3 establishes the absolute consistency of the postulate set for plane projective geometry as given in Section 4.5.

6.3.8 Let S be a set of elements and F a dyadic relation satisfying the following postulates:

    P1. *If* a *and* b *are elements of* S *and if* b F a, *then* a $\not{F}$ b.

    P2. *If* a *is an element of* S, *then there is at least one element* b *of* S *such that* b F a.

    P3. *If* a *is an element of* S, *then there is at least one element* b *of* S *such that* a F b.

    P4. *If* a, b, c *are elements of* S *such that* b F a *and* c F b, *then* c F a.

    P5. *If* a *and* b *are elements of* S *such that* b F a, *then there is at least one element* c *of* S *such that* c F a *and* b F c.

Show that the statement, "If a is an element of S, then there is at least one element b of S, distinct from a, such that $b \not{F} a$ and $a \not{F} b$," is consistent with the above postulates.

(This set of postulates, augmented by the above statement, has been used in relativity theory, where the elements of S are interpreted as *instants* and F as meaning "follows."[25])

6.3.9 Let K be a set of elements and R a triadic relation, and consider the following eight statements: (1) If a, b, c are elements of K, then at least one of the six relations $R(abc)$, $R(bca)$, $R(cab)$, $R(cba)$, $R(bac)$, $R(acb)$ holds. (2) There are at least three elements x, y, z of K such that $R(xyz)$ holds. (3) The two relations $R(abc)$ and $R(acb)$ cannot both hold. (4) If $R(abc)$ holds, then a, b, c are distinct. (5) If $R(abc)$ holds, then $R(bca)$ holds. (6) If $R(xab)$ and $R(ayb)$ hold, then $R(xay)$

---

[25] A. A. Robb.

holds. (7) If $R(xab)$ and $R(ayb)$ hold, then $R(xyb)$ holds. (8) If $R(abc)$ holds and x is any other element of K, then either $R(abx)$ or $R(xbc)$ holds.

(a) Show that the above eight statements are consistent.

(b) Show that the following four sets of statements are equivalent to one another: (1), (3), (4), (5), (6); (1), (3), (4), (5), (7); (1), (3), (4), (5), (8); (2), (3), (4), (5), (8). (Each of these four sets constitutes a postulate set for *cyclic order*.)

6.3.10 Consider the following postulate set, in which *bee* and *hive* are primitive terms:

P1. *Every hive is a collection of bees.*

P2. *Any two distinct hives have one and only one bee in common.*

P3. *Every bee belongs to two and only two hives.*

P4. *There are exactly four hives.*

Show that this set of postulates is absolutely consistent.

6.3.11 Deduce the following theorems from the postulate set of Problem 6.3.10:
(a) *There are exactly six bees.*
(b) *There are exactly three bees in each hive.*
(c) *For each bee there is exactly one other bee not in the same hive with it.*

6.4.1 Show that the statement in quotation marks in Problem 6.3.8 is independent of the postulates of the problem.

6.4.2 (a) Show that Postulates P2, P3, P4 of the postulate set of Problem 6.3.10 are independent.
(b) Show that the postulate set of Problem 6.3.10 is categorical.

6.4.3 Show that the postulates for an equivalence relation, as given in Section 5.5, are independent.

6.4.4 Show that the postulate set of the simple branch of pure mathematics in Section 6.2 is categorical.

6.4.5 Clearly, if a postulate p of a consistent postulate set P contains a primitive term which is not among the primitive terms occurring in any of the other postulates of P, then p is independent.
(a) Using the above principle, show that the additional postulate, "All hives lie in the same apiary," where *apiary* is a primitive term, is independent of the four postulates of Problem 6.3.10.
(b) Also show the independence of the additional postulate by the method of models.

6.4.6 Which of the additional statements in parts (a) through (f) of Problem 6.3.4 are independent of the ten field postulates of Section 5.1?

6.4.7 Which of the following sets of postulates are categorical?
(a) The ten field postulates of Section 5.1 plus the additional postulate, "S contains an infinite number of elements."
(b) The ten field postulates of Section 5.1 plus the additional postulate, "S contains only a finite number of elements."
(c) The ten field postulates of Section 5.1 plus the additional postulate, "S contains exactly two elements."

6.4.8 Show that the postulates for an ordered field, as given in Section 5.1, form a noncategorical set.

6.4.9 Consider the following set of postulates about a certain collection S of primitive elements and certain primitive subcollections of S called *m-classes*:

P1. *If* a *and* b *are distinct elements of* S, *then there is one and only one m-class containing* a *and* b.

D1. Two *m*-classes having no element in common are called *conjugate m*-classes.

P2. *For every* m-*class there is one and only one conjugate* m-*class.*

P3. *There exists at least one* m-*class.*

P4. *Every* m-*class contains at least one element of* S.

P5. *Every* m-*class contains only a finite number of elements of* S.

Establish the following theorems:

(a) T1. *Every* m-*class contains at least two elements.*

(b) T2. S *contains at least four elements.*

(c) T3. S *contains at least six* m-*classes.*

(d) T4. *No* m-*class contains more than two elements.*

(e) Prove that the postulate set is categorical.

6.4.10 If, in the ten field postulates of Section 5.1, we replace P5 by P′5: *If* a, b, c *are in* S, *then* a $\oplus$ (b $\otimes$ c) = (a $\oplus$ b) $\otimes$ (a $\oplus$ c), is the resulting set of postulates consistent?

6.4.11 Construct all geometries satisfying the following postulates:

P1. *Space* S *is a set of* n *points,* n *a positive integer.*

P2. *A line is a non-null subset of* S.

P3. *Any two distinct lines have exactly one common point.*

P4. *Every point lies in exactly two distinct lines.*

6.5.1 Explain how Moore's definition of complete independence of a postulate set is a generalization of the definition of independence as given in Section 6.4.

6.5.2 Criticize the following possible definitions in Euclidean plane geometry:

(a) A triangle is said to be *equilateral* if its three sides are equal and if its three angles are equal.

(b) A triangle is *isosceles* if two of its sides are equal and if the angles opposite these sides are equal.

(c) A *rectangle* is a parallelogram having all four of its angles right angles.

(d) A triangle is a *right triangle* if one of its angles is a right angle and the other two angles are complementary.

(e) The *altitude* on the base of an isosceles triangle is the line perpendicular to the base and bisecting the vertex angle of the triangle.

(f) A *diameter* of a circle is any straight line drawn through the center, terminated in both directions by the circumference of the circle, and bisecting the circle.

(g) A *diameter* of a circle is a maximum chord of the circle passing through the center of the circle.

(h) Two triangles are said to be *in perspective* if the joins of corresponding vertices of the two triangles are concurrent and the intersections of corresponding sides of the two triangles are collinear.

6.5.3 Show that the concept of isomorphism of interpretations of a postulate set is an equivalence relation.

6.5.4 Compare a branch of mathematics and its two constituents—a postulate set and a logic—with a vector represented as the sum of two variable components.

6.5.5 Let P be a postulate set composed of n postulates P1, P2, . . . , Pn. Suppose there is an interpretation that verifies P1 but not P2, another interpretation that verifies P1, P2, but not P3, . . . , another interpretation that verifies P1, P2, . . . , P(n − 1), but not Pn. We then say the postulates P, in the order given, are *type-I sequentially independent*. Does type-I sequential independence imply independence?

6.5.6 Let P be a postulate set composed of n postulates P1, P2, . . . , Pn. Suppose there is an interpretation that verifies P1 but none of the remaining postulates, another interpretation that verifies P1 and P2 but none of the remaining postulates, ..., another interpretation that verifies P1, P2, ..., P(n − 1), but not Pn. We then say the postulates P, in the order given, are *type-II sequentially independent*. Does type-II sequential independence imply independence?

6.5.7 A primitive term of a logical theory T is said to be *independent* of the remaining primitive terms of T if no definition of the concerned primitive term in terms of the remaining primitive terms can be given which is provable in T. In 1899 Alessandro Padoa gave the following method for establishing the independence of a primitive term:

Find two interpretations of T such that the meanings assigned to the concerned primitive term are different in two interpretations, but the meanings assigned to the remaining primitive terms of T are the same in the two interpretations.

(a) Show that Padoa's method serves its purpose.

(b) Consider the theory T given by the following:

Primitive terms: A set S of elements $a, b, c, \ldots$ and two dyadic relations R and D.

Postulates:

P1. If $a\,R\,b$, then $b\,\check{R}\,a$.

P2. If $a\,R\,b$ and $b\,R\,c$, then $a\,R\,c$.

P3. For each a there exists a b such that $b\,R\,a$.

P4. If $a\,D\,b$, then $b\,\check{D}\,a$.

Establish the independence of the primitive relation D.

6.5.8 A postulate P of a consistent postulate set S, it will be recalled, is *independent* in S if the system $(S − P) + $ not-P has an interpretation, and S itself is *independent* if each of its postulates is independent. Also, S is *completely independent* if for every proper subset M of S the system $(S − M) + $ not-M has an interpretation, where by not-M we mean the set composed of the denials of all the members of M. Now in the above definition of an independent postulate P, it is only necessary to have an interpretation for which P does not *always* hold. Sometimes it is possible to construct an interpretation for which all the other postulates hold and P *never* holds. In this case we say that postulate P is *very independent* in S, and S is said to be *very independent* if each of its postulates is very independent. Similarly, S is said to be *absolutely independent* if for every proper subset M of S there is an interpretation for which $S − M$ holds but the postulates of M *never* hold.

(a) Show that the following is a very independent, indeed an absolutely independent, postulate set for a dyadic relation R to be an equivalence relation.

P1. R is reflexive.

P2. R is symmetric.

P3. R is distinctly transitive. (See Problem 5.5.12.)

(b) Show that the postulate set for a group as given in Section 5.3 is not very independent.

(c) Show that the parallel postulate of Euclidean geometry is very independent.

THE REAL NUMBER SYSTEM

Mathematicians have divided the great bulk of existing mathematics into three large categories—geometry, algebra, and analysis. A given branch of mathematics cannot always be placed unfalteringly into one of these categories, for the categories lack clear-cut definitions. This is perhaps to be expected, since there is no agreement even as to what mathematics itself is. Roughly, however, the classification is determined somewhat as follows. The study of space inaugurated by the ancient Greeks, and so elegantly summarized by Euclid in his *Elements*, is called geometry, together with the multitude of variations, generalizations, and associated studies that have since been created. Among the variations are such studies as the Lobachevskian and the Riemannian non-Euclidean geometries, and the non-Archimedean and non-Desarguesian geometries; among the generalizations are the higher dimensional geometries, Riemann's broad classes of geometries, and instances of Klein's idea of a geometry as the invariant theory of a definite transformation group; and among the associated studies are the finite geometries, geometries of elements other than points, and various abstract theories of sets of elements, usually called points, with sets of relations in which these points are involved. The abstract symbolic study of ordinary arithmetic is called algebra, with its many variations, generalizations, and associated studies. Thus such studies as those of groups, rings, and fields are called algebra, as are the studies of such hypercomplex numbers as quaternions and matrices. Finally, analysis consists of those branches of mathematics allied to and arising from the calculus. Such studies as the calculus itself, the theory of differential equations, function theory, the theory of infinite series, generalized theories of integration, the calculus of variations, and so on, come under this third category. The idea of a limit plays an important part in each of these latter branches of mathematics; generally the presence of the limit concept distinguishes analysis from algebra.

It must be confessed that the classification of bodies of mathematics into the three categories of geometry, algebra, and analysis is largely based on more or less sentimental and traditional grounds, and that the boundaries between the categories are becoming less and less well defined. Some topics of mathematics generate little agreement regarding the category into which they should be placed. Thus topology, which in its early development was recognized as geometric and was classifiable within Klein's program, has since developed into a body of mathematics that perhaps can be placed into any one of the three categories; indeed, it seems quite possible that in the future topology may be characterized as constituting an independent fourth division of mathematics. Nevertheless, in spite of such uncertain and borderline cases, the classification of most of mathematics into the three categories of geometry, algebra, and analysis is reasonably sound, for most branches of mathematics have ultimate parenthood in either Euclidean geometry, symbolized arithmetic, or the limit processes of the calculus.

Earlier chapters have demonstrated the importance of the two categories of geometry and algebra insofar as the foundations and the fundamental concepts of mathematics are concerned, and how, in particular, these two categories influenced the development of the modern axiomatic method, until today each category has become a great bundle of associated postulational systems. Has the third category, that of analysis, similarly contributed to the foundations and the fundamental concepts of mathematics, and has it, too, become deeply penetrated by the postulational method? We shall see that the story of analysis is somewhat different from that of geometry and algebra.

The vital contributions of geometry and algebra to the foundations and fundamental concepts of mathematics arose when it was discovered that there are geometries and algebras other than the traditional ones. Thus it was the discovery of non-Euclidean Lobachevskian geometry and the discovery of noncommutative quaternion algebra that initiated a profound study of the foundations and basic concepts of these fields of mathematics. A corresponding study of the foundations and basic concepts of the third category of mathematics came about not by a similar discovery of a new and radically different kind of analysis but instead by the gradual emergence of contradictions within the discipline itself. These contradictions revealed the existence of some grave underlying defects, the removal of which required a careful and penetrating reexamination of the foundations on which the development rests. We shall see that the effort to establish a rigorous foundation to analysis proved to have perhaps an even more profound and revolutionary effect on mathematics as a whole than did the corresponding task in geometry and algebra, owing largely to the fact that analysis had to come to serious grips with the old and difficult concept of infinity.

Of the many remarkable mathematical discoveries that were made in the seventeenth century, and that rendered that century outstandingly productive in the development of the subject, unquestionably the most remarkable was the invention of the calculus, toward the end of the century, by Isaac Newton (1642–1727) and Gottfried Wilhelm Leibniz (1646–1716). The new tool proved to be almost unbelievably powerful in its astonishingly successful disposal of hosts of problems that had been baffling and quite unassailable in earlier days. Its general methods were able to cope with such matters as lengths of curved arcs, planar areas bounded by quite arbitrary curves, surface areas and volumes of all sorts of solids, intricate maximization and minimization problems, all kinds of problems involving related rates of change, geometrical questions about tangents and normals, asymptotes, envelopes, and curvature, and physical questions about work, energy, power, pressure, centers of gravity, inertia, and gravitational attraction. This wide and amazing applicability of the new subject attracted mathematical researchers of the day, and papers were turned out in great profusion with seemingly little concern regarding the very unsatisfactory foundations of the subject. It was much more exciting to apply the marvelous new tool than to examine its logical soundness, for, after all, the processes employed justified themselves to the researchers in view of the fact that they worked.

Although for almost 100 years after the invention of the calculus by Newton and Leibniz little serious work was done to strengthen logically the underpinning of the rapidly growing superstructure of the calculus, it must not be supposed that there was no criticism of the existing weak base. Long controversies were carried on by some mathematicians, and even the two founders themselves were dissatisfied with their accounts of the fundamental concepts of the subject. One of the ablest criticisms of the faulty foundations came from a nonmathematician, the eminent metaphysician Bishop George Berkeley (1685–1753), who insisted that the development of the calculus involved the logical fallacy of a shift in the hypothesis. Let us clarify this particular criticism by considering one of Newton's approaches to what is now called differentiation.

In his *Quadrature of Curves* of 1704, Newton determines the derivative (or fluxion, as he called it) of x^3 as follows. We here paraphrase Newton's treatment:

In the same time that x, by growing, becomes $x + o$, the power x^3 becomes $(x + o)^3$, or

$$x^3 + 3x^2o + 3xo^2 + o^3,$$

and the growths, or increments,

$$o \quad \text{and} \quad 3x^2o + 3xo^2 + o^3$$

are to each other as

$$1 \quad \text{to} \quad 3x^2 + 3xo + o^2.$$

Now let the increments vanish, and their last proportion will be 1 to $3x^2$, whence the rate of change of x^3 with respect to x is $3x^2$.

The shift of hypothesis to which Bishop Berkeley objected is evident; in part of the argument o is assumed to be nonvanishing, while in another part it is taken to be zero. Replies were made to Bishop Berkeley's criticism, but without a logically rigorous treatment of limits, the objection could not be well met. Alternative approaches proved to be no less confusing. Indeed, all the early explanations of the processes of the calculus are obscure, encumbered with difficulties and objections, and not easy to read. Some of the explanations border on the mystical and the seemingly absurd, as the statement by Johann Bernoulli that "a quantity which is increased or decreased by an infinitely small quantity is neither increased nor decreased."

When the theory of a mathematical operation is only poorly understood, the danger is that the operation will be applied in a blindly formal and perhaps illogical manner. The performer, not aware of possible limitations on the operation, is likely to use the operation in instances where it does not necessarily apply. Instructors of mathematics see mistakes of this sort made by their students almost every day. Thus one student of elementary algebra, firmly convinced that $a^0 = 1$ for all real numbers a, will set $0^0 = 1$, while another such student will assume that the equation $ax = b$ always has exactly one real solution for each pair of given real values a and b. Again, a student of trigonometry may think that the formula

$$\sqrt{1 - \sin^2 x} = \cos x$$

holds for all real x. A student of the calculus, not aware of improper integrals, may get an incorrect result by apparently correctly applying the rules of formal integration, or he may arrive at a paradoxical result by applying to a certain convergent infinite series some rule that holds only for absolutely convergent infinite series. This is essentially what happened in analysis during the century following the invention of the calculus. Attracted by the powerful applicability of the subject, and lacking a real understanding of the foundations on which the subject must rest, mathematicians manipulated analytical processes in an almost blind manner, often being guided only by a native intuition of what was felt must be valid. A gradual accumulation of absurdities was bound to result, until, as a natural reaction to the pell-mell employment of intuition and formal manipulation, some conscientious mathematicians felt impelled to attempt the difficult task of establishing a rigorous foundation under the subject.

The work of the great Swiss mathematician Leonhard Euler (1707–1783) represents the outstanding example of eighteenth-century formal manipulation in analysis. It was by purely formal devices that Euler discovered the remarkable formula

$$e^{ix} = \cos x + i \sin x,$$

which, for $x = \pi$, yields

$$e^{i\pi} + 1 = 0,$$

a relation connecting five of the most important numbers of mathematics. By formal manipulation Euler arrived at an enormous number of curious relations like

$$i \log_e i = -\frac{\pi}{2},$$

and succeeded in showing that any nonzero real number r has an infinite number of logarithms (for a given base), all imaginary if $r < 0$ and all imaginary but one if $r > 0$. The beta and gamma functions of advanced calculus, and many other topics in analysis, similarly originated with Euler. He was a most prolific writer on mathematics. His name is attached to practically every branch of the subject, and his amazing productivity was not in the least impaired when he had the misfortune to become totally blind. His remarkable mathematical intuition generally held him on the right path, but nevertheless on numerous instances Euler's formal manipulation led him into absurdities. For example, if the binomial theorem is applied formally to $(1 - 2)^{-1}$, we find

$$-1 = 1 + 2 + 4 + 8 + 16 + \cdots,$$

a result that caused Euler no wonderment! Also, by adding the two series

$$x + x^2 + \cdots = \frac{x}{1 - x} \quad \text{and} \quad 1 + \frac{1}{x} + \frac{1}{x^2} + \cdots = \frac{x}{x - 1},$$

Euler found that

$$\cdots + \frac{1}{x^2} + \frac{1}{x} + 1 + x + x^2 + \cdots = 0.$$

Seventeenth- and eighteenth-century mathematicians had little understanding of infinite series, and this field of analysis furnished many paradoxes. Thus consider the series

$$S = 1 - 1 + 1 - 1 + 1 - 1 + 1 - 1 + 1 - 1 + \cdots.$$

If we group the terms of this series in one way, we have

$$S = (1 - 1) + (1 - 1) + (1 - 1) + (1 - 1) + \cdots$$
$$= 0 + 0 + 0 + 0 + \cdots$$
$$= 0,$$

while, if we group the terms in another way, we have

$$S = 1 - (1 - 1) - (1 - 1) - (1 - 1) - (1 - 1) - \cdots$$
$$= 1 - 0 - 0 - 0 - 0 - \cdots$$
$$= 1.$$

Luigi Guido Grandi (1671–1742) argued that since the sums 0 and 1 are equally probable, the correct sum of the series is the average value $1/2$. This value, too, can be obtained in a purely formal manner, for we have

$$S = 1 - (1 - 1 + 1 - 1 + 1 - 1 + \cdots)$$
$$= 1 - S,$$

whence $2S = 1$, or $S = 1/2$.

The first suggestion of a real remedy for the unsatisfactory state of the foundations of analysis came from Jean-le-Rond d'Alembert (1717–1783), who very correctly observed in 1754 that a theory of limits was needed, but a sound development of this theory was not forthcoming until 1821. The earliest mathematician of the first rank actually to attempt a rigorization of the calculus was the Italian-born French mathematician Joseph Louis Lagrange (1736–1813). The attempt, based on representing a function by a Taylor's series expansion, was far from successful, for it ignored necessary matters of convergence and divergence. It was published in 1797 in Lagrange's monumental work, *Théorie des fonctions analytiques*. Lagrange was one of the leading mathematicians of the eighteenth century, and his work had a deep influence on later mathematical research; with Lagrange's work the long and difficult task of banishing intuition and blind formal manipulation from analysis had begun.

In the nineteenth century, the superstructure of analysis continued to rise, but on ever-deepening foundations. A debt is undoubtedly owed to Carl Friedrich Gauss, whom we have already met in connection with the development of non-Euclidean geometry, for Gauss, more than any other mathematician of his time, broke from intuitive ideas and set new high standards of mathematical rigor. Also, in a treatment of hypergeometric series made by Gauss in 1812, we encounter what is generally regarded as the first really adequate consideration of the convergence of an infinite series.

A great forward stride was made in 1821, when the French mathematician Augustin-Louis Cauchy (1789–1857) successfully executed d'Alembert's suggestion, by developing an acceptable theory of limits and then defining continuity, differentiability, and the definite integral in terms of the limit concept. Essentially these definitions are those found in the more carefully written of today's elementary textbooks on the calculus. The limit concept is certainly indispensable for the development of analysis, for convergence and divergence of infinite series also depend on this concept. Cauchy's rigor inspired other mathematicians to join the effort to rid analysis of formal manipulation and intuition.

The demand for an even deeper understanding of the foundations of analysis was strikingly brought out in 1874, when there was exhibited an example, earlier originated by the German mathematician Karl Weierstrass (1815–1897), of a continuous function having no derivative, or, what is the same thing, a continuous curve possessing no tangent at any of its points. This example dealt a severe blow to the employment of geometric intuition in analytical studies.

The theory of limits, on which the ideas of continuity and differentiability depend, had been built previously on a simple intuitive geometrical notion of the real number system. Indeed, the real number system had been taken more or less for granted, as it still is in most of our elementary calculus texts. It became apparent that the theory of limits, continuity, and differentiability depend on more recondite properties of the real number system than had been supposed. Other ties connecting analysis with properties of the real number system became evident when Riemann found that Cauchy had unnecessarily restricted his definition of a definite integral; Riemann showed that definite integrals as limits of sums exist even when the integrands are discontinuous. In elementary calculus these integrals are referred to as *improper integrals* and require some care in their handling. Riemann also produced a function that is continuous for all irrational values of the variable but discontinuous for all rational values. Examples such as these made it increasingly apparent that Cauchy had not struck the true bottom of the difficulties in the way of a sound foundation for analysis; beneath everything still lay properties of the real number system that required urgent understanding. Accordingly, Weierstrass advocated a program wherein the real number system itself should first be rigorized; then all the basic concepts of analysis should be derived from this number system. This remarkable program, known as the *arithmetization of analysis,* proved to be difficult and intricate, but was ultimately realized by Weierstrass and his followers, so that today it can be fairly said that classical analysis has been firmly established on the real number system as a foundation.

Weierstrass died in 1897, just 100 years after the first publication, in 1797 by Lagrange, of an attempt to rigorize the calculus. Weierstrass was a very influential teacher, and his meticulously prepared lectures and his program for arithmetizing analysis established a new ideal for analysts. Much of the remainder of this book is devoted to fundamental ideas in mathematics brought out by late nineteenth-century and early twentieth-century studies of the real number system. The next few sections examine the real number system itself.

7.2   The Postulational Approach to the Real Number System

As was pointed out in the last section, Weierstrass and his followers, through their program of arithmetizing analysis, established the real number system as the foundation for the whole of classical analysis. Euclidean geometry, through its analytical interpretation as considered in Section 4.4, can also be made to rest on the real number system, and mathematicians have shown that most other branches of geometry are consistent if Euclidean geometry is consistent. Again, since the real number system, or some part of it, can serve for interpreting so many branches of algebra, it appears that the consistency of a good deal of algebra can also be made to depend on that of the real number system. We have seen that a branch of mathematics that possesses an interpretation involving only a finite number of primitive elements may be shown to be absolutely consistent, whereas relative consistency is the best that can be hoped for, by models, when the branch of mathematics possesses no finite interpretation. In these latter situations, like that of Euclidean geometry, the real number system has frequently come to the mathematicians' assistance, so that today essentially all existing mathematics is consistent if the real number system is consistent. Herein lies the tremendous importance of the real number system for the foundations of mathematics. It is no wonder that the real number system has received intensive study in recent times and that this study has, as we shall see in the next two chapters, unearthed some new and even deeper foundational problems. For the present we shall concern ourselves with two current methods of approaching the study of the real number system, devoting this section to the postulational approach and the remaining sections of the chapter to the so-called definitional, or genetical, approach.

If the bulk of mathematics is to be derived from the real number system, it is desirable that the real number system itself, which mathematicians had long taken intuitively for granted, be developed in a logical fashion. One way of attempting this is by the postulational method. We might choose the real numbers, together with some operations on them, as primitive terms, and then hope to set down as postulates a number of the intuitive properties of the real numbers and the primitive operations that should be sufficient to characterize the system. It was observed in Section 5.1 that the real number system is an example of an ordered field. But the real number system is not characterized by this description, for the rational number system, which is not isomorphic with the real number system, also is an ordered field. It might seem, however, that our intuitive notion of the real number system could perhaps be characterized by the postulates of an ordered field augmented by some additional restrictions. Such is indeed the case, although the characterization also holds for systems isomorphic with the real number system, and the addition of only one more property to those constituting the postulates for an ordered field will suffice for the purpose. We refer to the additional property as the postulate of continuity, and we define a *complete ordered field* to be an ordered field that further satisfies the postulate of continuity.

Reproduced here, with a few simplifications, is the postulate set for an ordered field that was given earlier in Section 5.1. We retain, for pedagogical reasons, the two redundant postulates P1 and P9; for a derivation of them from the other postulates of a field, see Appendix, Section A.3.

An *ordered field* is a set S of elements, with two binary operations on the set, here denoted by $\oplus$ and $\otimes$, satisfying the following postulates:

P1.   *If* a *and* b *are in* S, *then* $a \oplus b = b \oplus a$.

P2.   *If* a *and* b *are in* S, *then* $a \otimes b = b \otimes a$.

P3.   *If* a, b, c *are in* S, *then* $(a \oplus b) \oplus c = a \oplus (b \oplus c)$.

P4.   *If* a, b, c *are in* S, *then* $(a \otimes b) \otimes c = a \otimes (b \otimes c)$.

P5.   *If* a, b, c *are in* S, *then* $a \otimes (b \oplus c) = (a \otimes b) \oplus (a \otimes c)$.

P6.   S *contains an element* z (zero) *such that for any element* a *of* S, $a \oplus z = a$.

P7.   S *contains an element* u (unity), *different from* z, *such that for any element* a *of* S, $a \otimes u = a$.

P8.   *For each element* a *in* S *there exists an element* ā *in* S *such that* $a \oplus \bar{a} = z$.

P9.   *If* a, b, c *are in* S, $c \neq z$, *and* $a \otimes c = b \otimes c$, *then* $a = b$.

P10. *For each element* $a \neq z$ *in* S, *there exists an element* a^{-1} *in* S *such that* $a \otimes a^{-1} = u$.

P11. *There exists a subset* P, *not containing* z, *of the set* S *such that if* $a \neq z$, *then one and only one of* a *and* ā *is in* P.

P12. *If* a *and* b *are in* P, *then* $a \oplus b$ *and* $a \otimes b$ *are in* P.

Definition 1   The elements of P are known as the *positive* elements of S, all other nonzero elements of S are known as *negative* elements of S.

Definition 2   If a and b are elements of S and if $a \oplus \bar{b}$ is positive, then we write $a \ominus b$ and $b \oslash a$, and we say that a is *greater than* b and that b is *less than* a.

Definition 3   An element a of S is called an *upper bound* of a nonempty collection M of elements of S if, for each element m of M, either $m \oslash a$ or $m = a$.

Definition 4   An element a of S is called a *least upper bound* of a non-empty collection M of elements of S if a is an upper bound of M and if $a \oslash b$ whenever b is any other upper bound of M.

We now define a *complete ordered field* (sometimes called a *linear continuum*) to be an ordered field that further satisfies

P13 (the postulate of continuity). *If a nonempty collection* M *of elements of* S *has an upper bound, then it has a least upper bound.*[1]

Our intuitive notion of the real number system satisfies all the postulates of a complete ordered field. Now mathematicians have shown that any two possible interpretations of a complete ordered field must be isomorphic with each other; in other words, mathematicians have shown that if we assume the postulate set for a complete ordered field to be consistent, then this postulate set is also categorical. Of course, the matter of consistency of the postulate set still remains a fundamental question. We cannot offer our intuitive concept of the real number system as an interpretation acceptable for establishing consistency of the postulate set for a complete ordered field any more than we can offer our intuitive concept of points and lines as an acceptable interpretation of the postulate set for Euclidean geometry, for an intuitive concept may involve inconsistencies. The question of consistency of the postulate set for a complete ordered field is obviously an important one, for on the settlement of this question rests the consistency of the great bulk of mathematics. We return to this matter in later sections.

Since, apart from isomorphisms, there can exist at most one interpretation of a complete ordered field, such a possible interpretation can be defined to be the *real number system.* Accordingly, we shall henceforth think of the elements of a complete ordered field as the familiar set of all real numbers, and we shall employ the symbols $+$, $\times$, $>$, $<$ (without circles around them) in their familiar senses.

Part of the significance of the postulate of continuity can be seen from the fact that the rational number system, which satisfies all the postulates of an ordered field, does not satisfy this additional postulate. Consider, for example, the collection M of all positive rational numbers whose squares are less than 2. The rational number 3 is certainly an upper bound of this collection M, but there is no rational number that is a least upper bound of M. The only number that could be a least upper bound of M is $\sqrt{2}$, and this, we know, is not a rational number.

The full significance of the postulate of continuity, however, only appears in the use made of it to derive many important theorems of the calculus.[2] We can illustrate this use by deriving with its aid the so-called Archimedean law of real numbers and Dedekind's theorem of continuity. We assume some of the simpler properties of the real number system that the system possesses by virtue of the fact that it is an ordered field. In particular, we shall assume that the familiar integers are embedded among the real numbers and that "between" any two

---

[1] Essentially this set of postulates, augmented with many redundancies, was given by E. V. Huntington in 1911. See E. V. Huntington [10].

[2] An interesting venture is the construction of an elementary calculus textbook based upon the above thirteen postulates for the real number system. To save time, one can assume the needed consequences of the first twelve postulates. This should cause the student no difficulty; since the first twelve postulates are those of ordinary algebra, he should already be familiar with the needed consequences, though not with their derivations. The basic concepts of the calculus depend upon the remaining postulate (the postulate of continuity) and these concepts are far different from those of algebra; these concepts should be derived. That such a venture is perfectly feasible is attested by the fact that it has already been successfully accomplished. See E. G. Begle.

distinct real numbers there lies at least one other real number. It would take us too far afield to derive all these needed properties.

The Archimedean Law of Real Numbers *If* a *and* b *are any two positive real numbers, there exists a positive integer* n *such that* na $>$ b.

Suppose the law is not true, and that $na \leq b$ for all positive integers n. Denote by M the collection of all numbers of the form na. Then, by hypothesis, b is an upper bound of M, and hence, by the postulate of continuity, M has a least upper bound k. That is, $pa \leq k$ for all positive integers p, and there is no number less than k for which this is true. It follows that $(n + 1)a = na + a \leq k$ for all positive integers n, or that $na \leq k - a$ for all positive integers n. But this is a contradiction of the fact that k is a least upper bound of M. The desired result now follows by *reductio ad absurdum*.

Dedekind's Theorem of Continuity *If all the real numbers are separated into two collections, which we denote by* L *and* R, *in such a way that* (1) *every real number is either in* L *or in* R, (2) L *and* R *each contain at least one real number,* (3) *every real number in* L *is less than every real number in* R, *then there is a real number* c *such that all real numbers less than* c *are in* L *and all real numbers greater than* c *are in* R.

Since the collection L of real numbers has an upper bound—namely, any real number of the collection R—it has a least upper bound c. Let m be any real number less than c; then m must lie in L, for otherwise m would lie in R and thereby be an upper bound of L that is less than the least upper bound c. On the other hand, let n be any real number greater than c; then n must lie in R, since no number in L exceeds c. This completes the proof of the theorem.

The reader has undoubtedly recognized in the Archimedean law of real numbers and in Dedekind's theorem of continuity arithmetized forms of the two important geometrical statements respectively referred to earlier in the book as the postulate of Archimedes and as Dedekind's postulate of continuity. A very interesting fact is that not only can Dedekind's theorem of continuity be deduced from the postulate of continuity (by employing properties of an ordered field), but the postulate of continuity can similarly be deduced from the Dedekind theorem of continuity. In other words, the Dedekind theorem of continuity and the postulate of continuity are equivalent statements, and either one can be used as Postulate P13 in the list of postulates for the real number system. Let us now complete the present section by deducing the postulate of continuity from Dedekind's theorem of continuity.

Let M be a nonempty collection of real numbers having an upper bound k. Choose L as the collection of all real numbers that are less than at least one number of M, and let R consist of all the rest of the real numbers. We now show that L and R satisfy the three conditions of the Dedekind continuity theorem.

1. The collections L and R contain all the real numbers; this follows from the definition of collection R.

2. L is not empty, for, inasmuch as M contains at least one number m, the number $m - 1$ belongs to L. R is not empty, for it contains k.

3. Every number of L is less than every number of R. For suppose there is a number b_1 of R that is less than a number a_1 of L. Since there exists a number m of M greater than a_1, we have $b_1 < a_1 < m$, and b_1 must belong to L. This contradicts the hypothesis that b_1 is a number of R.

Thus the three conditions of the Dedekind continuity theorem hold, and the theorem guarantees the existence of a real number c such that all real numbers less than c are in L and all real numbers greater than c are in R. Now suppose there is a number m_1 of M such that $m_1 > c$. Then between c and m_1 there would be a real number d that would be less than m_1 and that would therefore belong to L, a situation that is impossible, since every real number greater than c belongs to R. It follows that every number of M is less than, or at most equal to, c, and that c is therefore an upper bound of M. Next, let e be any real number less than c. Then e belongs to L, and, by the definition of collection L, there exists a number m of M such that $m > e$. Hence no number less than c is an upper bound of M. We have now shown that any nonempty collection of real numbers having an upper bound also has a least upper bound; this is the postulate of continuity.

7.3   The Natural Numbers and the Principle of Mathematical Induction

From the postulate set for the real number system one can, by simply making appropriate definitions in terms of the real numbers (as was indicated in Section 4.4), arrive at an interpretation of Euclidean geometry and thus make the consistency of Euclidean geometry depend on that of the real number system. This suggests that perhaps a similar undertaking can be carried out relative to the real number system itself. That is, perhaps we can start from some postulate set more basic than that for the real number system and, by merely making appropriate definitions, arrive at an interpretation of the real number system. Thus the consistency of the real number system would depend on the consistency of the more fundamental postulate set. This turns out to be possible. In the remaining sections of this chapter, therefore, we sketch a manner in which the real number system may be arrived at by definition from a postulate set for the much simpler and more basic system of natural numbers, the numbers of counting. In this way we make the consistency of the great bulk of mathematics depend on that of the very fundamental system of natural numbers. The success of carrying through the program is credited to late nineteenth-century researches by Peano, Dedekind, and Cantor, and its accomplishment has given the mathematician a considerable feeling of security concerning the consistency of most of mathematics. This attitude follows from the fact that the natural number system seems to have an intuitive simplicity lacking in most other mathematical systems, and the natural numbers have been very extensively handled over a long period of time without producing any known inner contradictions.

Let us in this section consider a postulate set for the system of natural numbers, and then in the following sections show how, by definitions in terms of the natural numbers, we can introduce first the set of all integers, next the set of all rational numbers, and then the set of all real numbers. We here choose for our postulate set for the system of natural numbers one that bears marked similarity to some of our earlier postulate sets.

For our primitive, or undefined, terms we take a set N of elements called *natural numbers*, together with two binary operations on the set, called *addition* and *multiplication* and denoted by $+$ and $\times$, satisfying the following ten postulates.

N1.   *If* a *and* b *are in* N, *then* a $+$ b $=$ b $+$ a.

N2.   *If* a *and* b *are in* N, *then* a $\times$ b $=$ b $\times$ a.

N3.   *If* a, b, c *are in* N, *then* (a $+$ b) $+$ c $=$ a $+$ (b $+$ c).

N4.   *If* a, b, c *are in* N, *then* (a $\times$ b) $\times$ c $=$ a $\times$ (b $\times$ c).

N5.   *If* a, b, c *are in* N, *then* a $\times$ (b $+$ c) $=$ (a $\times$ b) $+$ (a $\times$ c).

N6.   *There exists a natural number* 1 *such that* a $\times$ 1 $=$ a *for all* a *in* N.

N7.   *If* a, b, c *are in* N *and if* c $+$ a $=$ c $+$ b, *then* a $=$ b.

N8.   *If* a, b, c *are in* N *and if* c $\times$ a $=$ c $\times$ b, *then* a $=$ b.

N9.   *For given* a *and* b *in* N, *one and only one of the following holds:*
a $=$ b, a $+$ x $=$ b, a $=$ b $+$ y, *where* x *and* y *are in* N.

N10. *If* M *is a set of natural numbers such that* (1) M *contains the natural number* 1, (2) M *contains the natural number* k $+$ 1 *whenever it contains the natural number* k, *then* M *contains all the natural numbers.*

The first five postulates are exact counterparts of the first five postulates for the real number system; they postulate the commutativity and associativity of the two binary operations of addition and multiplication and the distributivity of multiplication over addition. N6 postulates the existence of a multiplicative identity. N7 and N8 grant the cancellation laws for addition and multiplication. These two postulates and Postulates N1, N2, N4 are actually redundant, inasmuch as they all can be derived from the remaining Postulates N3, N5, N6, N9, and N10, but all are retained here to simplify the student's experience in the subsequent deduction of theorems. The reader can find derivations of N1, N2, N4, N7, and N8 from the remaining five postulates in the Appendix, Section A.3. Postulate N9 introduces the idea of order among the natural numbers, for if we have the situation $a + x = b$, we write $a < b$ or $b > a$, and say a is *less than b* or b is *greater than a*.

Postulate N10 is an extremely interesting postulate and merits more extended comment. It is known as the *postulate of finite induction,* and it leads to some theorems that are useful for establishing certain propositions that involve all natural numbers. One of these theorems, often referred to as the *principle of mathematical induction,* is the following.

Theorem 1   *Let* P(n) *be a proposition that is defined for every natural number* n. *If* P(1) *is true, and if* P(k $+$ 1) *is true whenever* P(k) *is true, then* P(n) *is true for all natural numbers* n.

The proof of the theorem follows immediately from the postulate of finite induction; for consider the set M of natural numbers for which the proposition $P(n)$ is true. By hypothesis M contains the natural number 1 and also the natural number $k + 1$ whenever it contains the natural number k. Hence, by the postulate of finite induction, M contains all the natural numbers.

This theorem is referred to as the principle of mathematical induction but really has nothing to do with induction in the technical sense of the word. To reason inductively, it will be recalled from our discussion in Section 1.2, is to draw a general conclusion from specific data; for example, one may conclude that the sun will rise in the east tomorrow morning because it has risen there so many mornings in the past. In such reasoning, the conclusion is not one that *must* follow but is only one that *probably* follows. Reasoning of this sort is clearly inadmissible in a logical development. Induction in the usual sense often is employed in order to *guess* a general proposition $P(n)$ in the first place—that is, $P(n)$ may be conjectured from observing several instances, like $P(1)$, $P(2)$, and $P(3)$, of $P(n)$—but the rigid establishment of $P(n)$ by the so-called principle of mathematical induction as just explained is, of course, a purely deductive procedure. The reader is undoubtedly familiar with the principle of mathematical induction from his work in elementary algebra and is already aware that it yields a powerful procedure for establishing certain propositions.[3]

It is worth pointing out merely as a matter of interest that the postulate of finite induction is a consequence of the postulates for the real number system given in the preceding section. There we would *define* the natural numbers as the elements of the smallest subset N of the real numbers that are such that $a + 1$ is in N if and only if $a = 0$ or a is in N. The statement of the postulate of finite induction is then an immediate consequence of this definition. Thus, whereas the statement N10 is taken as a postulate in our development of the natural number system, it occupies the position of a theorem in our development of the real number system.

The consequences of the postulate set for the system of natural numbers constitute a very interesting body of mathematics known as the *theory of numbers*; this branch of mathematics has provided many recreational opportunities for the amateur mathematician as well as a collection of conjectural propositions that mathematical genius has been unable to prove or disprove. The following twenty theorems can be regarded as the start of a development of this branch of mathematics. We begin by repeating the theorem known as the principle of mathematical induction and, following convention, abbreviate $a \times b$ by ab.

Some Consequences of the Postulates N1 Through N10

Theorem 1 (the principle of mathematical induction)   *Let* $P(n)$ *be a proposition that is defined for every natural number* n. *If* $P(1)$ *is true, and if* $P(k + 1)$ *is true whenever* $P(k)$ *is true, then* $P(n)$ *is true for all natural numbers* n.

---

[3] One of the earliest acceptable statements of the principle of mathematical induction appears in Blaise Pascal's *Traité du triangle arithmétique*, written in 1653 but not printed until 1665. The principle reminds one of the children's game in which toy soldiers are lined up so that if one falls it will knock over the next one. If no soldier is pushed over, or if some soldier is AWOL, then the complete line will not fall.

Theorem 2 *If* a + c = b + c, *then* a = b.
  For

$$c + a = a + c \qquad\qquad\qquad \text{(by N1)}$$
$$= b + c \qquad\qquad \text{(by hypothesis)}$$
$$= c + b, \qquad\qquad\qquad \text{(by N1)}$$

whence

$$a = b. \qquad\qquad\qquad \text{(by N7)}$$

Theorem 3 *If* ac = bc, *then* a = b.
  For

$$ca = ac \qquad\qquad\qquad \text{(by N2)}$$
$$= bc \qquad\qquad \text{(by hypothesis)}$$
$$= cb, \qquad\qquad\qquad \text{(by N2)}$$

whence

$$a = b. \qquad\qquad\qquad \text{(by N8)}$$

Theorem 4 *If* e *is a natural number such that* ae = a *for some natural number* a, *then* e = 1.
  For

$$ae = a \qquad\qquad \text{(by hypothesis)}$$
$$= a1, \qquad\qquad\qquad \text{(by N6)}$$

whence

$$e = 1. \qquad\qquad\qquad \text{(by N8)}$$

Definition 1 *If* $a + x = b$, we say that a *is* *less than* b or b *is* *greater than* a, and we write $a < b$ or $b > a$.

Theorem 5 *If* a *and* b *are any two natural numbers, then one and only one of the following situations holds:* a = b, a < b, a > b.

  This is an immediate consequence of N9 and Definition 1.

Theorem 6 *If* a < b *and* b < c, *then* a < c.

  Since, by hypothesis, $a < b$ and $b < c$, there exist natural numbers x and y such that

$$a + x = b, \qquad b + y = c. \qquad\qquad \text{(by Def. 1)}$$

Then

$$a + (x + y) = (a + x) + y \qquad\qquad \text{(by N3)}$$
$$= b + y \qquad\qquad \text{(substitution)}$$
$$= c. \qquad\qquad \text{(substitution)}$$

It follows that

$$a < c. \qquad\qquad \text{(by Def. 1)}$$

Theorem 7 *If* a < b, *then* c + a < c + b.

  Since, by hypothesis, $a < b$, there exists a natural number x such that

$$a + x = b. \qquad\qquad \text{(by Def. 1)}$$

Then

$$(c + a) + x = c + (a + x) \qquad \text{(by N3)}$$

$$= c + b. \qquad \text{(substitution)}$$

It follows that

$$c + a < c + b. \qquad \text{(by Def. 1)}$$

Theorem 8 *If* a < b *and* c < d, *then* a + c < b + d.

Since, by hypothesis, $a < b$ and $c < d$, there exist natural numbers x and y such that

$$a + x = b, \qquad c + y = d. \qquad \text{(by Def. 1)}$$

Then

$$(a + c) + (x + y) = a + [c + (x + y)] \qquad \text{(by N3)}$$

$$= a + [(x + y) + c] \qquad \text{(by N1)}$$

$$= a + [x + (y + c)] \qquad \text{(by N3)}$$

$$= a + [x + (c + y)] \qquad \text{(by N1)}$$

$$= a + (x + d) \qquad \text{(substitution)}$$

$$= (a + x) + d \qquad \text{(by N3)}$$

$$= b + d. \qquad \text{(substitution)}$$

It follows that

$$a + c < b + d. \qquad \text{(by Def. 1)}$$

Theorem 9 *If* a < b, *then* ca < cb.

Since, by hypothesis, $a < b$, there exists a natural number x such that

$$a + x = b. \qquad \text{(by Def. 1)}$$

Then

$$ca + cx = c(a + x) \qquad \text{(by N5)}$$

$$= cb. \qquad \text{(substitution)}$$

It follows that

$$ca < cb. \qquad \text{(by Def. 1)}$$

Definition 2 If a and k are any natural numbers, we define

$$a^1 = a, \qquad a^{k+1} = a^k a.$$

Remark Definition 2 is said to be a *definition by induction.*

Theorem 10 $1^n = 1$ *for all natural numbers* n.

I. By Definition 2, $1^1 = 1$.

II. Suppose k is a natural number such that $1^k = 1$. Then

$$1^{k+1} = 1^k 1 \qquad \text{(by Def. 2)}$$

$$= 1^k \qquad \text{(by N6)}$$

$$= 1. \qquad \text{(by supposition)}$$

Hence the theorem follows, by the principle of mathematical induction (Theorem 1).

Theorem 11   *If* a, b, n *are any natural numbers, then* $(ab)^n = a^n b^n$.

I. By Definition 2, $(ab)^1 = ab = a^1 b^1$.

II. Suppose k is a natural number such that $(ab)^k = a^k b^k$. Then

$$
\begin{aligned}
(ab)^{k+1} &= (ab)^k(ab) & \text{(by Def. 2)}\\
&= (a^k b^k)(ab) & \text{(by supposition)}\\
&= [(a^k b^k)a]b & \text{(by N4)}\\
&= [a^k(b^k a)]b & \text{(by N4)}\\
&= [a^k(ab^k)]b & \text{(by N2)}\\
&= [(a^k a)b^k]b & \text{(by N4)}\\
&= (a^{k+1}b^k)b & \text{(by Def. 2)}\\
&= a^{k+1}(b^k b) & \text{(by N4)}\\
&= a^{k+1}b^{k+1}. & \text{(by Def. 2)}
\end{aligned}
$$

Hence we have the theorem, by the principle of mathematical induction (Theorem 1).

Theorem 12   *If* a, m, n *are any natural numbers, then* $a^m a^n = a^{m+n}$.

I. For any fixed m,

$$
\begin{aligned}
a^m a^1 &= a^m a & \text{(by Def. 2)}\\
&= a^{m+1}. & \text{(by Def. 2)}
\end{aligned}
$$

II. Suppose k is a natural number such that $a^m a^k = a^{m+k}$. Then

$$
\begin{aligned}
a^m a^{k+1} &= a^m(a^k a) & \text{(by Def. 2)}\\
&= (a^m a^k)a & \text{(by N4)}\\
&= a^{m+k}a & \text{(by supposition)}\\
&= a^{(m+k)+1} & \text{(by Def. 2)}\\
&= a^{m+(k+1)}. & \text{(by N3)}
\end{aligned}
$$

Hence the theorem follows, by the principle of mathematical induction (Theorem 1).

Theorem 13   *If* a, m, n *are any natural numbers, then* $(a^m)^n = a^{mn}$.

I. For any fixed m,

$$
\begin{aligned}
(a^m)^1 &= a^m & \text{(by Def. 2)}\\
&= a^{m1}. & \text{(by N6)}
\end{aligned}
$$

II. Suppose k is a natural number such that $(a^m)^k = a^{mk}$. Then

$$
\begin{aligned}
(a^m)^{k+1} &= (a^m)^k(a^m) & \text{(by Def. 2)}\\
&= a^{mk}a^m & \text{(by supposition)}
\end{aligned}
$$

$$= a^{mk+m} \qquad\qquad \text{(by Th. 12)}$$

$$= a^{m(k+1)}. \qquad\qquad \text{(by N6 and N5)}$$

Hence we have the theorem, by the principle of mathematical induction (Theorem 1).

Theorem 14. $1 \leqq n$ *for all natural numbers* n.

I. Clearly $1 \leqq 1$.

II. Suppose k is a natural number such that $1 \leqq k$. If $1 = k$, then

$$1 + 1 = 1 + k, \qquad\qquad \text{(substitution)}$$

and

$$1 < 1 + k. \qquad\qquad \text{(by Def. 1)}$$

If $1 < k$, then

$$1 + 1 < 1 + k, \qquad\qquad \text{(by Th. 7)}$$

whence there exists a natural number x such that

$$(1 + 1) + x = 1 + k. \qquad\qquad \text{(by Def. 1)}$$

Therefore

$$1 + (1 + x) = 1 + k, \qquad\qquad \text{(by N3)}$$

and

$$1 < 1 + k. \qquad\qquad \text{(by Def. 1)}$$

Thus, in any case, $1 < 1 + k$, and the theorem follows by the principle of mathematical induction (Theorem 1).

Theorem 15 *If* a *is a natural number, then there is no natural number* n *such that* $a < n < a + 1$.

Suppose there is a natural number n such that $a < n < a + 1$. Then there exist natural numbers x and y such that

$$a + x = n, \qquad n + y = a + 1. \qquad\qquad \text{(by Def. 1)}$$

Then

$$a + (x + y) = (a + x) + y \qquad\qquad \text{(by N3)}$$

$$= n + y \qquad\qquad \text{(substitution)}$$

$$= a + 1, \qquad\qquad \text{(substitution)}$$

or

$$x + y = 1. \qquad\qquad \text{(by N7)}$$

But

$$x \geqq 1, \qquad y \geqq 1. \qquad\qquad \text{(by Th. 14)}$$

Therefore

$$x + y \geqq 1 + 1 \qquad\qquad \text{(by Th. 8)}$$

$$\neq 1. \qquad\qquad \text{(by Def. 1 and Th. 5)}$$

But this is a contradiction, for (by Theorem 5) we cannot have both $x + y = 1$ and $x + y \neq 1$. Hence the theorem follows, by *reductio ad absurdum.*

Theorem 16 *If* $h < k + 1$, *then* $h \leq k$.

Suppose $h > k$. Then we have $k < h < k + 1$, which contradicts Theorem 15. The theorem thus follows, by *reductio ad absurdum.*

Theorem 17 (the principle of the smallest natural number). *In any nonempty set M of natural numbers there is always a smallest natural number* (that is, there is always a natural number in M which is less than any other natural number in M).

Suppose M has no smallest natural number. Then 1 is not in M, for if it were it would have to be the smallest natural number in M (by Theorem 14). Let T be the set of all natural numbers that are less than every natural number in M. Clearly M and T have no members in common.

I. We have seen that 1 is in T.

II. Suppose k is a natural number in T. Let h be any natural number less than $k + 1$. Then, by Theorem 16, $h \leq k$. It follows, by the definition of set T, that h cannot be in M. Hence, if $k + 1$ is in M, then $k + 1$ is the smallest natural number in M (by Theorem 15). Since, by supposition, M has no smallest natural number, it follows that $k + 1$ is not in M and that all natural numbers in M must be greater than $k + 1$. Thus $k + 1$ is in T.

It now follows, by the postulate of finite induction (Postulate N10), that T contains all the natural numbers. But this is impossible, since M is nonempty and M and T have no members in common. Hence the theorem follows by *reductio ad absurdum.*

Theorem 18 (a second principle of mathematical induction) *Let* P(n) *be a proposition that is defined for every natural number* n. *If* P(1) *is true, and if, for each natural number* m, P(m) *is true whenever* P(k) *is true for all natural numbers* k < m, *then* P(n) *is true for all natural numbers* n.

Let M be the set of natural numbers for which $P(n)$ is false, and suppose M is not empty. Then, by Theorem 17, M contains a smallest natural number m. Note that $m \neq 1$, for by hypothesis $P(1)$ is true. Hence, for all natural numbers $k < m$, $P(k)$ is true. By hypothesis we must then have $P(m)$ true. But this contradicts the fact that m is in M. It follows that M must be empty, or that $P(n)$ is true for all natural numbers n.

Theorem 19 (a third principle of mathematical induction) *Let* P(n) *be a proposition that is defined for every natural number* n. *If* P(n) *is true for all* n ≤ h, *where* h *is some fixed natural number, and if, for each natural number* m, P(m + h) *is true whenever* P(k) *is true for all natural numbers* k *such that* m ≤ k < m + h, *then* P(n) *is true for all natural numbers* n.

Let M be the set of natural numbers for which $P(n)$ is false, and suppose M is not empty. Then, by Theorem 17, M contains a smallest natural number s. Note that $s > h$, for by hypothesis $P(n)$ is true for all $n \leq h$. Since $s > h$, there exists a natural number m such that $s = m + h$. Hence $P(k)$ is true for all natural numbers $k < m + h$. In particular, then, $P(k)$ is true for all natural numbers k such that $m \leq k < m + h$. By hypothesis we must then have $P(m + h)$ true. But this contradicts the fact that $m + h = s$ is in M. It follows that M must be empty, or that $P(n)$ is true for all natural numbers n.

Note that, by virtue of Theorem 15, Theorem 1 is the special case of this theorem where $h = 1$.

As already revealed, the natural number system occupies a very basic position in the foundations of mathematics, and its development from the Postulates N1 through N10 is not difficult. There is much to recommend that this postulate set, rather than a postulate set for Euclidean geometry, say, be the first significant postulate set that a student of mathematics undertakes to study. This is because of the inherent simplicity of the postulate set and the inexperienced student's already long and wide acquaintance with elementary properties of the natural numbers. It is recommended that, at the start, the redundant postulates N1, N2, N4, N7, and N8 be retained. Perhaps the reader would care to try deductive work by adding some simple theorems to the list given above.

There are postulate sets for the natural number system other than the one we have considered, but in general these other sets are somewhat more sophisticated and less easy for a beginner to develop. Very popular with mathematicians is the following alternative set given by Peano.

For primitive terms Peano chose *natural number, successor,* and *1.* About these primitive terms he postulated:

N'1. *1 is a natural number.*

N'2. *For each natural number* x *there exists exactly one natural number, called the successor of* x, *which will be denoted by* x'.

N'3. *1 is not the successor of any natural number.*

N'4. *If* x' = y', *then* x = y.

N'5 (the postulate of finite induction). *If* M *is a set of natural numbers such that* (1) M *contains 1,* (2) M *contains* x' *whenever it contains* x, *then* M *contains all the natural numbers.*

If we interpret x' to mean $x + 1$, then Peano's postulates can easily be shown to be implied by our earlier postulate set. With appropriate definitions of addition and multiplication, it also can be shown that the earlier postulate set is implied by Peano's set. Thus the two postulate sets are equivalent. It can be shown that each set, if consistent, is categorical. Peano's set is closely allied to our conception of the natural numbers as an ordered sequence of elements starting with some first element. Many careful and detailed developments of the Peano postulate set exist in the literature, together with a definitional approach to the real number system from it.[4]

7.4 The Integers and the Rational Numbers

As promised, we now indicate a program wherein the real number system is obtained from the natural number system in a purely definitional way, without the introduction of any further postulates. Moreover, the same procedure actually is extended to the more general complex number system. The program is carried out in stages. In this section we show how, from the natural numbers, one can obtain the set of all integers (positive, zero, and negative) and then, after this is done, the set of all rational numbers. In the following section we then obtain the set of all real numbers and, finally, the set of all complex numbers. Because of limited space, our treatment will necessarily be very sketchy, but then

---

[4] See, for example, E. G. H. Landau or C. C. MacDuffee.

our aim is merely to indicate the procedure rather than to execute it in all its details. For a complete development, beginning with Peano's postulate set for the natural numbers, the interested reader may consult Landau's scholarly and very readable *Foundations of Analysis*.[5]

We continue the numbering of definitions and theorems that was started in the previous section. Each stage of the program is preceded by a brief paragraph explaining the motivation of the ideas that follow. Without this motivation, some of the definitions might seem highly artificial. It must be understood, however, that the discussion of motivation is quite informal and is not an actual part of the demonstration; in fact, in the discussion some concepts are anticipated even before they are introduced logically. The proofs of the theorems, in large part, are left for the reader to supply. We assume that the familiar properties of the natural numbers have been developed and are at our disposal.

The Integers

[**Motivation**   Any integer, be it positive, zero, or negative, can be represented as the difference $m - n$ of two natural numbers m and n. If $m > n$, then $m - n$ is a positive integer; if $m = n$, then $m - n$ is the zero integer; if $m < n$, then $m - n$ is a negative integer. This suggests the idea of logically introducing the integers as ordered pairs (m, n) of natural numbers, where by (m, n) we actually have in mind the difference $m - n$, although this fact does not enter into any demonstrations. With this interpretation of an ordered pair of natural numbers we are led to say that $(a, b) = (c, d)$ if and only if $a - b = c - d$, or $a + d = b + c$. Similarly, since $(a - b) + (c - d) = (a + c) - (b + d)$, we are led to define $(a, b) + (c, d)$ to be the ordered pair of natural numbers $(a + c, b + d)$, and since $(a - b)(c - d) = (ac + bd) - (ad + bc)$, we are led to define $(a, b)(c, d)$ to be the ordered pair of natural numbers $(ac + bd, ad + bc)$.]

Definition 3   We shall denote the set of all ordered pairs (m, n) of natural numbers m and n by I.

Definition 4   If (a, b) and (c, d) are two members of I, we say that $(a, b) = (c, d)$ if and only if $a + d = b + c$.

Remark   It is to be noted that the definition of equality of members of I is not the ordinary concept of equality as an identity of symbols.

Theorem 20   *Equality of members of* I *is an equivalence relation; that is,* (1) (a, b) = (a, b), (2) *if* (a, b) = (c, d), *then* (c, d) = (a, b), (3) *if* (a, b) = (c, d) *and* (c, d) = (e, f), *then* (a, b) = (e, f).

    The reflexive and symmetric properties are easily established. To establish the transitive property we note that, if $(a, b) = (c, d)$ and $(c, d) = (e, f)$, then $a + d = b + c$ and $c + f = d + e$. Consequently $a + d + c + f = b + c + d + e$, or $a + f = b + e$, whence $(a, b) = (e, f)$.

Theorem 21   *For any natural number* x, (x + a, x + b) = (a, b).

Definition 5   We define the *sum* of any two members (a, b) and (c, d) of I,

---

[5] This is E. G. H. Landau.

taken in this order, to be the ordered pair of natural numbers $(a + c, b + d)$, and we write

$$(a, b) + (c, d) = (a + c, b + d).$$

This binary operation on members of I will be called *addition*.

Theorem 22 I *is closed under addition, which is commutative, associative, and well defined.*

The closure, commutativity, and associativity properties follow immediately from the definition of addition and from the corresponding properties of natural numbers. Since equality of elements in I is not an identity equality, addition in I, if it is to be of use to us, should be a well-defined operation in I. That is, if $(a, b) = (a', b')$ and $(c, d) = (c', d')$, we should have that $(a', b') + (c', d') = (a, b) + (c, d)$. We show that this is so as follows:

$$(a', b') + (c', d') = (a' + c', b' + d') \qquad \text{(by Def. 5)}$$

$$= (a + c + a' + c', a + c + b' + d')$$
$$\text{(by Th. 21)}$$

$$= (a + c + a' + c', a' + b + c' + d')$$
$$\text{(by Def. 4)}$$

$$= (a + c, b + d) \qquad \text{(by Th. 21)}$$

$$= (a, b) + (c, d). \qquad \text{(by Def. 5)}$$

Definition 6 We define the *product* of any two members (a, b) and (c, d) of I, taken in this order, to be the ordered pair of natural numbers $(ac + bd, ad + bc)$, and we write

$$(a, b)(c, d) = (ac + bd, ad + bc).$$

This binary operation on members of I will be called *multiplication*.

Theorem 23 I *is closed under multiplication, which is commutative, associative, well defined, and distributive over addition.*

We shall show that multiplication is well defined and is distributive over addition; the other properties are easily established. Suppose, then, that $(a, b) = (a', b')$ and $(c, d) = (c', d')$. Then

$$(a', b')(c', d') = (a'c' + b'd', a'd' + b'c') \qquad \text{(by Def. 6)}$$

$$= (ac' + bd' + a'c' + b'd', ac' + bd' + a'd' + b'c')$$
$$\text{(by Th. 21)}$$

$$= (ac' + bd' + a'c' + b'd', c'(a + b') + d'(b + a'))$$

$$= (ac' + bd' + a'c' + b'd', c'(a' + b) + d'(b' + a))$$
$$\text{(by Def. 4)}$$

$$= (ac' + bd' + a'c' + b'd', c'a' + c'b + d'b' + d'a)$$

$$= (ac' + bd', c'b + d'a) \qquad \text{(by Th. 21)}$$

$$= (ac + bd + ac' + bd', ac + bd + c'b + d'a) \qquad \text{(by Th. 21)}$$

$$= (ac + bd + ac' + bd', a(c + d') + b(d + c'))$$

$$= (ac + bd + ac' + bd', a(c' + d) + b(d' + c)) \qquad \text{(by Def. 4)}$$

$$= (ac + bd + ac' + bd', ac' + ad + bd' + bc)$$

$$= (ac + bd, ad + bc) \qquad \text{(by Th. 21)}$$

$$= (a, b)(c, d) \qquad \text{(by Def. 6)}$$

and multiplication is well defined. To establish the distributivity property we note that

$$(a, b)[(c, d) + (e, f)]$$

$$= (a, b)(c + e, d + f) \qquad \text{(by Def. 5)}$$

$$= (a(c + e) + b(d + f), a(d + f) + b(c + e)) \qquad \text{(by Def. 6)}$$

$$= ((ac + bd) + (ae + bf), (ad + bc) + (af + be))$$

$$= (ac + bd, ad + bc) + (ae + bf, af + be) \qquad \text{(by Def. 5)}$$

$$= (a, b)(c, d) + (a, b)(e, f). \qquad \text{(by Def. 6)}$$

Definition 7 The elements of I, subject to Definitions 4, 5, 6, will be called *integers*.

Definition 8 The integers (m, n), where $m > n$, will be called *positive integers*.

Theorem 24 *Every positive integer has the form* $(x + a, a)$.

Definition 9 If two sets of symbols $a, b, c \ldots$ and $\alpha, \beta, \gamma, \ldots$, where for each set an equality, an addition, and a multiplication have been defined, are in one-to-one correspondence, say

$$a \leftrightarrow \alpha, \qquad b \leftrightarrow \beta, \qquad c \leftrightarrow \gamma, \qquad \ldots,$$

such that (1) $a = b$ if and only if $\alpha = \beta$, (2) $a + b \leftrightarrow \alpha + \beta$, (3) $ab \leftrightarrow \alpha\beta$, then we say that *the two sets are isomorphic relative to the operations of addition and multiplication.*

Theorem 25 *The positive integers* $(x + a, a)$ *and the natural numbers* x *are isomorphic relative to the operations of addition and multiplication.*

For, under the correspondence $(x + a, a) \leftrightarrow x$, we have

1. $(x + a, a) = (y + b, b)$ if and only if $x = y$,
2. $(x + a, a) + (y + b, b) = (x + y + a + b, a + b)$, whence
$$(x + a, a) + (y + b, b) \leftrightarrow x + y,$$
3. $(x + a, a)(y + b, b) = (xy + xb + ay + ab + ab, xb + ab + ay + ab)$, whence
$$(x + a, a)(y + b, b) \leftrightarrow xy.$$

Remark Because of the isomorphism established in Theorem 25, the natural numbers are embedded within the integers, and the integers thus form an extension of the set of natural numbers.

Definition 10   Henceforth we shall use the notation x for the positive integer $(x + a, a)$.

Theorem 26 $(b, c) = (m, m)$ *if and only if* $b = c$.

Theorem 27   *For every integer* (x, y), *we have* $(x, y) + (m, m) = (x, y)$ *and* $(x, y)(m, m) = (m, m)$.

For

$$(x, y) + (m, m) = (x + m, y + m) \qquad \text{(by Def. 5)}$$

$$= (x, y), \qquad \text{(by Th. 21)}$$

and

$$(x, y)(m, m) = (xm + ym, xm + ym) \qquad \text{(by Def. 6)}$$

$$= (m, m). \qquad \text{(by Th. 26)}$$

Definition 11   Henceforth we shall use the notation 0 for the integer (m, m), and shall call this integer the *zero integer*.

Theorem 28 $(m, n) + (n, m) = 0$.

Definition 12   We shall denote (m, n) by $-(n, m)$.

Definition 13   The integers (m, n), where $m < n$, will be called *negative integers*.

Theorem 29   *Every negative integer has the form* $(a, a + x)$.

Theorem 30 $(a, a + x) = -x$.

For

$$(a, a + x) = -(x + a, a) \qquad \text{(by Def. 12)}$$

$$= -x. \qquad \text{(by Def. 10)}$$

The familiar properties of the integers may now be developed. Since we lack space actually to do this, the next stage of our program freely assumes those properties that may be needed.

The Rational Numbers

[**Motivation**   Any rational number (positive, zero, or negative) can be expressed as the quotient m/n of two integers m and n, with $n \neq 0$; in fact, the word *rational* has its origin in this fact. This suggests the idea of logically introducing the rational numbers as ordered pairs $\langle m, n \rangle$ of integers, with $n \neq 0$, where we actually have in mind the quotient m/n. With this interpretation of an ordered pair of integers, in which the second integer is nonzero, we are led to say $\langle a, b \rangle = \langle c, d \rangle$ if and only if $a/b = c/d$, or $ad = bc$. Similarly, since $a/b + c/d = (ad + bc)/bd$, we are led to define $\langle a, b \rangle + \langle c, d \rangle$ to be the ordered pair of integers $\langle ad + bc, bd \rangle$, and since $(a/b)(c/d) = ac/bd$, we are led to define $\langle a, b \rangle \langle c, d \rangle$ to be the ordered pair of integers $\langle ac, bd \rangle$.]

Definition 14   We shall denote the set of all ordered pairs $\langle m, n \rangle$ of integers m and n, with $n \neq 0$, by Q.

Definition 15   If $\langle a, b \rangle$ and $\langle c, d \rangle$ are two members of Q, we say that $\langle a, b \rangle = \langle c, d \rangle$ if and only if $ad = bc$.

Remark It is to be noted that the definition of equality of members of Q is not the ordinary concept of equality as an identity of symbols.

Theorem 31 *Equality of members of* Q *is an equivalence relation; that is,* (1) $\langle a, b \rangle = \langle a, b \rangle$, (2) *if* $\langle a, b \rangle = \langle c, d \rangle$, *then* $\langle c, d \rangle = \langle a, b \rangle$, (3) *if* $\langle a, b \rangle = \langle c, d \rangle$ *and* $\langle c, d \rangle = \langle e, f \rangle$, *then* $\langle a, b \rangle = \langle e, f \rangle$.

Definition 16 We define the *sum* of any two members $\langle a, b \rangle$ and $\langle c, d \rangle$ of Q, taken in this order, to be the ordered pair of integers $\langle ad + bc, bd \rangle$, and we write

$$\langle a, b \rangle + \langle c, d \rangle = \langle ad + bc, bd \rangle.$$

This binary operation on members of Q will be called *addition*.

Theorem 32 Q *is closed under addition, which is commutative, associative, and well defined.*

Definition 17 We define the *product* of any two members $\langle a, b \rangle$ and $\langle c, d \rangle$ of Q, taken in this order, to be the ordered pair of integers $\langle ac, bd \rangle$, and we write

$$\langle a, b \rangle \langle c, d \rangle = \langle ac, bd \rangle.$$

This binary operation on members of Q will be called *multiplication*.

Theorem 33 Q *is closed under multiplication, which is commutative, associative, well defined, and distributive over addition.*

Definition 18 The elements of Q, subject to Definitions 15, 16, 17, will be called *rational numbers*.

Theorem 34 *The rational numbers* $\langle x, 1 \rangle$ *and the integers* x *are isomorphic relative to the operations of addition and multiplication.*

Remark Because of the isomorphism established in Theorem 34, the integers are embedded within the rational numbers, and the rational numbers thus form an extension of the set of integers.

Definition 19 Henceforth we shall use the notation x for the rational number $\langle x, 1 \rangle$ and shall call rational numbers of the form $\langle a, b \rangle$, where $a \neq mb$, *fractions*.

The familiar properties of the rational numbers may now be developed. Again, since we lack space actually to deduce these properties, the next stage of our program freely assumes those properties that may be needed. In particular, we assume the order properties connecting the rational numbers.

7.5 The Real Numbers and the Complex Numbers

In the first part of this section we continue our program by building up the real number system from the rational numbers. This, which is undoubtedly the most difficult of the stages in building up the complex number system from the natural numbers, may be accomplished in several ingenious ways. We select a procedure due to Richard Dedekind—a procedure that ties in with some of our earlier work and with the algebraic methods that we have so far been employing in the step-by-step extension of the natural number system. A second procedure, devised by

Georg Cantor and based on the concept of a regular sequence of rational numbers, can be found in many textbooks on the theory of functions of real variables.[6] A third method, based on the decimal representation of real numbers, is favored by some mathematicians,[7] while a fourth method, involving sequences of nested intervals, is preferred by others.[8]

The Real Numbers

The following is only an outline of the present stage of the program. For proofs of the theorems cited the reader may, for example, consult Goffman.

[**Motivation** Dedekind was led to his procedure for introducing the real numbers when he found it difficult in his lectures to give a clear definition of continuity, a property that is possessed by the ordered set of real numbers but that is not possessed by the ordered set of rational numbers. After considerable cogitation on the difficulty, Dedekind perceived that the essence of the continuity of a straight line lies in the property that if all the points of the line are divided into two classes, such that every point in the first class lies to the left of every point in the second class, then there exists one and only one point of the line that produces this severance of the line into the two classes. Arithmetizing this idea led to the consideration of cuts in the ordered set of rational numbers, and to a definition of real numbers that guarantees closure of the real number set under the operation of forming cuts within it, thus assuring the continuity of this set. Dedekind was one of the great nineteenth-century pioneers in the logical and philosophical study of the foundations of analysis, and his two essays, *Stetigkeit und irrationale Zahlen* (*Continuity and Irrational Numbers*) of 1872 and *Was sind und was sollen die Zahlen?* (*What Are and What Should Be Numbers?*) of 1887, profoundly influenced subsequent studies in the foundations of mathematics.]

Definition 20 A *Dedekind cut* $(A|B)$ *in the set of rational numbers* is a partition, or separation, of these numbers into two classes A and B such that (1) every rational number is either in A or in B, (2) A and B each contain at least one rational number, (3) every element of A is less than every element of B.

Types of Dedekind Cuts There are three possible types of Dedekind cuts $(A|B)$ in the set of rational numbers, for we may have the situations where (a) A has a largest element a and B has no smallest element, (b) A has no largest element and B has a smallest element b, (c) A has no largest element and B has no smallest element.

A seeming fourth possibility, where A has a largest element a and B has a smallest element b, cannot occur, for between any two distinct rational numbers a and b there is always a rational number c, and this number c could not lie in either A or B, which contradicts the formation of the sets A and B.

---

[6] Cf. J. Pierpont [1], vol. I., chap. 2, or C. Goffman, chap. 3.

[7] Cf. J. F. Ritt or H. Levi.

[8] For an elementary exposition of this method, see R. Courant and H. Robbins, chap. 2.

An example of a Dedekind cut of type (a) is the cut $(A \,|\, B)$ in which A consists of all the rational numbers less than or equal to 2, while B contains all the remaining rational numbers. Here A has a largest element—namely, 2—and B has no smallest element.

An example of a Dedekind cut of type (b) is the cut $(A \,|\, B)$ in which A consists of all the rational numbers less than 2, while B contains all the remaining rational numbers. Here A has no largest element, but B has a smallest element, namely 2.

An example of a Dedekind cut of type (c) is the cut $(A \,|\, B)$ in which A consists of all the nonpositive rational numbers and those positive rational numbers whose squares are less than 2, while B contains all the remaining rational numbers. Here A has no largest element and B has no smallest element.

Definition 21   A Dedekind cut of the rational numbers that is of type (a) or type (b) is called a *rational cut*. To each rational number r there correspond two cuts, one of type (a) and one of type (b). In order to have a one-to-one correspondence between the rational cuts and the rational numbers, we shall consider only rational cuts of type (b).

Definition 22   We shall denote the set of all Dedekind cuts $(A \,|\, B)$ in the set of rational numbers that are of type (b) or type (c) by R.

Definition 23   If $(A \,|\, B)$ and $(C \,|\, D)$ are members of R, we say that $(A \,|\, B) = (C \,|\, D)$ if and only if sets A and C are identical.

Remark   Note that, by definition, the equality of members of R is an identity equality.

Theorem 35   *Let* $(A_1 \,|\, B_1), (A_2 \,|\, B_2)$ *be two members of* R. *Let* A *be the set of all rational numbers of the form* x + y, *where* x *is in* A_1 *and* y *is in* A_2, *and let* B *be the set of all remaining rational numbers. Then* (A $|$ B) *is a member of* R.

Definition 24   We define the *sum* of any two members $(A_1 \,|\, B_1), (A_2 \,|\, B_2)$ of R to be the member $(A \,|\, B)$ of Theorem 35. We write

$$(A_1 \,|\, B_1) + (A_2 \,|\, B_2) = (A \,|\, B),$$

and refer to this binary operation on members of R as *addition*.

Theorem 36   *Addition of members of* R *is commutative and associative.*

Theorem 37   *Let* (A $|$ B) *be a member of* R *and let* (M $|$ P) *denote the member of* R *in which* M *is the set of all negative rational numbers and* P *is the set of all the remaining rational numbers. Then there exists a unique member of* R, *which we shall denote by* (A′ $|$ B′), *such that* (A $|$ B) + (A′ $|$ B′) = (M $|$ P).

Remark   The member $(M \,|\, P)$ plays the role of zero, or the additive identity, and $(A' \,|\, B')$ is then the additive inverse of $(A \,|\, B)$.

Definition 25   Member $(A \,|\, B)$ of R is said to be *positive* if A contains positive rational numbers; $(A \,|\, B)$ is said to be *negative* if B contains negative rational numbers.

Theorem 38   *If* (A $|$ B) *is positive, then* (A′ $|$ B′) *is negative; if* (A $|$ B) *is negative, then* (A′ $|$ B′) *is positive.*

Theorem 39   *Let* $(A_1 \,|\, B_1), (A_2 \,|\, B_2)$ *be two positive members of* R. *Let* A *be the set of all nonpositive rational numbers and those positive rational numbers of the form* xy, *where* x > 0 *is in* A_1 *and* y > 0 *is in* A_2. *Let* B *be the set of all remaining rational numbers. Then* (A $|$ B) *is a member of* R.

Definition 26 Let $(A_1 | B_1)$, $(A_2 | B_2)$ be any two members of R. If $(A_1 | B_1)$ and $(A_2 | B_2)$ are both positive, we define their *product*, $(A_1 | B_1)(A_2 | B_2)$, to be the member $(A | B)$ of Theorem 39. If $(A_1 | B_1)$ and $(A_2 | B_2)$ are both negative, we define their product to be $(A_1' | B_1')(A_2' | B_2')$. If $(A_1 | B_1)$ is positive and $(A_2 | B_2)$ is negative, we define their product to be $(C' | D')$, where $(C | D) = (A_1 | B_1)(A_2' | B_2')$. If $(A_1 | B_1)$ is negative and $(A_2 | B_2)$ is positive, we define their product to be $(C' | D')$, where $(C | D) = (A_1' | B_1')(A_2 | B_2)$. If either $(A_1 | B_1)$ or $(A_2 | B_2)$ is the member $(M | P)$, we define their product to be the member $(M | P)$. This binary operation on members of R will be called *multiplication*.

Theorem 40 *Multiplication of members of* R *is commutative, associative, and distributive over addition.*

Definition 27 The members of R, subject to Definitions 23, 24, 26, will be called *real numbers*.

Theorem 41 *The rational cuts* (A | B) *of type* (b), *in which* r *represents the smallest element of* B, *and the rational numbers* r, *are isomorphic relative to the operations of addition and multiplication.*

Remark Because of the isomorphism of Theorem 41, the rational numbers are embedded within the real numbers, and the real numbers thus form an extension of the set of rational numbers.

Definition 28 The members of R that are rational cuts [those of type (b)] are known as *rational real numbers* and those members of R that are not rational cuts [those of type (c)] are known as *irrational real numbers*.

All the postulates of a complete ordered field, as listed in Section 7.2, can now be shown to be satisfied if we interpret the elements of the field as the Dedekind cuts $(A | B)$ of R, the binary operations $\oplus$ and $\otimes$ as addition and multiplication of members of R, z as the rational cut $(M | P)$ corresponding to 0, a as $(A | B)$ and $\bar{a}$ as $(A' | B')$, u as the rational cut corresponding to 1, and the positive and negative elements of the field as the positive and negative members of R. It follows, then, that *the real numbers, defined as members of* R, *constitute a complete ordered field* and hence, because of the categoricalness of the postulate set for a complete ordered field, are, to within an isomorphism, the same as the real numbers considered in Section 7.2.

Since we were able to extend the set of rational numbers to the set of real numbers by making Dedekind cuts in the set of rational numbers, one might wonder if we can similarly extend the set of real numbers by making Dedekind cuts in this set. In view of Dedekind's theorem of continuity (see Section 7.2), the answer to this question is *no*; that is, no further numbers can be created by the method of Dedekind cuts.

The Complex Numbers

The last stage in our program involves the building of the complex number system on the real number system by the introduction of suitable definitions.

[**Motivation** We have already, in Section 5.2, given some consideration to Hamilton's procedure of defining the complex numbers in terms of ordered pairs of real numbers. For the sake of completeness, we record this procedure once more. Now any complex number is customarily written in the form $m + in$, where m and n are real numbers and i is the imaginary unit. This suggests the idea of logically introducing the complex numbers as ordered pairs $[m, n]$ of real numbers, where by $[m, n]$ we have in mind the representation $m + in$. With this interpretation of an ordered pair of real numbers we are led to say $[a, b] = [c, d]$ if and only if $a + ib = c + id$, or if and only if $a = c$ and $b = d$. Similarly, since $(a + ib) + (c + id) = (a + c) + i(b + d)$, we are led to define $[a, b] + [c, d]$ to be the ordered pair of real numbers $[a + c, b + d]$, and since $(a + ib)(c + id) = (ac - bd) + i(ad + bc)$, we are led to define $[a, b][c, d]$ to be the ordered pair of real numbers $[ac - bd, ad + bc]$.]

Definition 29 We shall denote the set of all ordered pairs $[m, n]$ of real numbers by C.

Definition 30 If $[a, b]$ and $[c, d]$ are two members of C, we say that $[a, b] = [c, d]$ if and only if $a = c$ and $b = d$.

Remark Note that, by definition, the equality of members of C is an identity equality.

Definition 31 We define the *sum* of any two members $[a, b]$ and $[c, d]$ of C, taken in this order, to be the ordered pair of real numbers $[a + c, b + d]$, and we write

$$[a, b] + [c, d] = [a + c, b + d].$$

This binary operation on members of C will be called *addition*.

Theorem 42 C *is closed under addition, which is commutative and associative.*

Definition 32 We define the *product* of any two members $[a, b]$ and $[c, d]$ of C, taken in this order, to be the ordered pair of real numbers $[ac - bd, ad + bc]$, and we write

$$[a, b][c, d] = [ac - bd, ad + bc].$$

This binary operation on members of C will be called *multiplication*.

Theorem 43 C *is closed under multiplication, which is commutative, associative, and distributive over addition.*

Definition 33 The elements of C, subject to Definitions 30, 31, 32, will be called *complex numbers*.

Theorem 44 *The complex numbers* $[a, 0]$ *and the real numbers* a *are isomorphic relative to the operations of addition and multiplication.*

Remark Because of the isomorphism of Theorem 44, the real numbers are embedded within the complex numbers, and the complex numbers thus form an extension of the set of real numbers.

Definition 34 Henceforth we shall use the notation a for the complex number $[a, 0]$.

Definition 35 We shall denote the complex number $[0, 1]$ by i.

Theorem 45 $i^2 = -1$.

Theorem 46 $[0, b] = ib$.

Definition 36 Complex numbers of the form $[0, b]$ shall be called *pure imaginary numbers*, and those of the form $[a, b]$, where $a \neq 0$ and $b \neq 0$, shall be called *mixed imaginary numbers*.

Theorem 47 $[a, b] = a + ib$.

Theorem 48 *When we use the notation* a + ib *for* [a, b], *we may carry out addition and multiplication as with real polynomials in* i, *provided we replace* i^2, *wherever it occurs, by* -1.

With this we bring to a close our sketch of the program by which the real and the complex number systems are obtained, without the need of any further postulates, from the basic system of natural numbers. The accomplishment of this program is an amazing achievement. Since we have seen that the consistency of the great bulk of mathematics can be made to rest on that of the real number system, it is now apparent that the consistency of the great bulk of mathematics actually can be made to rest on that of the much simpler system of natural numbers. In some sense, then, the ancient Pythagorean belief that everything depends on the whole numbers is justified, and we see meaning in Leopold Kronecker's often quoted remark, "Die ganzen Zahlen hat Gott gemacht, alles andere ist Menschenwerk."[9]

The program that we followed in reaching the complex number system from the natural number system can be represented by the scheme

$$N \to I \to Q \to R \to C,$$

where N represents the system of natural numbers, I the system of integers, Q the system of rational numbers, R the system of real numbers, and C the system of complex numbers. Various other developments are possible, of which a commonly followed program is[10]

$$N \to Q(+) \to R(+) \to R \to C,$$

where $Q(+)$ represents the system of positive rational numbers and $R(+)$ the system of positive real numbers. It is interesting that this alternative development reflects the historical growth of our number system. For mankind first employed the whole, or natural, numbers for counting, then the positive fractions for purposes of measurement; next he introduced positive real numbers to cover incommensurable situations (like the side and diagonal of a square), then acknowledged negative numbers, and, finally, accepted imaginary numbers. On the other hand, the development of the program as we pursued it reflects the successive algebraic need for new numbers. Thus, if a and b are natural numbers, the equation $x + a = b$ will always have a solution only if we extend the number system to include all integers. Again, if $a \neq 0$ and b are integers, the equation $ax = b$ will always have a solution only if we extend the number system to include all rational numbers. These numbers, though, are not sufficient to yield a solution to the equation $x^2 = 2$, and so the real numbers are introduced. Finally, the equation $x^2 = -2$ has no solution until the complex numbers are introduced. But here we reach an end, for it has been shown that any polynomial equation having coefficients in the complex number field has a solution within this field— this is the famous so-called *fundamental theorem of* (classical) *algebra*.

The complex number system need not be obtained by definition from the real number system; it, too, can be given by an appropriate categorical set of

---

[9] "God made the whole numbers, all the rest is the work of man."

[10] This is the program pursued in E. G. H. Landau.

postulates. We shall not list such a set of postulates here,[11] however, but conclude with the following schema, which exhibits a classification of the various kinds of numbers discussed in this chapter:

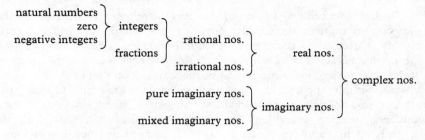

PROBLEMS

7.1.1   **(a)**   Under what conditions on the real number x is

$$\sqrt{1 - \sin^2 x} = \cos x?$$

     (b)   Under what conditions does the equation $ax = b$, where a and b are real numbers, have exactly one real solution? no real solution? infinitely many real solutions?

7.1.2   **(a)**   If b is a real number, we *define* b^5 to be an abbreviation for $(b)(b)(b)(b)(b)$, and, more generally, if n is any positive integer, we *define* b^n to be an abbreviation for $(b)(b)\cdots(b)$, where there are n factors in the product. In the symbol b^n, b is referred to as the *base* and n as the *exponent*. Show that the following so-called *laws of exponents*, in which a and b represent real numbers and m and n represent positive integers, are merely commonsense observations based on the above definition:
     (1) $(b^m)(b^n) = b^{m+n}$.
     (2) $\dfrac{b^m}{b^n} = b^{m-n}$, if $m > n$ and $b \neq 0$.
     (3) $(b^m)^n = b^{mn}$.
     (4) $(ab)^n = a^n b^n$.

     (b)   The extension of the notion of exponents to include all rational numbers as exponents is accomplished by definition and is motivated by a desire to preserve the above laws of exponents. (1) Show that if we wish to preserve law 2 for the situation where $m = n$, we are forced, for $b \neq 0$, to define $b^0 = 1$. (2) Show that if we wish to preserve law 2 for the situation where $n = m + 1$, we are forced, for $b \neq 0$, to define b^{-1} to mean $1/b$. (3) Show that if we wish to preserve law 3, we are forced to define $b^{1/n}$, where n is a positive integer, to mean an nth root of b.

     (c)   In (b) we have seen how one would try to formulate a set of definitions that might permit the universal application of the four exponential laws to numbers written in exponential form, where the bases of these numbers are real and the exponents are rational. Each definition was motivated by the effort to preserve some one of the four laws. It could well be that it is

---

[11] Such a postulate set for the complex number system may be found in E. V. Huntington [10].

impossible in this way to obtain suitable definitions that would satisfy *all* the exponential laws. Show that this is actually the case by (1) evaluating $[(-3)^2]^{1/2}$ in two ways, (2) evaluating $[(-4)(-4)]^{1/2}$ in two ways. Many students of elementary algebra are under the delusion that the four laws of exponents hold universally for exponential numbers having real bases and rational exponents; indeed, many textbooks of elementary algebra make this erroneous statement. The examples above show that this is not so. It can be shown, however, that *the four exponential laws do hold for real bases and rational exponents if we exclude the situations where the base is negative and the exponent is a reduced fraction with an even denominator.* Without an understanding of the limitations involved in the use of the exponential laws for rational exponents, a student can become involved in paradoxical results.

7.1.3 Explain the following paradoxes involving square root radicals:

(a) Since $\sqrt{a}\,\sqrt{b} = \sqrt{ab}$, we have

$$\sqrt{-1}\,\sqrt{-1} = \sqrt{(-1)(-1)} = \sqrt{1} = 1.$$

But, by definition, $\sqrt{-1}\,\sqrt{-1} = -1$. Hence $-1 = +1$.

(b) We have, successively,

$$\sqrt{-1} = \sqrt{-1},$$

$$\sqrt{\frac{1}{-1}} = \sqrt{\frac{-1}{1}},$$

$$\frac{\sqrt{1}}{\sqrt{-1}} = \frac{\sqrt{-1}}{\sqrt{1}},$$

$$\sqrt{1}\,\sqrt{1} = \sqrt{-1}\,\sqrt{-1},$$

$$1 = -1.$$

(c) Consider the following identity, which holds for all values of x and y:

$$\sqrt{x-y} = i\sqrt{y-x}.$$

Setting $x = a$, $y = b$, where $a \neq b$, we find

$$\sqrt{a-b} = i\sqrt{b-a}.$$

Now setting $x = b$, $y = a$, we find

$$\sqrt{b-a} = i\sqrt{a-b}.$$

Multiplying the last two equations, member by member, we get

$$\sqrt{a-b}\,\sqrt{b-a} = i^2\sqrt{b-a}\,\sqrt{a-b}.$$

Dividing both sides by $\sqrt{a-b}\,\sqrt{b-a}$, we finally get

$$1 = i^2, \text{ or } 1 = -1.$$

7.1.4 In elementary algebra one often learns the "axioms":

A1. *If equals are added to equals, the results are equal.*

A2. *If equals are subtracted from equals, the results are equal.*

A3. *If equals are multiplied by equals, the results are equal.*

A4. *If equals are divided by equals, the results are equal.*

A5. *Equal roots of equals are equal.*

Explain the following paradoxes, which use only these axioms:

(a) Let us solve the simple equation

$$x - 1 = 2$$

in the following roundabout way. Multiplying both sides by $x - 5$, we get, by A3,

$$x^2 - 6x + 5 = 2x - 10.$$

Now subtracting $x - 7$ from both sides, we get, by A2,

$$x^2 - 7x + 12 = x - 3.$$

Next, dividing both sides by $x - 3$, we get, by A4,

$$x - 4 = 1.$$

Finally, adding 4 to both sides, we get, by A1,

$$x = 5.$$

But this answer is not correct, for the value $x = 5$ does not satisfy the original equation.

(b) Let a and b be *any* two numbers and denote their sum by $2c$. Then

$$a + b = 2c.$$

Multiplying both sides by $a - b$, we get, by A3,

$$a^2 - b^2 = 2ac - 2bc.$$

Adding $b^2 + c^2 - 2ac$ to both sides, we get, by A1,

$$a^2 - 2ac + c^2 = b^2 - 2bc + c^2$$

or

$$(a - c)^2 = (b - c)^2.$$

Taking the square root of each side, we have, by A5,

$$a - c = b - c.$$

Adding c to both sides, we get, by A1,

$$a = b.$$

But a and b were chosen as *any* two numbers. Therefore *any* two numbers are equal to each other.

7.1.5 Find the sum of the roots of the equation

$$\frac{x + 3}{x^2 - 1} + \frac{x - 3}{x^2 - x} + \frac{x + 2}{x^2 + x} = 0.$$

7.1.6 Most students of elementary algebra will agree to the following theorem: "If two fractions are equal and have equal numerators, then they also have equal denominators." Now consider the following problem. We wish to solve the equation

$$\frac{x+5}{x-7} - 5 = \frac{4x-40}{13-x}.$$

Combining the terms on the left side, we find

$$\frac{(x+5)-5(x-7)}{x-7} = \frac{4x-40}{13-x}$$

or

$$\frac{4x-40}{7-x} = \frac{4x-40}{13-x}.$$

By the above theorem, it follows that $7 - x = 13 - x$, or, on adding x to both sides, that $7 = 13$. What is wrong?

7.1.7 Explain the following paradox: Let $\frac{a}{b}$ and $\frac{c}{d}$ be any two equal fractions. Then

$$\frac{a}{b} - 1 = \frac{c}{d} - 1 \quad \text{and} \quad \frac{a}{b} + 1 = \frac{c}{d} + 1.$$

That is

$$\frac{a-b}{b} = \frac{c-d}{d} \quad \text{and} \quad \frac{a+b}{b} = \frac{c+d}{d}.$$

It follows that

$$\frac{a-b}{a+b} = \frac{c-d}{c+d},$$

$$(a-b)(c+d) = (a+b)(c-d),$$

$$ac - bc + ad - bd = ac + bc - ad - bd,$$

$$ad - bc = -(ad - bc),$$

$$1 = -1.$$

7.1.8 Explain the following paradox:
Certainly

$$3 > 2.$$

Multiplying both sides by $\log (1/2)$, we find

$$3 \log (1/2) > 2 \log (1/2)$$

or

$$\log (1/2)^3 > \log (1/2)^2$$

whence

$$(1/2)^3 > (1/2)^2 \text{ or } 1/8 > 1/4.$$

7.1.9 Explain the following paradox:
Clearly $(-1)^2 = (+1)^2$. Taking the logarithm of each side, we have $\log (-1)^2 = \log (1)^2$. Therefore $2 \log (-1) = 2 \log 1$, or $-1 = 1$.

7.1.10 (*for students who have studied calculus*) Resolve the following paradoxes:

(a) By standard procedure we find,

$$\int_{-1}^{1} \frac{dx}{x^2} = \left[-\frac{1}{x} \right]_{-1}^{1} = -1 - 1 = -2.$$

But the function $y = 1/x^2$ is never negative; hence the above "evaluation" cannot be correct.

(b) Let e denote the eccentricity of the ellipse $x^2/a^2 + y^2/b^2 = 1$. It is well known that the length r of the radius vector drawn from the left-hand focus of the ellipse to any point $P:(x, y)$ on the curve is given by $r = a + ex$. Now $dr/dx = e$. Since there are no values of x for which dr/dx vanishes, it follows that r has no maximum or minimum. But the only closed curve for which the radius vector has no maximum or minimum is a circle. It follows that every ellipse is a circle.

(c) Consider the isosceles triangle ABC of Figure 7.1, in which base $AB = 12$ and altitude $CD = 3$. Surely there is a point P on CD such that

$$S = PC + PA + PB$$

is a minimum. Let us try to locate this point P. Denote DP by x. Then $PC = 3 - x$ and $PA = PB = (x^2 + 36)^{1/2}$. Therefore

$$S = 3 - x + 2(x^2 + 36)^{1/2},$$

and

$$\frac{dS}{dx} = -1 + 2x(x^2 + 36)^{-1/2}.$$

Setting $dS/dx = 0$, we find $x = 2\sqrt{3} > 3$, and P lies outside the triangle on DC produced. Hence there is no point on the segment CD for which S is a minimum.

(d) Consider the integral

$$I = \int \sin x \cos x \, dx.$$

Then we have

$$I = \int \sin x \, (\cos x \, dx) = \int \sin x \, d(\sin x) = \frac{\sin^2 x}{2}.$$

Also

$$I = \int \cos x \, (\sin x \, dx) = -\int \cos x \, d(\cos x) = -\frac{\cos^2 x}{2}.$$

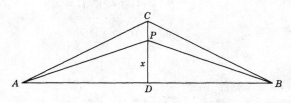

FIGURE 7.1

Therefore

$$\sin^2 x = -\cos^2 x,$$

or

$$\sin^2 x + \cos^2 x = 0.$$

But, for any x,

$$\sin^2 x + \cos^2 x = 1.$$

(e) Since

$$\int \frac{dx}{x} = \int \frac{-dx}{-x},$$

we have $\log x = \log(-x)$ or $x = -x$, whence $1 = -1$.

(f) Let S denote the sum of the *convergent* series

$$\frac{1}{(1)(3)} + \frac{1}{(3)(5)} + \frac{1}{(5)(7)} + \cdots.$$

Then

$$S = \left(\frac{1}{1} - \frac{2}{3}\right) + \left(\frac{2}{3} - \frac{3}{5}\right) + \left(\frac{3}{5} - \frac{4}{7}\right) + \cdots$$

$$= 1 - \frac{2}{3} + \frac{2}{3} - \frac{3}{5} + \frac{3}{5} - \frac{4}{7} + \cdots = 1,$$

since all terms after the first cancel out. Again

$$S = \frac{(1/1) - (1/3)}{2} + \frac{(1/3) - (1/5)}{2} + \frac{(1/5) - (1/7)}{2} + \cdots$$

$$= \frac{1}{2} - \frac{1}{6} + \frac{1}{6} - \frac{1}{10} + \frac{1}{10} - \frac{1}{14} + \cdots = \frac{1}{2},$$

since all terms after the first cancel out. It follows that $1 = 1/2$.

(g) Explain the paradoxical results concerning the series S in Section 7.1.

7.2.1 Give a definition of a lower bound and of a greatest lower bound of a nonempty collection M of real numbers.

7.2.2 Prove that any nonempty collection of real numbers can have at most one least upper bound.

7.2.3 Give an example of a nonempty collection M of real numbers that has

(a) both an upper and a lower bound;

(b) an upper bound but no lower bound;

(c) a lower bound but no upper bound;

(d) neither an upper nor a lower bound;

(e) a least upper bound that is in the collection M;

(f) a least upper bound that is not in the collection M.

7.2.4 Prove that if a nonempty collection M of real numbers has a lower bound, then it has a greatest lower bound.

7.2.5 Let M be a nonempty collection of real numbers, and let t be any fixed positive real number. Let N be the collection of numbers of the form tx, where x is in M. Show that if b is the least upper bound of M, then tb is the least upper bound of N.

7.2.6 Show that if a is a positive real number there is a positive integer n such that $1/n < a$.

7.2.7 Let M and N be two nonempty collections of real numbers having a and b, respectively, as least upper bounds. Let P be the collection of all numbers of the form $x + y$, where x is in M and y is in N. Show that $a + b$ is the least upper bound of P.

7.2.8   Let M be the collection of real numbers

$$x_n = (-1)^n[2 - 4/2^n], \quad n = 1, 2, \ldots.$$

Find the least upper bound and greatest lower bound of M. Do the same for the collection of real numbers

$$y_n = (-1)^n + 1/n, \quad n = 1, 2, \ldots.$$

7.2.9   Does the ordered system of all nonzero real numbers satisfy the postulate of continuity?

7.2.10   State the Archimedean law and Dedekind's theorem of continuity in the terminology and symbolism of a complete ordered field.

7.2.11   Show that for any real number r there is a smallest integer m such that $m > r$.

7.2.12   (a)   Show that between any two real numbers c and d, $c > d$, there lies at least one other real number.

      (b)   Show that between any two real numbers c and d, $c > d$, there lies a rational number m/n.

7.3.1   Deduce the following easy theorems about any natural numbers a, b, c, d:

      (a)   If $a > b$ and $b > c$, then $a > c$.

      (b)   If $a < b$, then $a + c < b + c$.

      (c)   If $c + a < c + b$, then $a < b$. (This is the converse of Theorem 7 of Section 7.3.)

      (d)   If $ca < cb$, then $a < b$. (This is the converse of Theorem 9 of Section 7.3.)

      (e)   If $a < b$ and $c < d$, then $ac < bd$.

      (f)   If $a > b$, then $a \geq b + 1$.

7.3.2   If n is an arbitrary natural number, prove, by means of the principle of mathematical induction, that

      (a) $1^2 + 2^2 + \cdots + n^2 = n(n + 1)(2n + 1)/6$.

      (b) $1^3 + 2^3 + \cdots + n^3 = [n(n + 1)/2]^2$.

      (c) $(1)(2) + (2)(3) + \cdots + n(n + 1) = n(n + 1)(n + 2)/3$.

      (d) $1/(1)(2) + 1/(2)(3) + \cdots + 1/n(n + 1) = n/(n + 1)$.

      (e) $8^n - 3^n$ is divisible by 5.

7.3.3   (*harder situations*) If n is an arbitrary natural number, prove, by means of the principle of mathematical induction, that

      (a) $1^5 + 2^5 + \cdots + n^5 = n^2(n + 1)^2(2n^2 + 2n - 1)/12$.

      (b) $1^2 + 4^2 + 7^2 + \cdots + (3n - 2)^2 = n(6n^2 - 3n - 1)/2$.

      (c) $(1 + a)^n \geq 1 + na$ if $a \geq -1$.

      (d) $n^3 + 1 > n^2 + n$ for $n \geq 2$.

      (e) $3^{2n+2} - 8n - 9$ is divisible by 64.

      (f) $9^n - 8n - 1$ is divisible by 64.

      (g) $7^{2n} - 48n - 1$ is divisible by 2304.

      (h) $a^n - b^n$ is divisible by $a - b$.

7.3.4   Establish the following geometrical theorems by mathematical induction:

      (a)   The sum of the interior angles of a convex n-gon is $(n - 2)180°$.

      (b)   The maximum number of lines determined by $n \geq 2$ points in a plane is $n(n - 1)/2$.

      (c)   The maximum number of points of intersection of $n \geq 2$ lines in a plane is $n(n - 1)/2$.

      (d)   The number of regions into which n straight lines divide a plane can never exceed 2^n.

7.3.5   If an amount of money P is invested at an annual interest rate r and interest is compounded annually, show that after n years the sum amounts to $P(1 + r)^n$.

7.3.6   By means of the second principle of mathematical induction prove that
   (a) for every natural number n, $n + 1$ is a prime or can be factored into primes;
   (b) the number of positive prime factors of $n \geq 2$ is less than $2 \log_e n$;
   (c) in every set of n distinct natural numbers there is a greatest.

7.3.7   Show that for every natural number n

$$u_n = \frac{(1 + \sqrt{5})^n - (1 - \sqrt{5})^n}{2^n \sqrt{5}}$$

is a natural number.

7.3.8   **(a)** Show that $n^2 - n + 41$ is a prime number for $n = 1, 2, \ldots, 40$. Is this sufficient to prove that $n^2 - n + 41$ is a prime number for all natural numbers n? Is $n^2 - n + 41$ a prime number for all natural numbers n?
   (b) Show that if $1 + 2 + \cdots + n = (n + 1/2)^2/2$ is true for $n = k \geq 1$, then it is true for $n = k + 1$. Is this sufficient to prove that the equality holds for all natural numbers n? Does the equality hold for all natural numbers n?

7.3.9   Find the fallacy in the following proof by mathematical induction:
   $P(n)$: *All numbers in a set of* n *numbers are equal to one another.*
   I. $P(1)$ is obviously true.
   II. Suppose k is a natural number for which $P(k)$ is true. Let $a_1, a_2 \ldots, a_k$, a_{k+1} be any set of $k + 1$ numbers. Then, by the supposition, $a_1 = a_2 = \cdots = a_k$ and $a_2 = \cdots = a_k = a_{k+1}$. Therefore $a_1 = a_2 = \cdots = a_k = a_{k+1}$, and $P(k + 1)$ is true.
   It follows that $P(n)$ is true for all natural numbers n.

7.3.10   Find the fallacy in the following proof by mathematical induction:
   $P(n)$: *If* a *and* b *are any two natural numbers such that* max $(a, b) = n$, *then* a = b. [Note: By max (a, b), when $a \neq b$, is meant the larger of the two numbers a and b. By max (a, a) is meant the number a. Thus max $(5, 7) = 7$, max $(8, 2) = 8$, max $(4, 4) = 4$.]
   I. $P(1)$ is obviously true.
   II. Suppose k is a natural number for which $P(k)$ is true. Let a and b be any two natural numbers such that max $(a, b) = k + 1$, and consider $\alpha = a - 1$, $\beta = b - 1$. Then max $(\alpha, \beta) = k$, whence, by the supposition, $\alpha = \beta$. Therefore $a = b$ and $P(k + 1)$ is true.
   It follows that $P(n)$ is true for all natural numbers n.

7.3.11   Comment on the following amusing "proof" that *every natural number is interesting*:
   Suppose it is not true that every natural number is interesting, and let M be the set of those natural numbers that are not interesting. Then M is nonempty. Therefore, by the principle of the least natural number, M contains a smallest natural number m. That is, m is the smallest natural number which is not interesting. But this is interesting!

7.3.12   Deduce Postulate N10 (the postulate of finite induction) from Postulates N1 through N9 together with the principle of the least natural number. (This will show that the postulate of finite induction and the principle of the least natural number are equivalent, and that either one can be used in place of Postulate N10 of the postulate set for the natural number system as given in Section 7.3.)

7.3.13   Show that Peano's postulate set for the natural number system is implied by the postulate set for the natural number system given in Section 7.3.

7.3.14   **(a)** Establish, as far as you can, type-I sequential independence of Postulates N1 through N10 (see Problem 6.5.5.)

(b) Establish, as far as you can, type-II sequential independence of the Peano postulates for the natural number system as listed in Section 7.3 (see Problem 6.5.6).

7.4.1 Give complete proofs of the following theorems of Section 7.4:
  (a) Theorem 20.   (e) Theorem 24.
  (b) Theorem 21.   (f) Theorem 26.
  (c) Theorem 22.   (g) Theorem 28.
  (d) Theorem 23.   (h) Theorem 29.

7.4.2 Prove the following theorems about integers:
  (a) If $(a, b) + (c, d) = (a, b) + (e, f)$, then $(c, d) = (e, f)$.
  (b) If $(x, y)(a, b) = (x, y)(c, d)$ and $(x, y) \neq (m, m)$, then $(a, b) = (c, d)$.
  (c) The equation $(a, b) + (x, y) = (c, d)$ has a unique solution.

7.4.3 Prove the following theorems about integers x, y, z:
  (a) $-(-y) = y$.
  (b) $(-x)(-y) = xy$.
  (c) $(-x)y = x(-y) = -(xy)$.

  Agreement We agree to write $x + (-y)$ as $x - y$.

  (d) $-(x + y) = -x - y$.
  (e) $x(y - z) = xy - xz$.
  (f) If $xy = 0$, then $x = 0$ or $y = 0$.

7.4.4 We may define inequalities between integers in terms of inequalities between natural numbers as follows. We say $(a, b) < (c, d)$, or $(c, d) > (a, b)$, if and only if $c + b > d + a$. On the basis of this definition, prove the following properties of the inequality signs for integers x, y, z:
  (a) If $x < y$ and $y < z$, then $x < z$.
  (b) If $x < y$, then $x + z < y + z$.
  (c) If $x < y$ and $z > 0$, then $xz < yz$.
  (d) If $x < y$ and $z < 0$, then $xz > yz$.

7.4.5 Give proofs of the following theorems of Section 7.4:
  (a) Theorem 31.
  (b) Theorem 32.
  (c) Theorem 33.

7.4.6 Prove the following theorems about rational numbers:
  (a) $\langle a, b \rangle + \langle 0, 1 \rangle = \langle a, b \rangle$.
  (b) $\langle a, b \rangle + \langle -a, b \rangle = \langle 0, 1 \rangle$.
  (c) $\langle a, b \rangle \langle m, m \rangle = \langle a, b \rangle$.
  (d) If $\langle a, b \rangle + \langle c, d \rangle = \langle a, b \rangle + \langle e, f \rangle$, then $\langle c, d \rangle = \langle e, f \rangle$.
  (e) If $\langle a, b \rangle \langle c, d \rangle = \langle a, b \rangle \langle e, f \rangle$, and $a \neq 0$, then $\langle c, d \rangle = \langle e, f \rangle$.
  (f) The equation $\langle a, b \rangle \langle x, y \rangle = \langle c, d \rangle$ has a unique solution if $a \neq 0$.

7.4.7 Let a^+ denote a positive integer (or zero) and let a^- denote the corresponding negative integer (or zero). Show that
  (a) $\langle a^-, b^- \rangle = \langle a^+, b^+ \rangle$.
  (b) $\langle a^+, b^- \rangle = \langle a^-, b^+ \rangle$.

  Agreement Of the two equal forms $\langle a^-, b^- \rangle$, $\langle a^+, b^+ \rangle$ we agree to use only the second, and of the two equal forms $\langle a^+, b^- \rangle$, $\langle a^-, b^+ \rangle$ we agree to use only the second.

  Definition We now say $\langle a, b \rangle > \langle c, d \rangle$ if and only if $ad > bc$.

  Show that
  (c) if $\langle a, b \rangle > \langle c, d \rangle$ and $\langle c, d \rangle > \langle e, f \rangle$, then $\langle a, b \rangle > \langle e, f \rangle$.
  (d) if $\langle a, b \rangle > \langle c, d \rangle$, then $\langle a, b \rangle + \langle e, f \rangle > \langle c, d \rangle + \langle e, f \rangle$.

(e) if $\langle a, b \rangle > \langle c, d \rangle$ and $\langle e, f \rangle > \langle g, h \rangle$, then $\langle a, b \rangle + \langle e, f \rangle > \langle c, d \rangle + \langle g, h \rangle$.

(f) if $\langle a, b \rangle > \langle c, d \rangle$ and $\langle e, f \rangle > \langle 0, b \rangle$, then $\langle a, b \rangle \langle e, f \rangle > \langle c, d \rangle \langle e, f \rangle$.

(g) if $\langle a, b \rangle > \langle c, d \rangle$ and $\langle 0, b \rangle > \langle e, f \rangle$, then $\langle c, d \rangle \langle e, f \rangle > \langle a, b \rangle \langle e, f \rangle$.

7.4.8 (a) Prove Theorem 34 of Section 7.4.

(b) Show that the order relation, $>$, for rational numbers (see Problem 7.4.7) is preserved under the isomorphism of Theorem 34.

7.5.1 Supply proofs for the following theorems of Section 7.5:

(a) Theorem 35.

(b) Theorem 36.

(c) Theorem 37.

(d) Theorem 38.

7.5.2 Let $(A_1 | B_1)$, $(A_2 | B_2)$ be any two members of R. Let A be the set of all rational numbers of the form xy, where x is in A_1 and y is in A_2, and let B be the set of all remaining rational numbers. Show why it would be inadequate to define the product of $(A_1 | B_1)$ and $(A_2 | B_2)$ to be $(A | B)$.

7.5.3 Show that for any member $(A | B)$ of R

(a) $(A | B) + (M | P) = (A | B)$,

(b) $(A | B)(M | P) = (M | P)$.

7.5.4

Definition We say $(A | B) > (C | D)$ if and only if the set of rational numbers C is a proper subset of the set of rational numbers A. Show that

(a) if $(A | B) > (C | D)$ and $(C | D) > (E | F)$, then $(A | B) > (E | F)$.

(b) if $(A | B) > (C | D)$, then $(A | B) + (E | F) > (C | D) + (E | F)$.

(c) if $(A | B) > (C | D)$ and $(E | F) > (G | H)$, then $(A | B) + (E | F) > (C | D) + (G | H)$.

(d) if $(A | B) > (C | D)$ and $(E | F) > (M | P)$, $(A | B)(E | F) > (C | D)(E | F)$.

(e) if $(A | B) > (C | D)$ and $(M | P) > (E | F)$, $(C | D)(E | F) > (A | B)(E | F)$.

7.5.5 Supply proofs for the following theorems of Section 7.5:

(a) Theorem 42.   (e) Theorem 46.

(b) Theorem 43.   (f) Theorem 47.

(c) Theorem 44.   (g) Theorem 48.

(d) Theorem 45.

7.5.6

Definition We call the complex number $[a, -b]$ the *complex conjugate* of the complex number $[a, b]$. Show that

(a) $[a, -b][a, b] = [a^2 + b^2, 0]$,

(b) the equation $[c, d][x, y] = [a, b]$ always has a solution if $c \neq 0$ or $d \neq 0$.

7.5.7 Show that

(a) $[r_1 \cos \theta_1, r_1 \sin \theta_1][r_2 \cos \theta_2, r_2 \sin \theta_2] = [r \cos \theta, r \sin \theta]$, where $r = r_1 r_2$ and $\theta = \theta_1 + \theta_2$.

(b) $[r \cos \theta, r \sin \theta]^n = [r^n \cos n\theta, r^n \sin n\theta]$, where n is a positive integer. (This is known as De Moivre's theorem.)

SETS

8.1 Sets and Their Basic Relations and Operations

The most important and most basic term to be found in modern mathematics and logic is that of *set*, or *class*. Although some modern studies make a technical distinction between set and class, we do not do so in this treatment; it is recognized, however, that mathematicians are inclined to use the word *set*, whereas logicians frequently refer to a *class*.

The modern mathematical theory of sets is one of the most remarkable creations of the human mind. Because of the unusual boldness of some of the ideas found in its study, and because of some of the singular methods of proof to which it has given rise, the theory of sets is indescribably fascinating. But above this, the theory has assumed tremendous importance for almost the whole of mathematics. It has enormously enriched, clarified, extended, and generalized many domains of mathematics, and its influence on the study of the foundations of mathematics has been profound. The theory of sets forms one of the connecting links between mathematics on the one hand and philosophy and logic on the other.

The reader can hardly have failed to notice the increasing use made in the past chapters of the word *set*. In each instance the set under consideration was some collection or aggregate of objects. In practice it is common to describe a set by itemizing the elements, or members, that comprise it or by specifying a property that each element of the set possesses and that objects that are not elements of the set do not possess.

It now is desirable that we discuss sets in general, and that we examine some of the relations that may exist between sets and some of the operations that may be performed on sets. We do this in an intuitive manner in the present section. In the following sections we shall consider the advisability of a postulational approach to the study of sets, and we shall touch on some of the matters mentioned above.

We shall, then, think of a set as simply a collection of definite distinct objects of our perception or thought. The objects that make up a set will be called *elements* of the set, and, for the time being, we put no restriction on the nature of these objects.[1] We may, for example, consider the set consisting of the men, Brown, Smith, and Jones, or the set consisting of a pencil, a chair, and a tree, or the set consisting of the integers 1, 2, and 3. Each of these sets contains three elements. Sets may, however, be of any size insofar as the number of elements is concerned. Thus we may speak of the set of all stars of magnitude not greater than six, or of the set of all natural numbers; the former set, though large, contains a finite number of elements, while the latter set contains infinitely many elements.[2] Another example of a set containing infinitely many elements is the set of all points on a given line segment. We even frequently find it convenient to consider a set containing no elements, such as the set of all kings of the United States, or the set of all even prime numbers greater than 10; such a set will be referred to as a *null*, or *empty*, *set*. If a is an element of set A, we write $a \in A$ (read, "a is an element of A"), and if a is not an element of set A, we write $a \notin A$ (read, "a is not an element of A"). We now introduce some definitions and some further notation. The first two definitions concern two important dyadic relations that may exist between pairs of sets.

Definition 1   Two sets A and B are said to be *equal*, and we write $A = B$, if and only if every element of A is an element of B and every element of B is an element of A.

It is important to note that the order in which we name the elements of a set is immaterial. Thus the sets[3] {Brown, Smith, Jones} and {Smith, Jones, Brown} are equal, whereas the sets {Brown, Smith, Jones} and {Black, Smith, Jones} are not equal. Equality of sets is an identical equality, although the identity of the elements in the two sets may at times be disguised, as in the two equal sets $A = \{$the smallest prime number, the first prime number after 15, the fourth prime number$\}$ and $B = \{$the positive root of $x^2 - 5x - 14 = 0$, $2^4 + 1$, the first even positive integer$\}$. We also see, by Definition 1, that all empty sets are equal. It is customary to denote this unique empty set by $\varnothing$.

Definition 2   A set A is said to be a *subset* of a set B if and only if every element of set A is an element of set B; we write $A \subset B$ (read, "A is included in B").

For example, the set {Brown, Smith, Jones} is included in the set {Black, Brown, Smith, Jones}. It is to be noted that every set is included in itself. The situation where $A \subset B$ and there is at least one element of B which is not an element of A is indicated by saying that A is a *proper* subset of B.[4] In the example above, the first set is a proper subset of the second one. By convention, the null set $\varnothing$ is regarded as a subset of every set, and is a proper subset of every set except itself. It is apparent that the inclusion relation is a transitive relation; that is, if $A \subset B$ and $B \subset C$, then $A \subset C$.

---

[1] See, however, Section 9.4.

[2] Explicit definitions of *finite* and *infinite sets* will be given in Section 8.3.

[3] Sets are frequently designated by exhibiting their elements within braces.

[4] Some authors use the notation $A \subseteq B$ to indicate that A is a subset of B, and reserve $A \subset B$ to mean that A is a proper subset of B.

The next two definitions concern two binary operations on sets.

Definition 3 The *union* of two sets A and B, written $A \cup B$, and sometimes read "A cup B," is the set of all elements which belong to A or to B or to both A and B.

Definition 4 The *intersection* of two sets A and B, written $A \cap B$, and sometimes read "A cap B," is the set of all elements which belong to both A and B.

Thus, if $A = \{$Brown, Smith, Jones$\}$ and $B = \{$Black, Smith, Brown$\}$, then $A \cup B = \{$Black, Brown, Smith, Jones$\}$, and $A \cap B = \{$Brown, Smith$\}$. It can be shown that each of these binary operations is commutative and associative, and each is distributive with respect to the other.

Possible confusion between the two symbols $\cup$ and $\cap$ may be eliminated by the mnemonic: "$\cup$nion and i$\cap$tersection."

In many discussions it is convenient to regard all the sets with which one may be concerned as subsets of an over-all embracing set I; I is then known as the *universe*. In such situations the following unary operation on sets is useful.

Definition 5 The *complement* of a set A relative to a universe I is the set of all elements of I that are not elements of A. When the universe I is clearly understood, we denote the complement of A by A'.[5]

For example, suppose the universe $I = \{$Black, Brown, Smith, Jones$\}$ and $A = \{$Brown, Smith$\}$. Then $A' = \{$Black, Jones$\}$. It is evident that in general the complement of a set depends strongly on the universe I; on the other hand, no matter, what set is chosen for I, it is easily seen that $I' = \varnothing$ and $\varnothing' = I$.

Arbitrary sets may be conveniently represented graphically by so-called Venn diagrams,[6] which are much like the Euler diagrams described in Section 1.2. In this representation, sets are pictured as subregions of some fixed region chosen to represent the universe I. The sets $A \cup B$, $A \cap B$, and A' are represented in Venn diagrams by the shaded regions in Figures 8.1, 8.2, and 8.3, respectively. Here I is the indicated square that includes A and B as subsets.

Many relationships among sets involve the operations of union, intersection, and complementation. Such relationships often may be neatly tested by means of

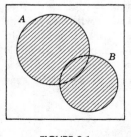

FIGURE 8.1

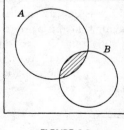

FIGURE 8.2

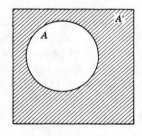

FIGURE 8.3

[5] Other notations for the complement include $\sim A$, $- A$, $I - A$, $C(A)$.

[6] John Venn employed the device in 1876 in a paper entitled "Boole's Logical System."

Venn diagrams. Consider, for example, the validity of the relationship $(A \cup B)' = A' \cap B'$. In the diagram of Figure 8.4, the region containing horizontal shading represents A', and the region containing vertical shading represents B'. It follows that the cross-hatched region then represents $A' \cap B'$, the intersection of A' and B'. But this cross-hatched region also represents the complement of the union of A and B. We have thus verified the relationship.

A test of a relationship among sets by Venn diagrams is not, of course, an acceptable proof of the concerned relationship, for, though the test is convincing to our common sense, it is merely an inductive procedure and is therefore technically impermissible as proof.

A proof of a relationship among sets may be accomplished by considering separately each possible case of an element e of the universe I. Thus, for the relationship considered above, there are four cases, for an element e of I may (1) lie in both A and B, (2) lie in A but not in B, (3) lie in B but not in A, (4) lie in neither A nor B. Let us consider the four cases in turn.

1. $e \in A$ and $e \in B$. Then $e \notin A'$ and $e \notin B'$, whence $e \notin A' \cap B'$. Also, $e \in A \cup B$, whence $e \notin (A \cup B)'$.

2. $e \in A$ but $e \notin B$. Then $e \notin A'$ and hence $e \notin A' \cap B'$. Also, $e \in A \cup B$, whence $e \notin (A \cup B)'$.

3. $e \in B$ but $e \notin A$. Then $e \notin B'$ and hence $e \notin A' \cap B'$. Also, $e \in A \cup B$, whence $e \notin (A \cup B)'$.

4. $e \notin A$ and $e \notin B$. Then $e \in A'$ and $e \in B'$, whence $e \in A' \cap B'$. Also, $e \notin A \cup B$, whence $e \in (A \cup B)'$.

It now follows that if e is an element of $(A \cup B)'$, then e is an element of $A' \cap B'$, and if e is an element of $A' \cap B'$, then e is an element of $(A \cup B)'$, whence $(A \cup B)' = A' \cap B'$. (This method of considering cases can be succinctly carried out by the use of "membership tables" as explained in Section A.4 of the Appendix and is then essentially the "truth table" method of logic described in Chapter 9.)

Another method of proving relationships among sets is to employ the definitions of *union*, *intersection*, and *complementation*. Thus, considering the same relationship as above, we note that "e is in $(A \cup B)'$" means "e is not in A or B," or, "e is in neither A nor B," and "e is in $A' \cap B'$" means "e is in A' and in

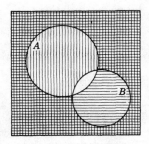

FIGURE 8.4

B'," or, "*e* is in neither *A* nor *B*." This approach assumes, as is customary in mathematical reasoning, the basic properties of the words *and*, *or*, and *not*.

There is still another method of establishing relationships among sets—namely, the postulational method. In this method, which mathematicians generally prefer, we assume certain relationships among sets as basic and then deduce all other relationships from these basic ones. This approach, when applied to a certain portion of set theory, gives rise to a remarkable branch of abstract mathematics known as Boolean algebra, to which we now turn.

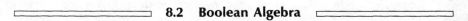

8.2   Boolean Algebra

Boolean algebra is a postulational abstraction of the theory of the subsets of a given set discussed in the last section. Since the algebra is developed by the axiomatic method, it is conceivable that there may be interpretations other than the one that originated the subject; this is indeed the case. Few branches of mathematics have received more diverse postulational treatments than has Boolean algebra, so that today a student has a choice of a wide variety of approaches to the study. The postulate set that we employ is a well-known set given by E. V. Huntington in 1904.[7]

Postulate Set for Boolean Algebra

A *Boolean algebra*[8] is a set B of elements $a, b, c, \ldots$, with two binary operations $\cup$ and $\cap$, called *cup* and *cap*, satisfying the following postulates.

B1. *Each of the operations $\cup$ and $\cap$ is commutative; that is,*

$$a \cup b = b \cup a \quad \textit{and} \quad a \cap b = b \cap a$$

for all elements a, b *of* B.

B2. *There exist distinct identity elements,* z *and* u, *relative to the operations* $\cup$ *and* $\cap$, *respectively; that is,*

$$a \cup z = a, \quad a \cap u = a, \quad z \neq u$$

for each element a *of* B.

B3. *Each operation is distributive relative to the other; that is,*

$$a \cup (b \cap c) = (a \cup b) \cap (a \cup c),$$
$$a \cap (b \cup c) = (a \cap b) \cup (a \cap c).$$

---

[7] E. V. Huntington [2].

[8] The term *Boolean algebra* has come to be used in the literature in two senses—namely, (1) as any interpretation of a certain postulate set and (2) as the resulting system of symbolic algebra following from this postulate set. The double usage causes no confusion, and the particular use intended is always clear from the context.

B4. *For each element* a *of* B *there exists an element* a′ *of* B *such that*

$$a \cup a' = u \quad and \quad a \cap a' = z.$$

It is easily verified that the subsets of a given set (as, for example, the subsets of a set of five chairs), under the binary operations of union and intersection and the unary operation of complementation, satisfy the postulates of a Boolean algebra. In this interpretation, the various subsets of the given set are identified with the elements $a, b, c, \ldots$ of the Boolean algebra, the empty set $\varnothing$ is identified with z, and the universe I (the given set) is identified with u. Since the postulate set possesses a concrete interpretation, it follows that the postulate set is absolutely consistent.

One of the first things one notices about the above postulate set is a perfect symmetry between the two operations of cup and cap. That is, if the operations $\cup$ and $\cap$, and the identity elements z and u, should be interchanged throughout the postulate set, we would obtain the postulate set all over again. There exists in Boolean algebra, then, a *principle of duality*, much like the principle of duality that we observed in plane projective geometry in Section 4.5. We state this principle of duality as the first theorem in our partial development of Boolean algebra.

Theorem 1 (principle of duality) *Any theorem of Boolean algebra remains valid if the operations* $\cup$ *and* $\cap$, *and the identity elements* z *and* u, *are interchanged throughout the statement of the theorem.*

This theorem is a theorem about theorems and accordingly comes under the subject of metamathematics. It permits us to state the theorems of Boolean algebra in dual pairs and guarantees that the proof of one statement of a dual pair is sufficient for the establishment of both statements. We now deduce a number of theorems of Boolean algebra, observing the fact that a separate demonstration is not necessary for the dual of a theorem that is proved.

Theorem 2 (the idempotent laws) $a \cup a = a; a \cap a = a.$

$$
\begin{array}{ll}
a = a \cup z & \text{(by B2)} \\
 = a \cup (a \cap a') & \text{(by B4)} \\
 = (a \cup a) \cap (a \cup a') & \text{(by B3)} \\
 = (a \cup a) \cap u & \text{(by B4)} \\
 = a \cup a. & \text{(by B2)}
\end{array}
$$

Theorem 3 $a \cup u = u; a \cap z = z.$

$$
\begin{array}{ll}
z = a \cap a' & \text{(by B4)} \\
 = a \cap (a' \cup z) & \text{(by B2)} \\
 = (a \cap a') \cup (a \cap z) & \text{(by B3)} \\
 = z \cup (a \cap z) & \text{(by B4)} \\
 = (a \cap z) \cup z & \text{(by B1)} \\
 = a \cap z. & \text{(by B2)}
\end{array}
$$

Theorem 4 (the absorption laws) $a \cap (a \cup b) = a;\ a \cup (a \cap b) = a.$

$$a = a \cup z \qquad\qquad\qquad \text{(by B2)}$$
$$= a \cup (b \cap z) \qquad\qquad \text{(by Th. 3)}$$
$$= (a \cup b) \cap (a \cup z) \qquad \text{(by B3)}$$
$$= (a \cup b) \cap a \qquad\qquad \text{(by B2)}$$
$$= a \cap (a \cup b). \qquad\qquad \text{(by B1)}$$

Theorem 5 (the associative laws) $a \cup (b \cup c) = (a \cup b) \cup c;\ a \cap (b \cap c) = (a \cap b) \cap c.$

Set $x = a \cup (b \cup c)$ and $y = (a \cup b) \cup c$. Then

$$a \cap x = (a \cap a) \cup [a \cap (b \cup c)] \qquad \text{(by B3)}$$
$$= a \cup [a \cap (b \cup c)] \qquad\qquad \text{(by Th. 2)}$$
$$= a, \qquad\qquad\qquad\qquad \text{(by Th.4)}$$

and

$$a \cap y = [a \cap (a \cup b)] \cup (a \cap c) \qquad \text{(by B3)}$$
$$= a \cup (a \cap c) \qquad\qquad\qquad \text{(by Th. 4)}$$
$$= a. \qquad\qquad\qquad\qquad \text{(by Th. 4)}$$

Therefore $a \cap x = a \cap y$. Now

$$a' \cap x = (a' \cap a) \cup [a' \cap (b \cup c)] \qquad \text{(by B3)}$$
$$= (a \cap a') \cup [a' \cap (b \cup c)] \qquad \text{(by B1)}$$
$$= z \cup [a' \cap (b \cup c)] \qquad\qquad \text{(by B4)}$$
$$= [a' \cap (b \cup c)] \cup z \qquad\qquad \text{(by B1)}$$
$$= a' \cap (b \cup c), \qquad\qquad\qquad \text{(by B2)}$$

and

$$a' \cap y = [a' \cap (a \cup b)] \cup (a' \cap c) \qquad\qquad \text{(by B3)}$$
$$= [(a' \cap a) \cup (a' \cap b)] \cup (a' \cap c) \qquad \text{(by B3)}$$
$$= [(a \cap a') \cup (a' \cap b)] \cup (a' \cap c) \qquad \text{(by B1)}$$
$$= [z \cup (a' \cap b)] \cup (a' \cap c) \qquad\qquad \text{(by B4)}$$
$$= [(a' \cap b) \cup z] \cup (a' \cap c) \qquad\qquad \text{(by B1)}$$
$$= (a' \cap b) \cup (a' \cap c) \qquad\qquad\qquad \text{(by B2)}$$
$$= a' \cap (b \cup c). \qquad\qquad\qquad\qquad \text{(by B3)}$$

Therefore $a' \cap x = a' \cap y$. Hence

$$(a \cap x) \cup (a' \cap x) = (a \cap y) \cup (a' \cap y), \qquad \text{(substitution)}$$
$$(x \cap a) \cup (x \cap a') = (y \cap a) \cup (y \cap a'), \qquad \text{(by B1)}$$
$$x \cap (a \cup a') = y \cap (a \cup a'), \qquad \text{(by B3)}$$

$$x \cap u = y \cap u, \qquad \text{(by B4)}$$

$$x = y. \qquad \text{(by B2)}$$

Theorem 6 a' *is unique.*

Let b and c be elements of B such that

$$a \cup b = u, \quad a \cap b = z; \quad a \cup c = u, \quad a \cap c = z.$$

Now

$$c = c \cap u \qquad \text{(by B2)}$$

$$= c \cap (a \cup b) \qquad \text{(substitution)}$$

$$= (c \cap a) \cup (c \cap b) \qquad \text{(by B3)}$$

$$= (a \cap c) \cup (b \cap c) \qquad \text{(by B1)}$$

$$= z \cup (b \cap c) \qquad \text{(substitution)}$$

$$= (b \cap c) \cup z \qquad \text{(by B1)}$$

$$= b \cap c, \qquad \text{(by B2)}$$

and, in the same way,

$$b = c \cap b$$

$$= b \cap c. \qquad \text{(by B1)}$$

Therefore $b = c$, and the theorem follows by reference to the definition of a'.

Theorem 7 (De Morgan's laws) $(a \cup b)' = a' \cap b'; (a \cap b)' = a' \cup b'.$

$$(a \cup b) \cup (a' \cap b') = [(a \cup b) \cup a'] \cap [(a \cup b) \cup b'] \qquad \text{(by B3)}$$

$$= [a' \cup (a \cup b)] \cap [(a \cup b) \cup b'] \qquad \text{(by B1)}$$

$$= [(a' \cup a) \cup b] \cap [a \cup (b \cup b')] \qquad \text{(by Th. 5)}$$

$$= [b \cup (a \cup a')] \cap [a \cup (b \cup b')] \qquad \text{(by B1)}$$

$$= (b \cup u) \cap (a \cup u) \qquad \text{(by B4)}$$

$$= u \cap u \qquad \text{(by Th. 3)}$$

$$= u, \qquad \text{(by B2)}$$

and

$$(a \cup b) \cap (a' \cap b') = (a' \cap b') \cap (a \cup b) \qquad \text{(by B1)}$$

$$= [(a' \cap b') \cap a] \cup [(a' \cap b') \cap b] \qquad \text{(by B3)}$$

$$= [a \cap (a' \cap b')] \cup [(a' \cap b') \cap b] \qquad \text{(by B1)}$$

$$= [(a \cap a') \cap b'] \cup [a' \cap (b' \cap b)] \qquad \text{(by Th. 5)}$$

$$= [b' \cap (a \cap a')] \cup [a' \cap (b \cap b')] \qquad \text{(by B1)}$$

$$= (b' \cap z) \cup (a' \cap z) \qquad \text{(by B4)}$$

$$= z \cup z \qquad \text{(by Th. 3)}$$

$$= z. \qquad \text{(by B2)}$$

Therefore, by Theorem 6, $(a' \cap b') = (a \cup b)'$.

As yet nothing has been said about a relation in Boolean algebra corresponding to the relation of inclusion, that plays such a fundamental role in the theory of sets. We now introduce such a relation.

Definition 1   If a and b are elements of B, we write $a \subset b$ if and only if $a \cup b = b$, and we say "a is contained in b."

Theorem 8 $a \subset a$.

For, by Theorem 2, we have $a \cup a = a$. Hence, by Definition 1, $a \subset a$.

Theorem 9   *If* $a \subset b$ *and* $b \subset a$, *then* $a = b$.

For

$$a \cup b = b \quad \text{and} \quad b \cup a = a. \qquad \text{(by Def. 1)}$$

But

$$a \cup b = b \cup a. \qquad \text{(by B1)}$$

Hence

$$a = b. \qquad \text{(substitution)}$$

Theorem 10   *If* $a \subset b$ *and* $b \subset c$, *then* $a \subset c$.

For

$$a \cup b = b \quad \text{and} \quad b \cup c = c. \qquad \text{(by Def. 1)}$$

Therefore

$$a \cup c = a \cup (b \cup c) \qquad \text{(substitution)}$$
$$= (a \cup b) \cup c \qquad \text{(by Th. 5)}$$
$$= b \cup c \qquad \text{(substitution)}$$
$$= c, \qquad \text{(substitution)}$$

and

$$a \subset c. \qquad \text{(by Def. 1)}$$

Theorem 11 $z \subset a \subset u$ *for each element* a *of* B.

For

$$z \cup a = a \cup z \qquad \text{(by B1)}$$
$$= a. \qquad \text{(by B2)}$$

Therefore

$$z \subset a. \qquad \text{(by Def. 1)}$$

Also

$$a \cup u = u, \qquad \text{(by Th. 3)}$$

whence

$$a \subset u. \qquad \text{(by Def. 1)}$$

Theorem 12   *If* $a \subset x$ *and* $b \subset x$, *then* $(a \cup b) \subset x$.

For

$$a \cup x = x \quad \text{and} \quad b \cup x = x. \qquad \text{(by Def. 1)}$$

Then

$$(a \cup b) \cup x = a \cup (b \cup x) \qquad \text{(by Th. 5)}$$
$$= a \cup x \qquad \text{(substitution)}$$
$$= x, \qquad \text{(substitution)}$$

whence

$$(a \cup b) \subset x. \qquad \text{(by Def. 1)}$$

We observe, from Theorems 8, 9, and 10, that the relation $\subset$ in Boolean algebra is reflexive, antisymmetric, and transitive (see Section 5.5 and Problem 5.5.11). The elements of a set S are said to be *partially ordered* with respect to a dyadic relation R if R possesses these three properties on S. It follows that the elements of a Boolean algebra are partially ordered with respect to the relation $\subset$. Theorem 11 gives universal bounds, relative to this relation, for all elements of B, and Theorem 12 establishes that $a \cup b$ is the least upper bound of a and b.

Of course, the above collection of twelve theorems is not exhaustive; in fact, mathematicians have deduced a tremendous number of theorems of Boolean algebra. The twelve theorems should serve, however, to give some idea of the nature of this algebra, which has, on the one hand, much in common with the familiar algebra of high school and yet, on the other hand, possesses many striking differences. The subject has a number of valuable applications. In the next chapter, we examine its application to certain portions of logic. Recently the algebra has found practical importance in the analysis and design of electrical networks[9] and in the development of a study known as the theory of information.

Boolean algebra is named after the English logician and mathematician George Boole (1815–1864), who was the first to give an algebraic formulation to the theory of classes in logic. Since the algebra originated in an attempt to establish a symbolic, or mathematical, logic, we withhold a fuller account of the history of the algebra until the next chapter, which is devoted to the subject of logic.

We now turn to the application of the theory of sets to numbers; this work originated with Georg Cantor (1845–1918) toward the end of the nineteenth century and is fundamental insofar as the foundations of mathematics are concerned.

8.3   Sets and the Foundations of Mathematics

In Chapter 7 we saw how the great bulk of mathematics can be reached from the natural number system in a purely definitional way, and thus how the consistency of mathematics can be made to rest on the consistency of this basic number system. It is natural to wonder whether the starting point of the definitional development of mathematics cannot perhaps be pushed to an even deeper level. Mathematicians who attempt such a construction usually start with the theory of sets, the concepts of which are already involved in the postulational development of the natural number system. The success of this program enables one to define the natural numbers in terms of sets and hence to reduce the number of undefined terms that must be assumed in mathematics. The program is far too extensive for us to carry through in detail, but we can indicate the main ideas of the development. Of course, a rigorous treatment along these lines requires that we start from a systematic postulational development of set theory. There has been a considerable amount of research expended on this task of building up set

---

[9] See F. E. Hohn [1]. This paper contains a bibliography of applications of Boolean algebra to electronics.

theory on a suitable postulational base. We shall forego this initial part of the program and merely sketch how the natural numbers may be defined in terms of set concepts.

It is interesting, as a mere observation at this point, that logicians have endeavored to push down still further the starting level of the definitional development of mathematics and to derive the theory of sets, or classes, from a foundation in the logic of propositions and propositional functions. Undoubtedly the most notable attempt of this kind to date is the monumental *Principia mathematica* of Whitehead and Russell. More will be said about this remarkable work in the next chapter.

We now take up a description of how the natural numbers may be introduced by means of set concepts. Some definitions are in order.

Definition 1   Two sets A and B are said to be in *one-to-one correspondence* when we have a pairing of the elements of A with the elements of B such that each element of A corresponds to one and only one element of B and each element of B corresponds to one and only one element of A.

There exists, for example, a one-to-one correspondence between the set of all letters of the alphabet and the set of the first twenty-six positive integers, for we may make the pairing

$$
\begin{array}{cccccccc}
a & b & c & d & \cdots & x & y & z \\
| & | & | & | & & | & | & | \\
1 & 2 & 3 & 4 & \cdots & 24 & 25 & 26
\end{array}
$$

There are, of course, many other permissible pairings of these two sets, such as

$$
\begin{array}{cccccccc}
a & b & c & d & \cdots & x & y & z \\
| & | & | & | & & | & | & | \\
3 & 4 & 5 & 6 & \cdots & 26 & 1 & 2
\end{array}
$$

Another illustration is the one-to-one correspondence between the set of all positive integers and the set of all even positive integers, for we may make the pairing

$$
\begin{array}{cccccccc}
1 & 2 & 3 & 4 & \cdots & n & \cdots \\
| & | & | & | & & | & \\
2 & 4 & 6 & 8 & \cdots & 2n & \cdots
\end{array}
$$

Also, there exists a one-to-one correspondence between the set of points on a line segment AB and the set of points on any other line segment $A'B'$, for we may make the pairing suggested in Figure 8.5.

Definition 2   Two sets A and B are said to be *equivalent*, and we write $A \sim B$, if and only if they can be placed in one-to-one correspondence.

Thus the set of all letters of the alphabet and the set of the first twenty-six positive integers are equivalent, as are the set of all positive integers and the set of all even positive integers. It is an easy matter, which we leave to the reader, to show that $\sim$ is an equivalence relation in the technical sense; that is, the relation $\sim$ is reflexive, symmetric, and transitive.

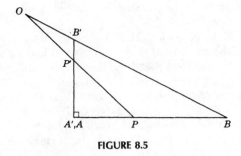

FIGURE 8.5

We now introduce the concept of number into the theory of sets by the following definition:

Definition 3   Two sets that are equivalent are said to have the *same cardinal number*. All sets that have the same cardinal number as the set $\{a\}$ are said to have *cardinal number one* (or to contain one element); all sets having the same cardinal number as the set $\{a, a'\}$ are said to have the *cardinal number two* (or to contain two elements); all sets having the same cardinal number as the set $\{a, a', a''\}$ are said to have *cardinal number three* (or to contain three elements); and so on. We shall denote the cardinal numbers one, two, three, ... by 1, 2, 3, ...[10]

Cardinal numbers thus appear as characters of sets. In other words, a triad of apples and a triad of pears have a property in common that we denote by *three*. Gottlob Frege in 1879 and Bertrand Russell in 1901, as have most other logicians who succeeded them, employed the idea of Definition 3 in basic fashion when they defined the cardinal number of a set S as the set of all sets equivalent to set S. Thus, according to this definition, *three* is the set of all sets each of which contains a triad of members.

Definition 4   A nonempty set is said to be *finite* if and only if its cardinal number is one of the cardinal numbers 1, 2, 3, ... A set that is not empty or finite is said to be *infinite*.

Definition 5   Let A be a set of cardinal number α and let B be a set of cardinal number β, and let us distinguish the elements of B from those of A so that sets A and B can be considered as having no elements in common; that is, $A \cap B = \varnothing$. Then, by $\alpha + \beta$, called the *sum* of α and β, we mean the cardinal number of the set $A \cup B$. This binary operation on cardinal numbers is called *addition*.

Theorem 1   *Addition of cardinal numbers is commutative and associative.*

These properties follow immediately from the commutativity and associativity of the union operation on sets.

Definition 6   The set C whose elements are all ordered pairs (a, b), where

---

[10]This definition presumes ability to recognize distinct objects.

$a \in A$ and $b \in B$, is called the *Cartesian product* of A and B, and is denoted by $A \times B$.

Theorem 2 *If* A, B, C *are sets, then* $A \times (B \cup C) = (A \times B) \cup (A \times C)$.

For let (a, d) be an element of $A \times (B \cup C)$. Then $a \in A$ and $d \in (B \cup C)$. Since $d \in (B \cup C)$, it follows that $d \in B$ or $d \in C$, whence (a, d) is an element of $A \times B$ or of $A \times C$, that is, of $(A \times B) \cup (A \times C)$. Conversely, suppose (a, d) is an element of $(A \times B) \cup (A \times C)$. Then $a \in A$, and $d \in B$ or $d \in C$. Since $d \in B$ or $d \in C$, it follows that $d \in (B \cup C)$, whence (a, d) is an element of $A \times (B \cup C)$. It now follows that $A \times (B \cup C) = (A \times B) \cup (A \times C)$.

Definition 7 Let A be a set of cardinal number α and let B be a set of cardinal number β. Then, by $\alpha\beta$, called the *product* of α and β, we mean the cardinal number of the set $A \times B$. This binary operation on cardinal numbers is called *multiplication*.

Theorem 3 *Multiplication of cardinal numbers is commutative, associative, and distributive with respect to addition of cardinal numbers.*

The commutative property holds, since the correspondence $(a, b) \leftrightarrow (b, a)$, where $a \in A$ and $b \in B$, shows that $A \times B \sim B \times A$. The associative property holds, since the correspondence $(a, (b, c)) \leftrightarrow ((a, b), c)$, where $a \in A$, $b \in B$, and $c \in C$, shows that $A \times (B \times C) \sim (A \times B) \times C$. Finally, the distributive property follows from Theorem 2, for if B and C have no elements in common and if α, β, γ are the cardinal numbers of A, B, C, then the cardinal number of $A \times (B \cup C)$ is $\alpha(\beta + \gamma)$, and the cardinal number of $(A \times B) \cup (A \times C)$ is $\alpha\beta + \alpha\gamma$.

Theorem 4 *For any cardinal number* α, $1\alpha = \alpha$.

The proof of this theorem is quite trivial.

Theorems 1, 3, and 4 hold for the cardinal numbers of both finite and infinite sets. These theorems cover the properties of Postulates N1 through N6 for the natural number system as given in Section 7.3. The properties of Postulates N7 through N10 do not hold for the cardinal numbers of infinite sets but can be shown to hold for those of finite sets, though we shall not undertake this task. It follows that the cardinal numbers $1, 2, 3, \ldots$, under the above operations of addition and multiplication, satisfy all the postulates for the natural number system. Since (assuming consistency of the postulate set) any two possible interpretations of this postulate set must be isomorphic, we may identify the set of cardinal numbers $1, 2, 3, \ldots$ with the set of natural numbers.[11]

8.4 Infinite Sets and Transfinite Numbers

In the last section we saw that the cardinal numbers of finite sets may be identified with the natural numbers. In this section we consider some of the properties of the cardinal numbers of infinite sets. These cardinal numbers are known as *transfinite numbers*, and their theory was first developed by Georg

---

[11] For a development of the real number system via the cardinal numbers of set theory, written at the college freshman level, see H. Levi; also see N. J. Lennes. For a very compact postulational treatment starting with set theory, see W. J. Thron, Sections 1 and 2.

Cantor in a remarkable series of articles beginning in 1872, and published, for the most part, in the German mathematics journals *Mathematische Annalen* and *Journal für Mathematik*. Prior to Cantor's study mathematicians accepted only one infinity, denoted by some symbol like ∞, and this symbol was employed indiscriminately to indicate the "number" of elements in such sets as the set of all natural numbers and the set of all real numbers. With Cantor's work, a whole new outlook was introduced, and a scale and arithmetic of infinities was achieved.

The basic principle that equivalent sets are to bear the same cardinal number presents us with many interesting and intriguing situations when the sets under consideration are infinite sets. Galileo Galilei observed as early as the latter part of the sixteenth century that the set of all positive integers can be placed in one-to-one correspondence with the set of all even positive integers, as was depicted in the last section. Hence, the same cardinal number should be assigned to each of these sets, and, from this point of view, we must say that there are as many even positive integers as there are positive integers in all! It is observed at once that the Euclidean postulate that states that the whole is greater than a part cannot be tolerated when cardinal numbers of infinite sets are under consideration. In fact, Dedekind, in about 1888, actually *defined* an infinite set to be one that is equivalent to some proper subset of itself. The simplicity of applying this definition in many circumstances is illustrated by Galileo's observation—namely, that the set of all positive integers is equivalent to a proper subset of itself. It can be shown[12] that this definition of Dedekind's is equivalent to the definition of infinite set given in Section 8.3.

We shall designate the cardinal number of the set of all natural numbers by d[13] and describe any set having this cardinal number as being *denumerable*. It follows that a set S is denumerable if and only if its elements can be written as an unending sequence $\{s_1, s_2, s_3, \ldots\}$. Since it is easily shown[14] that any infinite set contains a denumerable subset, it follows that d is the "smallest" transfinite number.

Cantor, in one of his earliest papers on set theory, proved the denumerability of two important sets that scarcely seem at first glance to possess this property.

The first set is the set of all rational numbers. This set has the important property of being *dense*. By this is meant that between any two distinct rational numbers there exists another rational number—in fact, infinitely many other rational numbers. For example, between 0 and 1 lie the rational numbers

$$\frac{1}{2}, \frac{2}{3}, \frac{3}{4}, \frac{4}{5}, \frac{5}{6}, \ldots, \frac{n}{n+1}, \ldots;$$

between 0 and 1/2 lie the rational numbers

$$\frac{1}{3}, \frac{2}{5}, \frac{3}{7}, \frac{4}{9}, \frac{5}{11}, \ldots, \frac{n}{2n+1}, \ldots;$$

---

[12] By employing the axiom of choice. See the Appendix, Section A.5.

[13] Cantor designated this cardinal number by the Hebrew letter aleph with the subscript zero, that is, by $\aleph_0$.

[14] Again, by employing the axiom of choice. See the Appendix, Section A.5.

between 0 and 1/4 lie the rational numbers

$$\frac{1}{5}, \frac{2}{9}, \frac{3}{13}, \frac{4}{17}, \frac{5}{21}, \ldots, \frac{n}{4n+1}, \ldots;$$

and so on. Because of this property one might well expect the transfinite number of the set of all rational numbers to be greater than d.[15] Cantor showed that this is *not* the case and that, on the contrary, the set of all rational numbers is denumerable. His proof is interesting and runs as follows.

Theorem 1   *The set of all rational numbers is denumerable.*

Consider the array

$$
\begin{array}{cccc}
1 \to 2 & 3 \to 4 & \cdots \\
\swarrow \nearrow \swarrow \\
\dfrac{1}{2} & \dfrac{2}{2} & \dfrac{3}{2} & \dfrac{4}{2} & \cdots \\
\downarrow \nearrow \swarrow \\
\dfrac{1}{3} & \dfrac{2}{3} & \dfrac{3}{3} & \dfrac{4}{3} & \cdots \\
\swarrow \\
\dfrac{1}{4} & \dfrac{2}{4} & \dfrac{3}{4} & \dfrac{4}{4} & \cdots \\
\downarrow \\
\cdot \quad \cdot \quad \cdot \quad \cdot & \cdots
\end{array}
$$

in which the first row contains, in order of magnitude, all the natural numbers (that is, all positive fractions with denominator 1), the second row contains, in order of magnitude, all the positive fractions with denominator 2, the third row contains, in order of magnitude, all the positive fractions with denominator 3, etc. Obviously, every positive rational number appears in this array, and if we list the numbers in the order of succession indicated by the arrows, omitting numbers that have already appeared, we obtain an unending sequence

$$1, 2, 1/2, 1/3, 3, 4, 3/2, 2/3, 1/4, \ldots$$

in which each positive rational number appears once and only once. Denote this sequence by $\{r_1, r_2, r_3, \ldots\}$. Then the sequence $\{0, -r_1, r_1, -r_2, r_2, \ldots\}$ contains the set of all rational numbers, and the denumerability of this set is established.

The second set considered by Cantor is a seemingly much more extensive set of numbers than the set of rational numbers. We first make the following definition.

Definition 1   A complex number is said to be *algebraic* if it is a zero of some polynomial

$$f(x) = a_0 x^n + a_1 x^{n-1} + \cdots + a_{n-1} x + a_n,$$

---

[15] The cardinal number of a set A is said to be *greater than* the cardinal number of a set B if and only if B is equivalent to a proper subset of A, but A is equivalent to no proper subset of B.

where n is a nonnegative integer, $a_0 \neq 0$, and all the a_k's are integers. A complex number that is not algebraic is said to be *transcendental*.

It is quite clear that the algebraic numbers include, among others, all rational numbers and all roots of such numbers. Accordingly, the following theorem is somewhat astonishing:

Theorem 2 *The set of all algebraic numbers is denumerable.*

Let $f(x)$ be a polynomial of the kind described in Definition 1, where, without loss of generality, we may suppose $a_0 > 0$. Consider the so-called *height* of the polynomial, defined by

$$h = n + a_0 + |a_1| + |a_2| + \cdots + |a_{n-1}| + |a_n|.$$

Obviously h is an integer $\geqq 1$, and there are plainly only a finite number of polynomials of a given height h and therefore only a finite number of algebraic numbers arising from polynomials of a given height h. We may now list (theoretically speaking) all the algebraic numbers, refraining from repeating any number already listed, by first taking those arising from polynomials of height 1, then those arising from polynomials of height 2, then those arising from polynomials of height 3, and so on. We thus see that the set of all algebraic numbers can be listed in an unending sequence, whence the set is denumerable.

In view of the past two theorems, there remains the possibility that all infinite sets are denumerable. That this is not so was shown by Cantor in a striking proof of the following significant theorem:

Theorem 3 *The set of all real numbers in the interval $0 < \mathrm{x} < 1$ is nondenumerable.*

The proof is indirect and employs an unusual method known as the *Cantor diagonal process*. Let us, then, assume the set to be denumerable. Then we may list the numbers of the set in a sequence $\{p_1, p_2, p_3, \ldots\}$. Each of these numbers p_i can be written uniquely as a nonterminating decimal fraction; in this connection it is useful to recall that every rational number may be written as a "repeating decimal"; a number such as 0.3, for example, can be written as 0.29999 . . . . We can then display the sequence in the following array,

$$p_1 = 0.a_{11}a_{12}a_{13} \cdots$$
$$p_2 = 0.a_{21}a_{22}a_{23} \cdots$$
$$p_3 = 0.a_{31}a_{32}a_{33} \cdots$$
$$\ldots \ldots \ldots \ldots \ldots \ldots \ldots$$

where each symbol a_{ij} represents some one of the digits 0, 1, 2, 3, 4, 5, 6, 7, 8, 9. Now, in spite of any care that has been taken to list all the real numbers between 0 and 1, there is a number that could not have been listed. Such a number is $0.b_1 b_2 b_3 \ldots$, where, say, $b_k = 7$ if $a_{kk} \neq 7$ and $b_k = 3$ if $a_{kk} = 7$, for $k = 1, 2, 3, \ldots, n, \ldots$. This number clearly lies between 0 and 1, and it must differ from each number p_i, for it differs from p_1 in at least the first decimal place, from p_2 in at least the second decimal place, from p_3 in at least the third decimal place, and so on. Thus the original assumption that all the real numbers between 0 and 1 can be listed in a sequence is untenable, and the set must therefore be nondenumerable.

Cantor deduced the following remarkable consequence of Theorems 2 and 3:

Theorem 4   *Transcendental numbers exist.*

Since, by Theorem 3, the set of all real numbers between 0 and 1 is nondenumerable, it is easily demonstrated that the set of all complex numbers is also nondenumerable. But by Theorem 2, the set of all algebraic numbers is denumerable. It follows that there must exist complex numbers that are not algebraic, and the theorem is established.

Not all mathematicians are willing to accept the above proof of Theorem 4. The acceptability or nonacceptability of the proof hinges on what one believes mathematical existence to be, and some mathematicians feel that mathematical existence is established only when one of the objects whose existence is in question is actually constructed and exhibited. The above proof does not establish the existence of transcendental numbers by producing a specific example of such a number. There are many existence proofs in mathematics of this nonconstructive sort, where existence is presumably established by merely showing that the assumption of nonexistence leads to a contradiction. Most proofs of the fundamental theorem of algebra, for example, are formulated along such lines.

Because of the dissatisfaction of some mathematicians with nonconstructive existence proofs, a good deal of effort has been made to replace such proofs by those that actually yield one of the objects concerned. In the Appendix, Section A.5, is a proof of this kind of the existence of transcendental numbers.

The proof of the existence of transcendental numbers and the proof that some particular number is transcendental are two quite different matters, the latter often being a very difficult problem. Hermite, in 1873, proved that the number e, the base for natural logarithms, is transcendental, and Lindemann, in 1882, first established the transcendentality of the number π. Unfortunately, it is inconvenient for us to prove these interesting facts here. The difficulty of identifying a particular given number as algebraic or transcendental is illustrated by the fact that it is not yet known whether the number π^π is algebraic or transcendental. A recent gain along these lines was the establishment of the transcendental character of any number of the form a^b, where a is an algebraic number different from 0 or 1, and b is any irrational algebraic number. This result was a culmination of an almost thirty-year effort to prove that the so-called *Hilbert number*, $2^{\sqrt{2}}$, is transcendental.

Since the set of all real numbers in the interval $0 < x < 1$ is nondenumerable, the transfinite number of this set is greater than d. We shall denote it by c and refer to it as the *cardinal number of the continuum*. It has been generally believed that c is the next transfinite number after d—that is, that there is no set having a cardinal number greater than d but less than c. This belief is known as the *continuum hypothesis*, but, in spite of the most strenuous efforts, no proof has been found to establish it. Many consequences of the hypothesis have been deduced, and, in about 1940, the Austrian logician Kurt Gödel succeeded in showing that the continuum hypothesis is consistent with a famous postulate set of set theory provided these postulates themselves are consistent.[16] Gödel conjectured that the denial of the continuum hypothesis is also consistent with the postulates of set theory. This conjecture was established, in 1963, by Dr. Paul J. Cohen of Stanford University, thus proving that the continuum hypothesis is independent of the postulates of set theory and hence can never be deduced from those postulates.[17] The situation is analogous to that of the parallel postulate in Euclidean geometry.

It has been shown that the set of all single-valued functions $f(x)$ defined over the interval $0 < x < 1$ has a cardinal number greater than c, but whether this cardinal number is or is not the next after c is not known. Cantor's theory provides for an infinite sequence of transfinite numbers, and there are demonstrations that purport to show that an unlimited number of cardinal numbers greater than that of the continuum actually exist.

8.5 Sets and the Fundamental Concepts of Mathematics

Following the work of Cantor, interest in the theory of sets developed rapidly until today virtually every field of mathematics has felt the impact of the new discipline. Notions of space and the geometry of a space, for example, have been completely revolutionized by the theory of sets. Also, the basic concepts in analysis, such as those of limit, function, continuity, derivative, and integral, are now most aptly described in terms of set theory ideas. Most important, however, has been the opportunity for new mathematical developments undreamed of fifty years ago. Thus, in companionship with the new appreciation of postulational procedures in mathematics, abstract spaces have been born, general theories of dimension and measure have been created, and the branch of mathematics called topology has undergone a spectacular growth. In short, under the influence of set theory, a considerable unification of traditional mathematics has occurred, and new mathematics has been created at an almost explosive rate. The present section expands on the above remarks and, in particular, shows how the theory of sets has clarified and generalized some of the fundamental concepts of mathematics.

Let us first consider some notions of space and the geometry of a space. These concepts have undergone marked changes since the days of the ancient Greeks. For the Greeks there was only one space and one geometry; these were absolute concepts. The space was not thought of as a collection of points but rather as a realm, or locus, in which objects could be freely moved about and compared with one another. From this point of view, the basic relation in geometry was that of congruence or superimposability.

With the development of analytic geometry, space came to be regarded as a collection of points, and with the invention of the classical non-Euclidean

---

[16]K. Gödel [3].

[17]P. J. Cohen.

geometries, mathematicians accepted the situation that there is more than one conceivable space and hence more than one geometry. But space was still regarded as a locus in which figures could be compared with one another. The central idea became that of a group of congruent transformations of space into itself, and a geometry came to be regarded as the study of those properties of configurations of points which remain unchanged when the enclosing space is subjected to these transformations. We have seen, in Section 5.4, how this point of view was expanded by Felix Klein in his Erlanger Programm of 1872. In the Erlanger Programm, a geometry was defined as the invariant theory of a transformation group. This concept synthesized and generalized all earlier concepts of geometry and supplied a singularly neat classification of a large number of important geometries.

At the end of the last century there was developed the idea of a branch of mathematics as an abstract body of theorems deduced from a set of postulates, and each geometry became, from this point of view, a particular branch of mathematics. Postulate sets for a large variety of geometries were studied, but the Erlanger Programm was in no way upset, for a geometry could be regarded as a branch of mathematics which is the invariant theory of a transformation group.

In 1906, however, Maurice Fréchet inaugurated the study of abstract spaces, and very general geometries came into being that no longer necessarily fit into the neat Kleinian classification. A space became merely a set of objects, usually called points, together with a set of relations in which these points are involved, and a geometry became simply the theory of such a space. The set of relations to which the points are subjected is called the *structure* of the space, and this structure may or may not be explainable in terms of the invariant theory of a transformation group. Through set theory, then, geometry received a further generalization. Although abstract spaces were first formally introduced in 1906, the idea of a geometry as the study of a set of points with some superimposed structure was really already contained in remarks made by Riemann in his famous lecture of 1854. It is interesting that some of these new geometries have found valuable application in the Einstein theory of relativity and in other developments of modern physics.

Let us illustrate the notion of abstract spaces by briefly considering, in turn, a Hausdorff space, a metric space, and a topological space. We find that in the study of these spaces many of the basic concepts of analysis have reached a remarkable degree of generalization. Since the subject of abstract spaces is based on set theory ideas, we freely employ the concepts and notations of set theory that have already been developed in earlier sections.

Definition 1   A *Hausdorff space* is a set H of elements, called *points*, together with a collection of certain subsets of these points, called *neighborhoods*, satisfying the following four postulates:

H1. *For each point* x *of* H *there corresponds at least one neighborhood* N_x, *where the symbol* N_x *means that* $x \in N_x$.

H2. *For any two neighborhoods* N_x *and* N_x' *of* x, *there exists a third neighborhood* N_x'' *such that* $N_x'' \subset (N_x \cap N_x')$.

H3. *If* y *is a point of* H *such that* $y \in N_x$, *then there is a neighborhood* N_y *of* y *such that* $N_y \subset N_x$.

H4. *If* x ≠ y, *there exists an* N_x *and an* N_y *such that* $N_x \cap N_y = \varnothing$.

To assist in understanding the postulates for a Hausdorff space, let H be the set of all points on a straight line, and select for the neighborhoods of a point x of H the segments of the straight line that have x as midpoint. It is easy to see that the above postulates are satisfied by this interpretation of points and neighborhoods, and thus we have an example of a Hausdorff space. The arithmetical counterpart of this Hausdorff space is very important in the study of analysis.

Within a Hausdorff space it is possible to define the concept of limit point as follows. A point x of H will be called a *limit point* of a subset S of H provided every neighborhood of x contains at least one point of S distinct from x. There are some interesting consequences of this definition when taken in conjunction with the space given above. We have, for example, the following theorem.

Theorem 1 *Any neighborhood* N_x *of a limit point* x *of a set* S *contains an infinite number of points of* S.

Since x is a limit point of S, any neighborhood N_x of x contains a point y_1 of S, where $y_1 \neq x$. By Postulate H4, there then exist neighborhoods N_{y_1} of y_1 and N'_x of x such that $N'_x \cap N_{y_1} = \varnothing$. Again, by Postulate H2, there exists a neighborhood N''_x of x such that $N''_x \subset (N_x \cap N'_x)$. It follows that $y_1 \notin N''_x$. But since x is a limit point of S, N''_x, and hence N_x, contains a point y_2 of S, where $y_2 \neq x$ and $\neq y_1$. Continuing in this way, we find that N_x contains an infinite sequence of distinct points $y_1, y_2, y_3 \ldots$ of S, and the theorem is established.

This theorem brings out in striking fashion the fact that the concept of limit point is applicable only to infinite sets. Not every infinite subset of the points of a Hausdorff space necessarily has a limit point, but a theorem that we owe to Weierstrass states that if H is the set of all points on a straight line, and if by neighborhood of a point x of H we mean any segment of the straight line that has x as midpoint, then any infinite subset of points of H lying *within a finite segment* of H has at least one limit point. The arithmetized version of this theorem is very fundamental in analysis, and the theorem is readily extended to spaces other than the set of points on a straight line.

Among the various concepts related to a Hausdorff space, we might mention the following. A subset S of H is said to be *compact* if every infinite subset of points of S possesses a limit point that belongs to S. If S is a subset of H, by the *closure* of S, written $\bar{S}$, is meant the set consisting of all points of S and all limit points of S. A subset S of H is said to be *separable* provided it contains a finite or a denumerable subset E such that $S \subset \bar{E}$. A subset S of H is said to be *connected* provided that however S may be divided into nonempty subsets, one of these subsets will contain a limit point of the other. A subset S of H which is both compact and connected is called a *continuum*. If, in any of the above definitions, $S = H$, then the whole Hausdorff space is said to possess the corresponding property. All of these properties are generalizations of analogous properties possessed by the set of all real numbers x of a closed number interval I, where by a neighborhood of x is meant any number interval lying within I and containing x.

In many of the familiar geometries, such as Euclidean geometry, there is the notion of distance between two points. This idea has been generalized by set theory into the study of so-called metric spaces. These are the abstract spaces that Fréchet introduced in 1906.

Definition 2   A *metric space* is a set M of elements, called *points*, together with a real number $\rho(x, y)$, called the *distance function* or *metric* of the space, associated with each pair of points x and y of M, satisfying the following two postulates:

M1. $\rho(x, y) = 0$ *if and only if* $x = y$.

M2. $\rho(x, y) \leqq \rho(y, z) + \rho(z, x)$, *where* x, y, z *are any three, not necessarily distinct, points of* M. (This is referred to as the *triangle inequality*.)

Some important examples of metric spaces[18] follow:

1. The set of all real numbers in which the distance function is defined by $\rho(x, y) = |x - y|$. A simple geometric interpretation of this space is the set of all points on a straight line, with the ordinary concept of distance between two points.

2. The set of all ordered pairs of real numbers in which the distance function is defined by
$$\rho(p_1, p_2) = [(x_1 - x_2)^2 + (y_1 - y_2)^2]^{1/2},$$
where
$$p_1 = (x_1, y_1), \qquad p_2 = (x_2, y_2).$$
The student of plane analytic geometry will recognize as a geometric interpretation here the set of all points in the plane, together with the ordinary concept of distance between two points.

3. The set of all ordered pairs of real numbers in which the distance function is defined by
$$\rho(p_1, p_2) = |x_2 - x_1| + |y_2 - y_1|,$$
where
$$p_1 = (x_1, y_1), \qquad p_2 = (x_2, y_2).$$
By plotting on the Cartesian plane, the reader will readily see why this space is frequently referred to as *taxi-cab space*.

4. The set of all infinite sequences $x = \{x_1, x_2, \ldots\}$ of real numbers, for which the infinite series $\sum_{i=1}^{\infty} x_i^2$ is convergent, in which the distance function is defined by
$$\rho(x, y) = \left[\sum_{i=1}^{\infty} (x_i - y_i)^2 \right]^{1/2}.$$
This example of a metric space is known as *Hilbert space*.

We now list several theorems about metric spaces.

---

[18] We shall not *prove* that the spaces are metric spaces. A proof is easy for (1) and (3) but is more difficult for (2) and (4).

Theorem 2 $\rho(x, y) = \rho(y, x)$.

By Postulate M2 we have, for any points z and w,

$$\rho(x, y) \leqq \rho(y, z) + \rho(z, x) \quad \text{and} \quad \rho(y, x) \leqq \rho(x, w) + \rho(w, y).$$

Taking $z = x$ in the first of these inequalities and $w = y$ in the second one, we find (recalling Postulate M1)

$$\rho(x, y) \leqq \rho(y, x) \quad \text{and} \quad \rho(y, x) \leqq \rho(x, y).$$

It follows that $\rho(x, y) = \rho(y, x)$.

Theorem 3 $\rho(x, y) \geqq 0$.

By Postulate M2, for any point z,

$$\rho(x, z) \leqq \rho(z, y) + \rho(y, x).$$

Taking $z = x$ we have

$$\rho(x, x) \leqq \rho(x, y) + \rho(y, x).$$

But $\rho(x, x) = 0$ (by Postulate M1), and $\rho(y, x) = \rho(x, y)$ (by Theorem 2). It follows that $0 \leqq 2\rho(x, y)$, whence $\rho(x, y) \geqq 0$.

Theorem 4 *Any subset of a metric space is, for the same metric, a metric space.*

Theorem 5 *Any metric space can be made into a Hausdorff space.*

Let x be any point of a metric space M, and consider the subset of all points y of M such that $\rho(x, y) < r$, where r is a positive real number. Such a subset, which is nonempty inasmuch as it necessarily contains x, will be called an *open sphere* and will be denoted by $S(x, r)$. We now show that these open spheres, if regarded as neighborhoods, satisfy the four postulates of a Hausdorff space.

We omit the obvious verification of H1 and H2.

To verify H3, let $\rho(x, y) < r$, and put $R = r - \rho(x, y) > 0$. The triangle inequality states that $\rho(x, y') \leqq \rho(x, y) + \rho(y, y') = (r - R) + \rho(y, y') < r$, if $\rho(y, y') < R$. That is, $S(y, R) \subset S(x, r)$.

To verify H4, let x be distinct from y and set $r = \rho(x, y) > 0$. Then it is easy to show that the two open spheres $S(x, r/3)$ and $S(y, r/3)$ have no points in common.

The important inverse problem—namely, the discussion regarding to what extent any Hausdorff space can be made into a metric space—cannot be taken up here, and the interested reader must consult a more exhaustive treatment of the subject.

Some interesting definitions can be introduced in connection with metric spaces. Thus, to define a *limit point* x of a subset S of M by the use of the metric $\rho(x, y)$, it is sufficient to state that for any given positive number ε there exists a point y of S such that $y \neq x$ and $\rho(x, y) < \varepsilon$. Again, the *diameter* of a subset S of M may be defined as the least upper bound of $\rho(x, y)$, where x and y are points in S. Similarly, the *distance between two subsets* S_1 *and* S_2 of M may be defined as the greatest lower bound of $\rho(x, y)$, where x is a point in S_1 and y is a point in S_2.

An abstract space with a very simple and very basic structure is a so-called topological space.

Definition 3   A *topological space*[19] is a set T of elements, called *points*, with a collection of subsets of these points, called *open sets*, satisfying the following three postulates.

T1. T *and the null set $\emptyset$ are open sets.*

T2. *The union of any number of open sets is an open set.*

T3. *The intersection of any two open sets is an open set.*

When, for a given point set T, we devise a system of subsets of T satisfying the above postulates, we say that the set T has been *topologized,* or has received a *topology.* A given set T may be topologized in more than one way. In fact, it is an easy matter to show that if we select for the collection of open sets of T all of the subsets of T, then T is topologized; T is also topologized if we choose for the open sets of T only the two sets T and $\emptyset$. The first of these two topologies imposed on T is called the *discrete topology* of T; the second one is called the *trivial topology* of T. The notion of limit point can be introduced into a topological space by the definition: A point p of T is called a *limit point* of a subset S of T if every open set containing p (whether p belongs to S or not) contains at least one point of S different from p. The study of the consequences of the postulates for a topological space is called *general topology,* and the limit point concept plays an important role in this study. We shall content ourselves with the statement of only one theorem about topological spaces—a theorem that links topological and Hausdorff spaces.

Theorem 6   *A topological space for which there exist, for any two distinct points* x *and* y, *disjoint open sets* S_x *and* S_y *containing* x *and* y, *respectively, is a Hausdorff space with respect to the open sets as neighborhoods.*
     We leave it to the reader to show that the postulates for a Hausdorff space are satisfied if one chooses for neighborhoods the open sets of T.

Having given some consideration to the matter of individual sets possessing a superimposed structure, we shall now very briefly look at pairs of related sets. A considerable amount of mathematical theory is concerned with two related sets. This leads to the concept of function and, consequently, to the elaborate branch of mathematics known as function theory.

The concept of function, like the notions of space and geometry, has undergone a marked evolution, and every student of mathematics encounters various refinements of this evolution as his or her studies progress from the elementary courses of high school into the more advanced and sophisticated courses of the graduate college level.

The history of the notion of function furnishes another interesting example of the tendency of mathematicians to generalize and extend their concepts. The word *function*, in its Latin equivalent, seems to have been introduced by Leibniz (1646–1716) in 1694, at first as a term to denote any quantity connected with a curve, such as the coordinates of a point on the curve, the slope of the curve, and so on. Johann Bernoulli (1667–1748), by 1718, had come to regard a function as any expression made up of a variable and some constants, and Euler

---

[19] The definition of a topological space is not uniform in the literature.

(1707–1783), somewhat later, regarded a function as any equation or formula involving variables and constants. This latter idea is the notion of a function formed by most students of elementary mathematics courses. The familiar notation $f(x)$ was not used at first but entered about 1734 with A. C. Clairaut (1713–1765) and Euler. The Euler concept of function remained unchanged until Fourier (1768–1830) was led, in his investigations of heat flow, to consider so-called trigonometric series. These series involve a more general type of relationship between variables than had previously been studied, and, in an attempt to furnish a definition of function broad enough to encompass such relationships, Lejeune Dirichlet (1805–1859) arrived at the following formulation: A *variable* is a symbol that represents any one of a set of numbers; if two variables x and y are so related that whenever a value is assigned to x there is automatically assigned, by some rule or correspondence, a value to y, then we say y is a (single-valued) *function* of x. The variable x, to which values are assigned at will, is called the *independent variable*, and the variable y, whose values depend on those of x, is called the *dependent variable*. The permissible values that x may assume constitute the *domain of definition* of the function, and the values taken on by y constitute the *range of values* of the function.

The student of mathematics usually meets the Dirichlet definition of function in his introductory course in calculus. The definition is a very broad one and does not imply anything regarding the possibility of expressing the relationship between x and y by some kind of analytic expression; it stresses the basic idea of a relationship between two sets of numbers.

Set theory has naturally extended the concept of function to embrace relationships between any two sets of elements, whether the elements are numbers or anything else. Thus, in set theory, a function f is defined to be any set of ordered pairs of elements such that if $(a_1, b_1) \in f$, $(a_2, b_2) \in f$, and $a_1 = a_2$, then $b_1 = b_2$. The set A of all first elements of the ordered pairs is called the *domain (of definition)* of the function, and the set B of all second elements of the ordered pairs is called the *range (of values)* of the function. A functional relationship is thus nothing but a special kind of subset of the Cartesian product set $A \times B$. A one-to-one correspondence is, in its turn, a special kind of function—namely, a function f such that if $(a_1, b_1) \in f$, $(a_2, b_2) \in f$, and $b_1 = b_2$, then $a_1 = a_2$. If, for a functional relationship f, $(a, b) \in f$, we write $b = f(a)$, which we read, "b equals f at a."

The notion of function pervades much of mathematics, and since the early part of the present century various influential mathematicians have advocated the employment of this concept as the unifying and central principle in the organization of elementary mathematics courses. The concept seems to form a natural and effective guide for the selection and development of textual material. There is no doubt of the value of a mathematics student's early acquaintance with the function concept.

The property of continuity may immediately be associated with the concept of function if the study pertains to spaces A and B defined by a system of neighborhoods. In fact, a function f may be said to be *continuous* at a point a of A if for any neighborhood $N_b \subset B$, where b is the point of B associated with the given point a of A, there exists a neighborhood $N_a \subset A$ such that for any $z \in N_a$, $f(z) \in N_b$. Other definitions of continuity are possible.

8.1.1 Show that if $A \subset B$ and $B \subset A$, then $A = B$.

8.1.2 If a and b are real numbers and $a < b$, let

$$[a, b] \text{ represent the number interval } a \leqq x \leqq b,$$

$$(a, b] \text{ represent the number interval } a < x \leqq b,$$

$$[a, b) \text{ represent the number interval } a \leqq x < b,$$

$$(a, b) \text{ represent the number interval } a < x < b.$$

Find

(a) $[1, 3] \cup [3, 4]$.	**(h)** $[1, 3] \cap [2, 4]$.
(b) $[1, 3] \cup [2, 4]$.	**(i)** $[1, 3] \cap [3, 7]$.
(c) $[1, 7] \cup [2, 5]$.	**(j)** $[1, 3] \cap [5, 7]$.
(d) $[1, 3) \cup [3, 4]$.	**(k)** $[1, 3) \cap [2, 4]$.
(e) $(1, 3) \cup [3, 4]$.	**(l)** $[1, 3] \cap (2, 4]$.
(f) $(1, 3) \cup [3, 4)$.	**(m)** $[1, 3) \cap (2, 4)$.
(g) $(1, 3) \cup (3, 4)$.	**(n)** $(1, 3) \cap (3, 7)$.

8.1.3 Show that $\varnothing \cup A = A$ and $\varnothing \cap A = \varnothing$.

8.1.4 **(a)** If $B \subset A$, show that $B \cup A = A$ and $B \cap A = B$.

   (b) Show that $A \cup A = A$ and $A \cap A = A$.

8.1.5 Let A be the set of all points of a Cartesian plane such that $x \geqq 0$, and let B be the set of all points of the plane such that $x \leqq 0$. Find $A \cup B$ and $A \cap B$.

8.1.6 Let $I = [-50, 50]$. Find the following sets (see Problem 8.1.2 for meaning of notation):

(a) $[-10, 10]'$.	**(f)** $[-5, 0] \cap [5, 10]'$.
(b) $(5, 15)'$.	**(g)** $\{[-5, 0] \cup [5, 10]\}'$.
(c) $\varnothing'$.	**(h)** $\{(-5, 0) \cup (5, 10)'\}'$.
(d) $(-50, 50)'$.	**(i)** $\{[-5, 0] \cap [5, 10]\}'$.
(e) $[-5, 0] \cup [5, 10]'$.	**(j)** $\{(-5, 0) \cap (5, 10)'\}'$.

8.1.7 Show that

   (a) $A \cup A' = I$.

   (b) $A \cap A' = \varnothing$.

   (c) $(A')' = A$.

8.1.8 Show that $A \subset B$ if and only if $A \cap B' = \varnothing$.

8.1.9 Verify, by Venn diagrams,

   (a) the associative laws for $\cup$ and $\cap$,

   (b) the distributive law of $\cup$ with respect to $\cap$,

   (c) the distributive law of $\cap$ with respect to $\cup$.

8.1.10 Label each of the eight regions of Figure 8.6 by an appropriate expression involving A, B, and C which will exactly describe that region.

8.1.11 By shading the appropriate regions on a Venn diagram, determine which of the following relationships are true:

   (a) $(A' \cup B)' = A \cap B'$.

   (b) $A' \cup B' = (A \cup B)'$.

   (c) $A \cup (B \cap C)' = (A \cup B') \cap C'$.

   (d) $A \cup (A \cap B) = A$.

   (e) $A \cap (A \cup B) = A$.

8.1.12 **(a)** Show that a set A consisting of five elements contains 2^5 subsets (including itself and the null set).

   (b) Let A be a set with seven elements and B a set with five elements. What can be said about the number of elements in the sets $A \cap B$ and $A \cup B$?

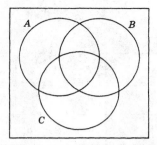

FIGURE 8.6

(c) Generalize (a) and (b).

8.1.13 **(a)** Let $D = A \cup B$ and $E = B \cup C$. Show that $D \cup C = A \cup E$.

   (b) Let $D = A \cap B$ and $E = B \cap C$. Show that $D \cap C = A \cap E$.

8.2.1 Establish the following theorems of Boolean algebra:

   (a) The identity elements z and u are unique.

   (b) $z' = u$ and $u' = z$.

   (c) $(a')' = a$.

8.2.2 Establish the following theorems of Boolean algebra:

   (a) $(a \cap b) \cup (b \cap c) \cup (c \cap a) = (a \cup b) \cap (b \cup c) \cap (c \cup a)$.

   (b) If $a \cap b = a \cap c$ and $a \cup b = a \cup c$, then $b = c$.

   (c) Show that the above two theorems are self-dual.

8.2.3 Prove *Poretsky's law*: For given x and t, $x = z$ if and only if $t = (x \cap t') \cup (x' \cap t)$.

8.2.4 Establish the following theorems of Boolean algebra:

   (a) $a \subset (a \cup b)$.

   (b) $(a \cap b) \subset a$.

   (c) If $x \subset a$ and $x \subset b$, then $x \subset (a \cap b)$. (This shows that $a \cap b$ is the greatest lower bound of a and b relative to the partial ordering relation $\subset$.)

   (d) If $a \subset b$, then $(a \cup c) \subset (b \cup c)$ and $(a \cap c) \subset (b \cap c)$.

8.2.5 Establish the following theorems of Boolean algebra:

   (a) $a \subset b$ if and only if $a \cap b' = z$.

   (b) $a \subset b$ if and only if $b' \subset a'$.

   (c) $a \subset b$ if and only if $a \cap b = a$.

   (d) $b \subset a'$ if and only if $a \cap b = z$.

   (e) $a' \subset b$ if and only if $a \cup b = u$.

8.2.6 Let $B = \{1, 2, 3, 5, 6, 10, 15, 30\}$ and let $a \cup b$ = the least common multiple of a and b, and $a \cap b$ = the greatest common divisor of a and b. Show that B, together with $\cup$ and $\cap$, constitutes a Boolean algebra.

8.2.7 Let $B = \{z, u\}$, with binary operations $\cup$ and $\cap$ defined by the following tables:

$\cup$	z	u
z	z	u
u	u	u

$\cap$	z	u
z	z	z
u	z	u

Show that B, with $\cup$ and $\cap$, constitutes a Boolean algebra.

8.2.8 Let $B = \{a, b, c, d\}$, with binary operations $\cup$ and $\cap$ defined by the following tables:

$\cup$	a	b	c	d
a	a	b	c	d
b	b	b	b	b
c	c	b	c	b
d	d	b	b	d

$\cap$	a	b	c	d
a	a	a	a	a
b	a	b	c	d
c	a	c	c	a
d	a	d	a	d

Show that B, with $\cup$ and $\cap$, constitutes a Boolean algebra.

8.2.9 (a) Define operations of addition and multiplication in a Boolean algebra by

$$a + b = (a \cap b') \cup (a' \cap b), \quad ab = a \cap b.$$

Show that B constitutes a commutative ring with respect to these operations of addition and multiplication. (For a definition of a commutative ring, see Section 5.2.)

(b) Show that in the ring of (a), $a^2 = a$ for each element a of B. (A ring having this latter property is called a *Boolean ring*.)

(c) Show that $a + a = z$ for each element a of B.

(d) Show that $ab(a + b) = z$ for all elements a, b of B.

8.2.10 Establish the independence of the postulates for Boolean algebra given in Section 8.2.

8.2.11 Referring to Definition 1 of Section 8.2, write $b \supset a$ (read "b contains a") if and only if $a \subset b$. Show that $b \supset a$ if and only if $b \cap a = a$, thus showing that $\subset$ and $\supset$ are dual symbols.

8.3.1 Whether an end point A (or B) of a line segment AB is to be considered as belonging or not belonging to the segment will be indicated by using a bracket or a parenthesis, respectively, about the letter A (or B). Using this notation, show that the segments $[AB], (AB], [AB), (AB)$, considered as sets of points, are equivalent to one another.

8.3.2 Show, starting with Figure 8.7, that the set of points composing any finite segment and the set composing any infinite segment are equivalent to each other.

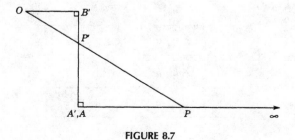

FIGURE 8.7

8.3.3 Show that equivalence of sets is reflexive, symmetric, and transitive.

8.3.4 (a) Let $A = \{1, 2, 3, 4\}$ and $B = \{a, b, c\}$. Form the sets $A \times B$ and $B \times A$.

(b) Take A and B as in (a) and let $C = \{x, y\}$. Exhibit the sets $A \times (B \cup C)$, $A \times (B \times C)$, $(A \times B) \times C$.

8.3.5 (a) Carry out the details of the proof of Theorem 1 of Section 8.3.

(b) Prove Theorem 4 of Section 8.3.

8.3.6 Show that the cancellation laws for addition and multiplication do not hold for cardinal numbers of infinite sets.

8.4.1 Devise an infinite sequence of subsets of the set of natural numbers such that each subset is a proper subset of its predecessor and yet has cardinal number d.

8.4.2 Explain how Figure 8.8 shows, with the aid of Dedekind's definition of infinite set, that the set of points on a line segment is an infinite set.

8.4.3 Prove that the assumed existence of at least one rational number between any two rational numbers implies the existence of a denumerable number of rational numbers between any two rational numbers.

8.4.4 (a) Show that every rational number is an algebraic number and hence that every real transcendental number is irrational.

(b) Is every irrational number a transcendental number?

(c) Is the imaginary unit i algebraic or transcendental?

(d) Show that $\pi/2$ is transcendental.

(e) Show that $\pi + 1$ is transcendental.

(f) Generalize (d) and (e).

(g) Show that any complex number which is a zero of a polynomial of the form

$$a_0 x^n + a_1 x^{n-1} + \cdots + a_{n-1} x + a_n,$$

where $a_0 \neq 0$ and all the a_k's are *rational numbers*, is an algebraic number.

8.4.5 (a) Prove that the union of a finite number of denumerable sets is a denumerable set.

(b) Prove that the union of a denumerable number of denumerable sets is a denumerable set.

8.4.6 (a) Show that 1 is the only polynomial of height 1.

(b) Show that x and 2 are the only polynomials of height 2.

(c) Show that $x^2, 2x, x + 1, x - 1$, and 3 are the only polynomials of height 3, and that they yield the distinct algebraic numbers $0, 1, -1$.

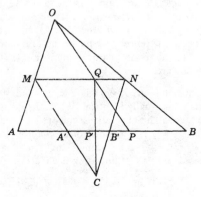

FIGURE 8.8

(d) Form all possible polynomials of height 4 and show that the only new real algebraic numbers contributed are $-2, -1/2, 1/2, 2$.

(e) Show that polynomials of height 5 contribute 12 more real algebraic numbers.

8.4.7 **(a)** Show that the set of all irrational numbers is nondenumerable.

(b) Show that the set of all transcendental numbers is nondenumerable.

8.4.8 **(a)** Show that if a circle has a center with at least one irrational coordinate, then there are at most two points on the circle with rational coordinates.

(b) Show that if a circle has a center with at least one transcendental coordinate, then there are at most two points on the circle with algebraic coordinates.

(c) Is it possible for a straight line or a circle in the Cartesian plane to contain only points having rational coordinates? algebraic coordinates?

8.4.9 **(a)** Complete the details of the following proof that the set of all points on a line segment AB is nondenumerable:

Take the length of AB to be 1 unit, and assume that the points on AB constitute a denumerable set. The points on AB can then be arranged in a sequence $\{P_1, P_2, P_3, \ldots\}$. Enclose point P_1 in an interval of length $1/10$, point P_2 in an interval of length $(1/10)^2$, point P_3 in an interval of length $(1/10)^3$, and so on. It follows that the unit interval AB is entirely covered by an infinite sequence of possibly overlapping subintervals of lengths $1/10$, $(1/10)^2$, $(1/10)^3, \ldots$. But the sum of the lengths of these subintervals is

$$\frac{1}{10} + \left(\frac{1}{10}\right)^2 + \left(\frac{1}{10}\right)^3 + \cdots = \frac{1}{9} < 1.$$

(b) By choosing the subintervals in (a) to be of lengths $\varepsilon/10, \varepsilon/10^2, \varepsilon/10^3, \ldots$, where ε is an arbitrarily small positive number, show that a denumerable set of points can be covered by a set of intervals the sum of whose lengths can be made as small as we please. (Using the terminology of measure theory, we say that a denumerable set of points has *zero* measure.)

8.4.10 Consider the following two arguments:

I. Theorem *Of all triangles inscribed in a circle, the equilateral is the greatest.*

(1) If ABC is a nonequilateral triangle inscribed in a circle, so that $AB \neq AC$, say, construct triangle XBC, where X is the intersection of the perpendicular bisector of BC with arc BAC. (2) Then triangle XBC > triangle ABC. (3) Hence, if we have a nonequilateral triangle inscribed in a circle, we can always construct a greater inscribed triangle. (4) Therefore, of all triangles inscribed in a circle, the equilateral is the greatest.

II. Theorem *Of all natural numbers, 1 is the greatest.*

(1) If m is a natural number other than 1, construct the natural number m^2. (2) Then $m^2 > m$. (3) Hence, if we have a natural number other than 1, we can always construct a greater natural number. (4) Therefore, of all natural numbers, 1 is the greatest.

Now the conclusion in argument I is true, and that in argument II is false. But the two arguments are formally identical. What, then, is wrong?

8.4.11 Let E_1 denote the set of all points on the segment $(0, 1]$ and E_2 the set of all points in the unit square $0 < x, y \leq 1$. A point Z of E_1 may be designated by an unending decimal $z = 0.z_1 z_2 z_3 \cdots$ lying between 0 and 1, and a point P of E_2 may be designated by an ordered pair of unending decimals

$$(x = 0.x_1 x_2 x_3 \ldots, y = 0.y_1 y_2 y_3 \ldots),$$

each decimal lying between 0 and 1. Suppose we let each z_i, x_i, y_i in these representations denote either a nonzero digit or a nonzero digit preceded by a possible block of zeros. For example, if $z = 0.73028007\ldots$, then $z_1 = 7$, $z_2 = 3$, $z_3 = 02$, $z_4 = 8$, $z_5 = 007$, $\ldots$. Show that a one-to-one correspondence may be set up between the points of E_1 and those of E_2 by associating with the point $0.z_1 z_2 z_3 \ldots$ of E_1 the point

$$(0.z_1 z_3 z_5 \ldots, \; 0.z_2 z_4 z_6 \ldots)$$

of E_2, and with the point

$$(0.x_1 x_2 x_3 \ldots, \; 0.y_1 y_2 y_3 \ldots)$$

of E_2 the point $0.x_1 y_1 x_2 y_2 x_3 y_3 \ldots$ of E_1. Thus show that the set of all points in a unit square has the cardinal number c. (This shows that the dimension of a manifold cannot be distinguished by the cardinal number of the manifold.)

8.4.12 **(a)** Show that any infinite set of mutually external closed intervals on a straight line is denumerable.

(b) Show that any infinite set of mutually external circles lying in a plane is denumerable.

8.4.13 Show that the set of all finite sequences of nonnegative integers is denumerable.

8.4.14 Let S denote the set of all single-valued functions of one variable x which assume positive integral values whenever x is a positive integer. Show that S is nondenumerable.

8.4.15 The postulate set P1, P2, P3, P4, of Section 6.2 is categorical (see Problem 6.4.4). Show that if P4 is replaced by P′4: K *consists of exactly* d *distinct elements*, then the set is no longer categorical.

8.5.1 **(a)** Show that the set of all points in the plane can be made into a Hausdorff space by selecting for neighborhoods of a point P the interiors of all circles having P as center.

(b) Show that the set of all points in the plane can be made into a Hausdorff space by selecting for neighborhoods of a point P the interiors of all squares having centers at P and having sides parallel to two given perpendicular lines of the plane.

(c) Show that any set of points can be made into a Hausdorff space if we select for neighborhoods the points themselves.

8.5.2 Let H be the Hausdorff space of Problem 8.5.1(a). Find the set of limit points of each of the following sets of points in H (here described, for convenience, in terms of a Cartesian coordinate system superimposed on the plane):

(a) All points (x, y) for which $x > 0$.

(b) All points (x, y) such that $x^2 + y^2 < 9$.

(c) The set of points $(1, 1)$, $(\frac{1}{2}, \frac{1}{2})$, $(\frac{1}{3}, \frac{1}{3})$, $\ldots$.

(d) All points (x, y) such that $y = \sin(1/x)$ and $0 < x \leqq 1$.

(e) The set of all points $(m/n, 1/n)$, where m is an integer and n is a positive integer.

8.5.3 Define a *circle* with center C and radius r in a metric space M as the set of all points X of M such that $\rho(C, X) = r$. Describe the appearance of a *circle* in each of the Examples 1, 2, 3 of metric spaces given in Section 8.5.

8.5.4 **(a)** Prove Theorem 4 of Section 8.5.

(b) Complete the proof of Theorem 5 of Section 8.5.

(c) Prove Theorem 6 of Section 8.5.

8.5.5 Show that if $\rho(x, y)$ is a metric for a set M of points, then we may also use as a metric for M

(a) $k\rho(x, y)$, where k is a positive real number,

(b) $\sqrt{\rho(x, y)}$,

(c) $\rho(x, y)/[1 + \rho(x, y)]$.

8.5.6 **(a)** Show that any set M of elements can be made into a metric space by setting $\rho(x, y) = 1$ if $x \neq y$, and $\rho(x, y) = 0$ if $x = y$.

  (b) Show that the set M of all ordered pairs of real numbers, in which the distance function is defined by $\rho(p_1, p_2) = \max\ (|x_2 - x_1|, |y_2 - y_1|)$, where $p_1 = (x_1, y_1)$, $p_2 = (x_2, y_2)$, is a metric space.

  (c) Show that the set M of all points on the circumference of a circle, in which the distance between two points is defined as the length of the shorter arc of the circle connecting the two points, is a metric space.

  (d) Generalize (c) to the set M of all points on the surface of a sphere.

8.5.7 **(a)** Let T be the three distinct points A, B, C, and let the open sets of T be $\varnothing, \{A\}$, $\{A, B\}, \{A, C\}, T$. Show that T is a topological space.

  (b) Show that T in (a) is not a Hausdorff space if the open sets of T are chosen as neighborhoods.

  (c) The example of (a) shows that the postulates for a topological space are consistent. Show, by appropriate models, that these postulates are independent.

8.5.8 Establish the consistency and independence of the postulates for a Hausdorff space.

8.5.9 Establish the consistency and independence of the postulates for a metric space.

8.5.10 (*for students who have studied calculus*) Show that the definition of continuity of a function appearing at the end of Section 8.5 is a generalization of the concept of continuity of a real function of a real variable as defined in the usual course in calculus.

8.5.11 Let A and B be two sets of elements. Show that we may define a subset R of $A \times B$ to be a dyadic relation between the elements of A and the elements of B.

8.5.12 In elementary mathematics, functions of a limited sort only are considered. Thus a typical definition of *function* in elementary mathematics might read as follows: A *function* is a rule that assigns one and only one number to each number in some collection of numbers.

  (a) Show that this definition is a special case of the more general definition of a function as given in Section 8.5.

  (b) What constitutes the *domain of definition* in the above elementary concept of a function?

  (c) What constitutes the *range of values* of a function as defined in elementary mathematics?

  (d) In what sense can the equation $F = (9/5)C + 32$ be regarded as a function in elementary mathematics?

  (e) In what sense can a bar graph showing the number of students failing a certain course in each of certain years be regarded as a function in elementary mathematics?

  (f) In what sense can a table of trigonometric tangents be regarded as a function in elementary mathematics?

  (g) Compare a function to a machine equipped with a hopper at the top and a spout at the bottom. Raw materials are fed into the machine through the hopper. The machine then converts these raw materials into a product that is expelled through the spout of the machine.

9

LOGIC AND PHILOSOPHY

We noted, in Chapter 6, that a mathematical theory results from the interplay of two factors, a set of postulates and a logic. The set of postulates constitutes the basis from which the theory starts, and the logic constitutes the rules by which such a basis may be expanded into a body of theorems. The mathematical theory, or system, comprises the totality of statements composed of both the postulates and the theorems.

Considerable attention has been paid in the preceding chapters to the first of the two factors, namely, the underlying postulate set, but little attention has been paid to the second factor, namely, the logic employed in deducing the theorems of the discourse from the postulational base. The purpose of the present chapter is partially to remedy this deficiency.

It would be a well nigh hopeless task to discuss modern considerations of logic by the use of only ordinary language. A symbolic language has become necessary in order to achieve the required exact scientific treatment of the subject. Because of the presence of such symbolism, the resulting treatment is known as *symbolic*, or *mathematical*, *logic*. In symbolic logic the various relations among propositions, classes, and so forth are represented by formulas whose meanings are free from the ambiguities so common to ordinary language. It becomes possible to develop the subject from a set of initial formulas in accordance with certain clearly prescribed rules of formal transformation, much like the development of a piece of common algebra. Also, and again as in the development of a piece of common algebra, the advantages of the symbolic language over ordinary language insofar as compactness and ease of comprehension are concerned are very great.

Leibniz is regarded as the first to consider seriously the desirability of a symbolic logic. One of his earliest works was an essay, *De arte combinatoria*, published in 1666, in which he indicated his belief in the possibility of a universal scientific language, expressed in an economical and workable symbolism for guidance in the reasoning process. Returning to these ideas between the years

1679 and 1690, Leibniz made considerable headway toward the creation of a symbolic logic, and he formulated a number of the concepts that are so important in modern studies.

An important renewal of interest in symbolic logic took place in 1847 when George Boole (1815–1864) published his little pamphlet entitled *The Mathematical Analysis of Logic, Being an Essay towards a Calculus of Deductive Reasoning*. Another paper followed in 1848, and finally, in 1854, Boole gave a notable exposition of his ideas in the work, *An Investigation into the Laws of Thought, on Which Are Founded the Mathematical Theories of Logic and Probability*.

Augustus De Morgan (1806–1871) was a contemporary of Boole, and his treatise on *Formal Logic; or, the Calculus of Inference, Necessary and Probable*, published in 1847, in some ways went considerably beyond Boole. Later De Morgan also made extended studies of the hitherto neglected logic of relations.

In the United States, outstanding work in the field was contributed by Charles Sanders Peirce (1839–1914), son of the distinguished Harvard mathematician Benjamin Peirce. Peirce rediscovered some of the principles enunciated by his predecessors. It is unfortunate that his work appeared somewhat out of the stream of normal development; only in comparatively recent times has the merit of much of Peirce's thought been properly appreciated.

The notions of Boole were given a remarkable completeness in the massive treatise by Ernst Schröder (1841–1902) entitled *Vorlesungen über die Algebra der Logic*, published during the period 1890–1895. In fact, modern logicians are inclined to characterize symbolic logic in the Boolean tradition by the term *Boole-Schröder algebra*. Considerable work is still being done in Boolean algebra, and many papers on the subject are to be found in present-day research journals.

A still more modern approach to symbolic logic originated with the work of the German logician Gottlob Frege (1848–1925) during the period 1879–1903 and with the studies of Peano referred to in Section 4.1. Peano's work was motivated by a desire to express all mathematics in terms of a logical calculus, and Frege's work stemmed from the need of a sounder foundation for mathematics. Frege's *Begriffsschrift* appeared in 1879, and his historically important *Grundgesetze der Arithmetik* in 1893–1903; the *Formulaire de mathématiques* of Peano and his co-workers began its appearance in 1894. The work started by Frege and Peano led directly to the very influential and monumental *Principia mathematica* (1910–1913) of Whitehead and Russell. The basic idea of this work is the identification of much of mathematics with logic by the deduction of the natural number system, and hence of the great bulk of existing mathematics, from a set of premises or postulates for logic itself. In 1934–1939 appeared the comprehensive *Grundlagen der Mathematik* of David Hilbert and Paul Bernays. This work, based on a series of papers and university lectures given by Hilbert, attempts to build up mathematics by the use of symbolic logic in a new way that renders possible the establishment of the consistency of mathematics.

At the present time elaborate studies in the field of symbolic logic are being pursued by many mathematicians, chiefly as a result of the impetus given to the work by the publication of the *Principia mathematica*. A periodical, known as the *Journal of Symbolic Logic*, was established in 1935 to publicize the writings of this group.

In this section and the next we shall endeavor to give a slight idea of the nature of symbolic logic, restricting ourselves to the so-called *propositional calculus* as developed by Whitehead and Russell. In the present section we introduce needed concepts and symbolism, and in the next section we outline a postulational approach to the subject. Of course, in such a limited space, our treatment must be brief and incomplete.

A *proposition* is to be understood as any statement concerning which it is meaningful to say that its content is true or false,[1] whether or not we know which of these terms actually applies. Examples of propositions are: "Spring is a season," "8 is a prime number," "The billionth digit in the decimal expansion of π is 7." The first proposition is true, the second false, and though the truth or falsity of the third is not known, we shall consider it meaningful to say that the third statement is true or false. Propositions will be denoted in the symbolic logic by the lower-case letters $m, n, p, q, r, \ldots$, and we shall agree that when we assert a proposition p, without qualification as to its truth or falsity, we mean to imply that p is true.

Propositions can be combined in various ways to form new propositions. For example, from the two propositions "Spring is a season," "8 is a prime number," we can construct the new propositions: "Spring is a season *and* 8 is a prime number," "Spring is a season *or* 8 is a prime number," "*If* spring is a season, *then* 8 is a prime number," and "Spring is a season *if and only if* 8 is a prime number." Finally, from the single proposition, "Spring is a season," we can construct the new proposition, "Spring is *not* a season"; that is, we can construct the proposition that denies the first proposition.

The above combinations of propositions are rendered verbally by the words *and, or, if-then, if and only if, not.* In symbolic logic these fundamental combinations of propositions are rendered by suitable symbolism. For this purpose we introduce the following five symbols:

1. $p \wedge q$ (read "p *and* q") represents the proposition that is true when and only when both p and q are true. A proposition of this form is called a *conjunction*.

2. $p \vee q$ (read "p *or* q") represents the proposition that is true when and only when at least one of the two propositions p, q is true. A proposition of this form is called a *disjunction*.

3. $p \rightarrow q$ (read "*if* p, *then* q") represents the proposition that is false when and only when p is true and q is false. A proposition of this form is called an *implication*.

4. $p \leftrightarrow q$ (read "p *if and only if* q") represents the proposition that is true when and only when p and q are both true or both false. The assertion, $p \leftrightarrow q$, means, therefore, that p and q have the same truth value. A proposition of this form is called an *equivalence*.

5. p' (read "*not* p") represents the denial, or contradiction, of p. That is, p' represents that proposition that is true when p is false and that is false when p is true. A proposition of this form is called a *negation*.[2]

---

[1] We shall not go into the epistemological question of the meaning of *true* and of *false*.

[2] The symbols for conjunction, disjunction, implication, equivalence, and negation are by no means standardized, and vary considerably among authors. Thus Whitehead and Russell write $p \cdot q, p \vee q, p \supset q, p \equiv q, \sim p$, and Hilbert writes $p \mathbin{\&} q, p \vee q, p \rightarrow q, p \sim q, \bar{p}$.

We observe that $\wedge$, $\vee$, $\rightarrow$, and $\leftrightarrow$ are symbols denoting binary operations performed on propositions, and that $'$ is a symbol denoting a unary operation performed on propositions.

It is to be noted that the logical meanings of *and*, *or*, *if-then*, and *if and only if* differ somewhat from the customary uses of these connectives in ordinary language.

For example, in ordinary language, *and* is commonly used to conjoin two propositions that are relevant to each other in what might be called connected discourse, as in a coherent description of a sequence of events, such as

He took the train and arrived in Boston.

In logic, however, *and* is used to connect any two propositions whatever, without regard to their possible relevance or bearing on one another, as in the example,

Spring is a season and 8 is a prime number.

Similarly, in ordinary language, the connective *or* appears in two senses, the exclusive sense of the Latin *aut* ("*p* or *q* but not both") and the inclusive sense of the Latin *vel* ("*p* or *q* or both"). It is in the latter sense, the sense of the familiar juristic barbarism "and/or," that *or* is used in logic,[3] and, again, no regard is paid to the possible relevancy or bearing of one proposition upon the other.

Particularly interesting is the logical use of *if-then*. In ordinary language, the compound statement, "If *p*, then *q*," is taken to indicate a relation of premise and conclusion or of cause and effect, as in the statement,

If it is raining, then we shall not go.

In logic, on the other hand, the implication $p \rightarrow q$ may connect any two propositions p and q whatever, and, by Definition 3 above, the implication is to be considered as false when and only when p is true and q is false. We thus have the interesting situation that the implication $p \rightarrow q$ is true whenever p is false and also whenever q is true. Thus all three of the following implications are to be considered true:

If 7 is a prime number, then 2 times 2 is 4,

If 8 is a prime number, then 2 times 2 is 4,

If 8 is a prime number, then 2 times 2 is 5.

Though somewhat startling, these results need not be unexpected if we remember that we are concerned here, not with the meanings of propositions, but only with their truth values. An implication that is based on a relation of premise and conclusion or of cause and effect depends on the structure of the component propositions rather than on their truth or falsity.[4]

---

[3] The meaning of the logical symbol $\vee$ may be kept in mind by associating it with the initial letter of the Latin word *vel*.

[4] The definition of implication in logic is a controversial matter, and other analyses of the nature of implication have been proposed. Implication as defined above is called *material implication*. C. I. Lewis has introduced a concept called *strict implication*, which seems more nearly to correspond to the relation holding when a conclusion is said to be deducible from premises, but as yet there is no definition of implication dependent on propositional structure that has won general acceptance among logicians. In any event, material implication is bound up in any idea of implication, and, no matter what definition of implication be adopted, at least we have a material implication.

Finally, the assertion, "*p* if and only if *q*," is not intended in logic to imply that the component propositions *p* and *q* have the same meaning or significance, but merely that the assertion is true when and only when *p* and *q* are both true or both false. Thus the two statements,

7 is a prime number if and only if 2 times 2 is 4,

8 is a prime number if and only if 2 times 2 is 5,

are both to be considered true.

Probably the neatest and most compact way of summarizing the logical meanings of conjunction, disjunction, implication, equivalence, and negation is by the use of *truth tables*. In these tables, the symbol *T* is employed to denote that the corresponding proposition is true, *F* that it is false.

Let us illustrate by forming the truth table for the operation of conjunction.

Conjunction

p	q	$p \wedge q$
T	T	T
T	F	F
F	T	F
F	F	F

This table is merely an expansion of the definition of conjunction given above. It says that $p \wedge q$ is true if p and q are each true, and that $p \wedge q$ is false in every other case. To form such a table one must list all possible combinations of T and F for the component propositions (p and q in this case) involved. The truth or falsity of each combination is then entered in the final column.

Truth tables for disjunction, implication, equivalence, and negation are as follows:

Disjunction

p	q	$p \vee q$
T	T	T
T	F	T
F	T	T
F	F	F

Implication

p	q	$p \rightarrow q$
T	T	T
T	F	F
F	T	T
F	F	T

Equivalence

p	q	$p \leftrightarrow q$
T	T	T
T	F	F
F	T	F
F	F	T

Negation

p	p'
T	F
F	T

The composite propositions considered above may be modified or combined into more elaborate composite propositions, such as

$$[p \wedge (p \to q)] \to q,$$

and the truth value of the new proposition for any combination of truth values of the fundamental propositions p and q may be determined by means of the truth tables for conjunction, disjunction, implication, equivalence, and negation. Thus, for the proposition just cited, we have

p	q	$p \to q$	$p \wedge (p \to q)$	$[p \wedge (p \to q)] \to q$
T	T	T	T	T
T	F	F	F	T
F	T	T	F	T
F	F	T	F	T

The first, second, and last columns constitute the truth table for the new proposition; the third and fourth columns serve merely as aids for obtaining the last column. This truth table happens to possess the curious feature of having nothing but T's in the last column. This means that the proposition concerned has the truth value T no matter what truth value p has and no matter what truth value q has. Propositions of this special sort are called *tautologies*, or *laws of logic*, and they play an essential role in the study of logic. The reader can easily show that the following propositions are all tautologies.

Some Laws of Logic

Law	Name of Law
$p \vee p'$	law of excluded middle
$(p \wedge p')'$	law of contradiction
$[(p \to q) \wedge (q \to r)] \to (p \to r)$	law of the syllogism
$p \leftrightarrow (p')'$	law of double negation
$(p \to q) \leftrightarrow (q' \to p')$	law of contraposition

If one should form the truth tables for the two propositions $p \to q$ and $q' \to p'$, it would be found that the tables agree with each other in point of truth and falsity. Two propositions m and n of this sort are said to be *logically equivalent*, and it follows that the proposition $m \leftrightarrow n$ is a tautology, or law of logic. Thus the proposition $(p \to q) \leftrightarrow (q' \to p')$ is a law of logic, as has been indicated in the above list of some of the laws of logic. The importance of knowing that two propositions are logically equivalent lies in the fact that we can alter the structure of one proposition into that of the logically equivalent proposition with full confidence that we are not thereby affecting the truth value of the original proposition. Thus the proposition $q' \to p'$ may, anywhere it appears, be replaced by the proposition $p \to q$. Similarly, $(p')'$ may, anywhere it appears, be replaced by p.

We leave to the reader the establishment of the logical equivalence of the following pairs of propositions:

Some Pairs of Logically Equivalent Propositions

$p \lor q$	$(p' \land q')'$
$p \to q$	$(p \land q')'$
$p \leftrightarrow q$	$(p \land q')' \land (q \land p')'$
$p \land q$	$(p' \lor q')'$
$p \to q$	$p' \lor q$
$p \leftrightarrow q$	$[(p' \lor q)' \lor (q' \lor p)']'$
$p \land q$	$(p \to q')'$
$p \lor q$	$p' \to q$
$p \leftrightarrow q$	$[(p \to q) \to (q \to p)']'$

These are interesting logical equivalences, for they show that conjunction, disjunction, implication, equivalence, and negation are, to some extent, inter-definable, and that some of the logical connectives are therefore dispensable. Thus the first three logical equivalences exhibit, respectively, ways of expressing disjunction, implication, and equivalence in terms of conjunction and negation; the middle three logical equivalences exhibit, respectively, how we may express conjunction, implication, and equivalence in terms of disjunction and negation; the last three logical equivalences exhibit, respectively, how we may render conjunction, disjunction, and equivalence in terms of implication and negation.

It can be shown that negation is indispensable for expressing combinations of propositions, and that not all the fundamental logical connectives can be rendered by equivalence and negation.

It is worth noting that disjunction can be expressed by implication alone, without the use of negation, for $p \lor q$ and $(p \to q) \to q$ are logically equivalent. It is not possible to do this for conjunction.

H. M. Sheffer showed, in 1913, that all the fundamental logical connectives can be rendered by means of a single logical symbol, which he called *stroke*. Sheffer's stroke combination is written as p/q, and means: "Not p or not q." The reader may easily show that p/p and p' are logically equivalent and that $(p/p)/(q/q)$ and $p \lor q$ are logically equivalent. Since disjunction and negation can be expressed by Sheffer's stroke symbol, it follows that the other fundamental logical connectives can also be expressed by means of this symbol.

9.2   The Calculus of Propositions

In the employment of the postulational method we deduce theorems from postulates or from earlier theorems. In making these deductions we engage in an argumentation from premises to conclusion, and our reasoning takes the implicative form of "if such and such, then so and so." We are not concerned with the truth or falsity of either our premises or our conclusion but are concerned only with the validity of the argument leading from the one to the other. That is, we insist that our argument be *formally* correct or that the involved implication be true regardless of the truth or falsity of the premises and the conclusion. It follows that the implication must be a tautologous one. Conversely, if the implication is tautologous, then the argument is valid. Thus all that is required for testing our reasoning for validity is to ascertain whether the involved implication is or is not a tautology. In short, propositional formulas that are tautologies represent what we call valid argument.

In view of the above, it is evident that a primary task of logic is to ferret out tautologies—that is, to find those combinations of component propositions p, q, r, . . . that are true independently of the truth values of the components themselves. These tautologies are the laws of the logical system, and they constitute the structure of what is to be regarded as correct formal reasoning. Now, in the method of truth tables, we possess an adequate and sure means of deciding whether a composite proposition is or is not a tautology. The application of this method will, however, result in a purely random and unsystematic catalogue of tautologies. The present section explains a method in which the tautologies are arrived at in an orderly and interconnected manner. We show that from a small and specially selected set of tautologies, the remaining tautologies can be obtained in accordance with certain precise rules. Furthermore, the new method will yield *only* tautologies, and we shall not have the task of sorting out the tautologies from the nontautologies, as must be done in the employment of the truth-table method. Since the new method obtains tautologies by a procedure of symbolic calculation, the new method is referred to as the *calculus of propositions*.

The approach to the calculus of propositions explained next is essentially that given by Whitehead and Russell in their *Principia mathematica*. A remarkable feature of the development is that it employs the postulational method. A small selection is made from the set of all tautologies to serve as the postulates of the development, and then several formal rules are given in accordance with which all other tautologies can be obtained from the selected few. The rules play the same role in the development of the propositional calculus that logical inference plays in the usual development of a branch of mathematics. Of course, logical inference, in the usual sense, cannot be used here, for this very logical inference now constitutes the object of our study.

We now list, and briefly comment on, the primitive terms, the postulates, and the rules for obtaining theorems in the calculus of propositions.

Primitive Terms

For the primitive, or undefined, terms of the calculus of propositions we choose: (1) a set P of elements p, q, r, . . . , called *propositions*, (2) a binary operation performed on the elements of P and denoted by $\vee$, and (3) a unary operation performed on the elements of P and denoted by $'$.

Because of the interdefinability of the logical connectives (pointed out at the end of Section 9.1), there is no need also to include among the primitive terms the binary operations $\wedge$, $\rightarrow$, and $\leftrightarrow$. Note that if p and q are members of P, then $p \vee q$ and p' also are members of P; that is, they also are propositions.

Postulates, or Primitive Tautologies

All tautologies are propositions, but a proposition may or may not be a tautology. From among the infinitely many tautologies, four are selected to serve as postulates, or as *primitive tautologies*. These primitive tautologies are here denoted by L1, L2, L3, L4, and for convenience in stating them we precede them by a definition.

Definition 1 $p \rightarrow q$ means $p' \vee q$.

L1. $(p \vee p) \rightarrow p$.

L2. $q \rightarrow (p \vee q)$.

L3. $(p \vee q) \rightarrow (q \vee p)$.

L4. $(q \rightarrow r) \rightarrow [(p \vee q) \rightarrow (p \vee r)]$.

In *Principia mathematica*, Postulate L1 is called the *principle of tautology*. It is clear that we could have asserted something stronger than just implication, for actually $(p \vee p) \leftrightarrow p$ is a tautology. Since, however, $p \rightarrow (p \vee p)$ can be proved, this half of the equivalence need not be assumed.

Postulate L2 is motivated by the fact that a disjunction is true when one of its components is true. In *Principia mathematica*, this postulate is called the *principle of addition*.

Postulate L3, called the *principle of permutation*, is motivated by the commutative nature of disjunction. Again, an equivalence could have been asserted here instead of just an implication, but then more would be assumed than is necessary.

Postulate L4 is called the *principle of summation* and is motivated by the fact that in an implication a disjunct may be added to both the antecedent and the consequent, and the implication will still hold. There is a similarity between this postulate and the familiar law of arithmetic of natural numbers: If $a > b$, then $a + c > b + c$.

Rules for Obtaining Theorems or Derived Tautologies

There are four rules by means of which we are permitted to obtain *derived tautologies* from given tautologies. These rules are

R1 (rule of substitution) *We may substitute any proposition for all occurrences of a proposition represented by a single letter in a given tautology and thus obtain a derived tautology.* For example, the substitution of the proposition $p \vee q$ for the proposition q in the tautology L2 gives the new tautology

$$(p \vee q) \rightarrow [p \vee (p \vee q)].$$

R2 (rule of definitional substitution) *We may replace any expression in a given tautology by another expression that is definitionally identical with it, and thus obtain a derived tautology.* For example, the replacement of $q \rightarrow r$ by $q' \vee r$ in the tautology L4 gives the new tautology

$$(q' \vee r) \rightarrow [(p \vee q) \rightarrow (p \vee r)].$$

R3 (rule of detachment, or rule of implication) *From two given tautologies* m *and* m $\rightarrow$ n, *the derived tautology* n *may be obtained.*

To facilitate the statement of the fourth rule we first introduce the following definition:

Definition 2 $p \wedge q$ means $(p' \vee q')'$.

We now state Rule R4.[5]

R4 (rule of adjunction) *From two given tautologies* m *and* n, *the derived tautology* m $\wedge$ n *may be obtained.*

With the above postulates and rules of procedure in mind, we now turn to some of the theorems of the calculus of propositions. Only enough demonstrations will be presented to give some idea of the workings of the calculus; for the rest, we shall be content with merely the statements of the theorems.

Theorem 1 $(q \rightarrow r) \rightarrow [(p \rightarrow q) \rightarrow (p \rightarrow r)]$.

Proof $(q \rightarrow r) \rightarrow [(p \vee q) \rightarrow (p \vee r)]$ (by Post. L4)

$(q \rightarrow r) \rightarrow [(p' \vee q) \rightarrow (p' \vee r)]$ (substituting p' for p)

$(q \rightarrow r) \rightarrow [(p \rightarrow q) \rightarrow (p \rightarrow r)]$ (by Def. 1)

(This theorem is one form of the transitivity of implication.)

Theorem 2 $p \rightarrow (p \vee p)$.

Proof $q \rightarrow (p \vee q)$ (by Post. L2)

$p \rightarrow (p \vee p)$ (substituting p for q)

Theorem 3 $p \rightarrow p$.

Proof $(q \rightarrow r) \rightarrow [(p \rightarrow q) \rightarrow (p \rightarrow r)]$ (by Th. 1)

$[(p \vee p) \rightarrow p] \rightarrow [\{p \rightarrow (p \vee p)\} \rightarrow (p \rightarrow p)]$

(substituting $p \vee p$ for q and p for r)

---

[5] Actually, Rule R4 can be established from the postulates and Rules R1, R2, R3, but to carry out the establishment is a lengthy and involved task. Since we shall need the rule, we here simply assume it.

$(p \lor p) \rightarrow p$ (by Post. L1)

$\{p \rightarrow (p \lor p)\} \rightarrow (p \rightarrow p)$ (by Rule R3)

$p \rightarrow (p \lor p)$ (by Th. 2)

$p \rightarrow p$ (by Rule R3)

(This theorem asserts that any proposition implies itself, or that implication is reflexive.)

Theorem 4 $p' \lor p$.

Proof $p \rightarrow p$ (by Th. 3)

$p' \lor p$ (by Def. 1)

Theorem 5 $p \lor p'$.

Proof $(p \lor q) \rightarrow (q \lor p)$ (by Post. L3)

$(p' \lor p) \rightarrow (p \lor p')$ (substituting p' for p and p for q)

$p' \lor p$ (by Th. 4)

$p \lor p'$ (by Rule R3)

(This is the *law of the excluded middle*, which says, "Either p is true or p is false," or "Either p is true or not-p is true.")

Theorem 6 $(p \rightarrow p') \rightarrow p'$.

Proof $(p \lor p) \rightarrow p$ (by Post. L1)

$(p' \lor p') \rightarrow p'$ (substituting p' for p)

$(p \rightarrow p') \rightarrow p'$ (by Def. 1)

(This theorem, which asserts that any proposition that implies its own falsity is false, is the principle of *reductio ad absurdum* in its simplest form. It is possible to prove more general forms of *reductio ad absurdum*, such as

$$(p \rightarrow q) \rightarrow [(p \rightarrow q') \rightarrow p'].$$

This latter theorem says that a proposition p which implies both q and q' is false. Later we shall state the most general form of the principle of *reductio ad absurdum*.)

Theorem 7 $p \rightarrow (p')'$.

Proof $p \lor p'$ (by Th. 5)

$p' \lor (p')'$ (substituting p' for p)

$p \rightarrow (p')'$ (by Def. 1)

Theorem 8 $p \lor \{(p')'\}'$.

Proof $(q \rightarrow r) \rightarrow [(p \lor q) \rightarrow (p \lor r)]$ (by Post. L4)

$[p' \rightarrow \{(p')'\}'] \rightarrow [(p \lor p') \rightarrow (p \lor \{(p')'\}')]$

(substituting p' for q and $\{(p')'\}'$ for r)

$$p \to (p')' \hspace{4cm} \text{(by Th. 7)}$$

$$p' \to \{(p')'\}' \hspace{3.5cm} \text{(substituting } p' \text{ for } p)$$

$$(p \vee p') \to (p \vee \{(p')'\}') \hspace{2cm} \text{(by Rule R3)}$$

$$p \vee p' \hspace{5cm} \text{(by Th. 5)}$$

$$p \vee \{(p')'\}' \hspace{4cm} \text{(by Rule R3)}$$

Theorem 9 $(p')' \to p.$

Proof $(p \vee q) \to (q \vee p)$ $\hspace{3cm}$ (by Post. L3)

$$(p \vee \{(p')'\}') \to (\{(p')'\}' \vee p) \hspace{1.5cm} \text{(substituting } \{(p')'\}' \text{ for } q)$$

$$p \vee \{(p')'\}' \hspace{4.5cm} \text{(by Th. 8)}$$

$$\{(p')'\}' \vee p \hspace{4.5cm} \text{(by Rule R3)}$$

$$(p')' \to p \hspace{5cm} \text{(by Def. 1)}$$

Theorem 10 $[p \to (p')'] \wedge [(p')' \to p].$

Proof $p \to (p')'$ $\hspace{5cm}$ (by Th. 7)

$$(p')' \to p \hspace{5cm} \text{(by Th. 9)}$$

$$[p \to (p')'] \wedge [(p')' \to p] \hspace{2.5cm} \text{(by Rule R4)}$$

Definition 3 $p \leftrightarrow q$ means $(p \to q) \wedge (q \to p).$
Theorem 11 $p \leftrightarrow (p')'.$

Proof $[p \to (p')'] \wedge [(p')' \to p]$ $\hspace{2.5cm}$ (by Th. 10)

$$p \leftrightarrow (p')' \hspace{5cm} \text{(by Def. 3)}$$

(This is the *law of double negation*.)

Since the aim of the above demonstrations is merely to present some idea of the nature of the calculus of propositions rather than to give a complete account of its workings, we shall simply state the following additional theorems:

Theorem 12 $(p \vee q) \leftrightarrow (p' \wedge q')' \leftrightarrow (p' \to q).$
Theorem 13 $(p \to q) \leftrightarrow (p' \vee q) \leftrightarrow (p \wedge q')'.$
Theorem 14 $(p \wedge q) \leftrightarrow (p' \vee q')' \leftrightarrow (p \to q')'.$

(Theorems 12, 13, and 14 show the logical equivalences between the conjunctive, disjunctive, and implicative types of propositions. The significance of these logical equivalences has already been discussed.)

Theorem 15 $(p' \to p) \to p.$

(This theorem sanctions a frequently used type of argument in which a proposition p is proved by showing that the assumption of the falsity of p implies p.)

Theorem 16 $(p \to q) \leftrightarrow (q' \to p').$

(This is the *law of contraposition*, and it permits us to prove $p \to q$ by proving $q' \to p'$.)

Theorem 17 $[(p \to q) \wedge q'] \to p'.$

(This theorem asserts that if p implies q, and q is false, then p is false.)

Theorem 18 $[(p \lor q) \land q'] \rightarrow p$.

(This theorem asserts that if *p* or *q* is true, and *q* is false, then *p* is true.)

Theorem 19 $[(p \rightarrow q) \land (q \rightarrow r)] \rightarrow (p \rightarrow r)$.

(This is the *law of the syllogism* and is another form of the transitivity of implication.)

Theorem 20 $[p \land \{[(p \land q') \rightarrow r] \land [(p \land q') \rightarrow r']\}] \rightarrow q$.

(This theorem, which asserts that if *p* is true, and if *p* and *q'* imply both *r* and *r'*, then *q* is true, is the principle of *reductio ad absurdum* in its most general form. It is frequently used in mathematical demonstrations, where *p* represents the entire set of postulates and *q* is the theorem we wish to prove by *reductio ad absurdum*. We assume *q'*, or that *q* is false. The proof then consists in showing that the original postulates together with the falsity of *q* imply a pair of contradictory propositions, *r* and *r'*. When this is established, we may accept *q* as true.)

Theorem 21 $q \rightarrow (p \rightarrow q)$.

Theorem 22 $p' \rightarrow (p \rightarrow q)$.

(Theorems 21 and 22 exhibit two features of material implication that we have considered earlier. Thus Theorem 21 asserts that a true proposition *q* is implied by any proposition *p*, and Theorem 22 asserts that a false proposition *p* implies any proposition *q*.)

Theorem 23 $(p \land p')'$.

(This is the *law of contradiction*, which asserts that it is false that both *p* and not-*p* are true.)

The above theorems are sufficient to illustrate the general nature of the calculus of propositions, and it should now be seen how the principles of the argument of compound propositions are derivable from this calculus. But though the calculus of propositions is adequate for the precise rendering of those logical connections in which the propositions appear as unanalyzed wholes, the calculus is not adequate for the purposes of logic in general, for there are logical inferences of the sort that depend not only on the propositions as wholes but also on the inner content of the propositions themselves. For example, the calculus of propositions cannot render the logical relation expressed in the classical syllogism:

All men are mortal.

Socrates is a man.

Therefore Socrates is mortal.

The reason is evident; the inference made in this syllogism depends on the relations of subject to predicate in the various sentences of the syllogism and not just on the propositions considered as unanalyzed wholes. The calculus of classes must next be introduced to take care of this type of logical structure. It would unduly lengthen our treatment to attempt here the detailed introduction and unfolding of this phase of logic, though we have given some idea of the calculus of classes in an earlier chapter. We accordingly leave our discussion of the development of symbolic logic at this point and finish the section with a few comments bearing on some of the matters already covered.

In the first edition of *Principia mathematica*, Whitehead and Russell employed *five* postulates for the development of the calculus of propositions. In addition to Postulates L1, L2, L3, L4 given above, they also employed the postulate

$$[p \vee (q \vee r)] \rightarrow [q \vee (p \vee r)].$$

Later, in 1926, P. Bernays[6] showed that this additional postulate is not independent of the other four but that it can, by the permitted rules, be derived from L1, L2, L3, L4. He also showed, on the other hand, that L1, L2, L3, L4 constitute an independent set of postulates.

As with other systems developed via the postulational method, the calculus of propositions can be evolved from more than one set of primitive terms and from more than one set of postulates. J. B. Rosser and W. V. Quine, for example, have given developments based on conjunction and negation as primitive operations. Under these circumstances the postulate set may now be taken as

Def. $p \rightarrow q$ means $(p \wedge q')'$.

L'1. $p \rightarrow (p \wedge p)$.

L'2. $(p \wedge q) \rightarrow p$.

L'3. $(p \rightarrow q) \rightarrow [(q \wedge r)' \rightarrow (r \wedge p)']$.

Rules R1, R2, R3, and R4 are still used for deriving the theorems. As far back as 1879, Frege chose implication and negation as primitive operations and offered a set of six postulates in terms of these operations. J. Lukasiewicz has shown that Frege's set of six postulates may be replaced by the following simpler system, consisting of only three postulates:

L"1. $p \rightarrow (q \rightarrow p)$.

L"2. $[p \rightarrow (q \rightarrow r)] \rightarrow [(p \rightarrow q) \rightarrow (p \rightarrow r)]$.

L"3. $(p' \rightarrow q') \rightarrow (q \rightarrow p)$.

Again, Rules R1, R2, R3, R4 are used for deriving theorems. In 1916, J. Nicod[7] showed that the calculus of propositions follows from Sheffer's stroke as a primitive operation, together with the single postulate

$$[p/(q/r)]/\{[s/(s/s)]/[(t/q)/\{(p/t)/(p/t)\}]\}.$$

The rule of implication is now replaced by the rule: *From two given tautologies* u *and* u/(v/w), *the derived tautology* w *may be obtained.* Other postulational approaches to the calculus of propositions, besides those we have mentioned, have been offered.

A remarkable feature of the calculus of propositions is that it actually is an example of a Boolean algebra. To see this, let us interpret the elements $a, b, c, \ldots$ of a Boolean algebra to mean propositions, let $\cup$ and $\cap$ mean $\vee$ and $\wedge$, let equality mean logical equivalence, and let u and z mean a tautologous proposition and the negative of a tautologous proposition, respectively. Under these

[6]P. Bernays [1].

[7]J. Nicod [1].

interpretations, the postulates of Boolean algebra (see Section 8.2) become the following theorems in the calculus of propositions:

B1. $(a \lor b) \leftrightarrow (b \lor a)$, $(a \land b) \leftrightarrow (b \land a)$.

B2. $(a \lor z) \leftrightarrow a$, $(a \land u) \leftrightarrow a$.

B3. $[a \lor (b \land c)] \leftrightarrow [(a \lor b) \land (a \lor c)]$,
     $[a \land (b \lor c)] \leftrightarrow [(a \land b) \lor (a \land c)]$.

B4. $(a \lor a') \leftrightarrow u$, $(a \land a') \leftrightarrow z$.

The reader may easily verify, by using the method of truth tables, for example, that these are indeed theorems in the calculus of propositions—that is, are tautologies. Since, for our interpretations, the postulates of Boolean algebra become tautologies, it follows that the calculus of propositions can be considered an example of a Boolean algebra.

This representation of the calculus of propositions as a Boolean algebra is of great importance. It means that any theorem of Boolean algebra leads to a corresponding theorem in the calculus of propositions. For example, the De Morgan laws of Boolean algebra,

$$(a \cup b)' = a' \cap b', \qquad (a \cap b)' = a' \cup b',$$

lead to the following theorems in the calculus of propositions:

$$(a \lor b)' \leftrightarrow (a' \land b'), \qquad (a \land b)' \leftrightarrow (a' \lor b').$$

Again, since there is a principle of duality in Boolean algebra, there is a corresponding principle of duality in the calculus of propositions; it is: *Any tautology of the form* m $\leftrightarrow$ n, *where* m *and* n *are expressed only by means of conjunctions, disjunctions, and negations, remains a tautology if the conjunctions and disjunctions are interchanged throughout the tautology.*

In view of the above fact that the calculus of propositions is an interpretation of Boolean algebra, one might wonder if the calculus of propositions can be derived from the calculus of classes. The answer to this is no, for the calculus of classes requires the logical concepts of implication, negation, and so on, for its development. Indeed, the calculus of propositions cannot be derived from any other system, for in order to set up any other system we need the properties of propositions. That is, the properties of propositions must be assumed first and thus cannot be derived from any other system. It follows that the calculus of propositions occupies a unique position among deductive systems in that it must be assumed as fundamental. The calculus of propositions constitutes the central core of all work involving deduction.

9.3   Other Logics

An interesting analogy (if it is not pushed too far) exists between the parallelogram law of forces and the postulational method. By the parallelogram law, two component forces are combined into a single resultant force. Different resultant forces are obtained by varying one or both of the component forces, although it is possible to obtain the same resultant force by taking different pairs of initial component forces. Now, just as the resultant force is determined by the two initial component forces, so (see Figure 9.1) is a mathematical theory determined by a set of postulates and a logic. That is, the set of statements constituting a mathematical theory results from the interplay of an initial set of statements, called the postulates, and another initial set of statements, called the logic or the rules of procedure. For some time mathematicians have been aware of the variability of the first set of initial statements, namely, the postulates, but until very recent times the second set of initial statements, namely, the logic, was universally thought to be fixed, absolute, and immutable. Indeed, this is still the prevailing view among most people, for it seems quite inconceivable, except to the very few students of the subject, that there could be any alternative to the laws of logic stated by Aristotle in the fourth century B.C. The general feeling is that these laws are in some way attributes of the structure of the universe and that they are inherent in the very nature of human reasoning. As with many other absolutes of the past, this one, too, has toppled, but only as late as 1921. The modern viewpoint can hardly be more neatly put than in the following words of the outstanding American logician, Alonzo Church.[8]

> We do not attach any character of uniqueness or absolute truth to any particular system of logic. The entities of formal logic are abstractions, invented because of their use in describing and systematizing facts of experience or observation, and their properties, determined in rough outline by this intended use, depend for their exact character on the arbitrary choice of the inventor. We may draw the analogy of a three dimensional geometry used in describing physical space, a case for which, we believe, the presence of such a situation is more commonly recognized. The entities of the geometry are clearly of abstract character, numbering as they do planes without thickness and points which cover no area in the plane, point sets containing an infinitude of points, lines of infinite length, and other things which cannot be reproduced in any physical experiment. Nevertheless the geometry can be applied to physical space in such a way that an extremely useful correspondence is set up between the theorems of the geometry and observable facts about material bodies in space. In building the geometry, the proposed application to physical space serves as a rough guide in determining what properties the abstract entities shall have, but does not assign these properties completely. Consequently there may be, and actually are, more than one geometry whose use is feasible in describing physical space. Similarly, there exist, undoubtedly, more than one formal system whose use as a logic is feasible, and of these

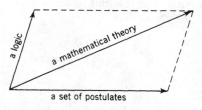

FIGURE 9.1

---

[8] A. Church [3], pp. 348, 349.

systems one may be more pleasing or more convenient than another, but it cannot be said that one is right and the other wrong.

It will be recalled that new geometries first came about through the denial of Euclid's parallel postulate and that new algebras first came about through the denial of the commutative law of multiplication. In similar fashion, the new so-called many-valued logics first came about by denying Aristotle's law of the excluded middle. According to this law, the disjunctive proposition $p \vee p'$ is a tautology, and a proposition p in Aristotelian logic is always either true or false. Because a proposition may possess any one of two possible truth values, namely truth or falsity, this logic is known as a two-valued logic. In 1921, in a two-page paper, J. Lukasiewicz considered a three-valued logic, or a logic in which a proposition p may possess any one of three possible truth values. Very shortly after, and independently of Lukasiewicz's work, E. L. Post considered m-valued logics, in which a proposition p may possess any one of m possible truth values, where m is any integer greater than 1. If m exceeds 2, the logic is said to be *many-valued*. Another study of m-valued logics was given in 1930 by Lukasiewicz and A. Tarski. Then, in 1932, the m-valued truth systems were extended by H. Reichenbach to an infinite-valued logic, in which a proposition p may assume any one of infinitely many possible truth values.[9]

Not all new logics are of the type just discussed. Thus A. Heyting has developed a symbolic two-valued logic to serve the intuitionist school of mathematicians; it differs from Aristotelian logic in that it does not universally accept the law of the excluded middle or the law of double negation. Like the many-valued logics, then, this special-purpose logic exhibits differences from Aristotelian laws. Such logics are known as *non-Aristotelian logics*. As a matter of fact, the symbolic two-valued logic of *Principia mathematica*, which was described in the last section, is really non-Aristotelian; it differs from Aristotelian logic in its meaning of implication.

Like the non-Euclidean geometries, the non-Aristotelian logics have proved not to be barren of application. Reichenbach actually devised his infinite-valued logic to serve as a basis for the mathematical theory of probability. And in 1933 F. Zwicky observed that many-valued logics can be applied to the quantum theory of modern physics. Many of the details of such an application have been supplied by Garrett Birkhoff, J. von Neumann, and H. Reichenbach. Lukasiewicz has employed three-valued logics to establish the independence of the postulates of familiar two-valued logic.[10] The part that non-Aristotelian logics may play in the future development of mathematics is uncertain but intriguing to contemplate; the application of Heyting's symbolic logic to intuitionistic mathematics indicates that the new logics may be mathematically valuable. In the next section we shall point out a possible use of these logics in the resolution of a modern crisis in the foundations of mathematics.

---

[9] As a matter of historical interest, in 1936 K. Michalski discovered that three-valued logics had actually been anticipated as early as the fourteenth century by the medieval schoolman, William of Occam. The possibility of a three-valued logic had also been considered by the philosopher Hegel and, in 1896, by Hugh MacColl. These speculations, however, had little effect on subsequent thought and so cannot be considered as decisive contributions.

[10] See J. Lukasiewicz.

Anything like a full treatment of many-valued logics would not be feasible here, but perhaps it is possible to give some idea of what such a logic is like. For simplicity we shall consider a three-valued logic and shall attempt to build up the logic by generalizing some of the concepts of the familiar two-valued logic considered in the last section.

We shall employ the method of truth tables, starting with a truth table for conjunction. We first of all repeat the truth table for conjunction for the two-valued logic, displaying the table in the more convenient form of a multiplication table as shown in Figure 9.2. This table is to be read as follows. Down the left-hand column appear the possible truth values for proposition p, and across the top row appear the possible truth values for the proposition q. Now, knowing the truth value of p and of q, one can find the truth value of $p \land q$ by looking in the table to the right of the value of p and below the value of q. Since, by definition, $p \land q$ is to be true when and only when both p and q are true, a T appears in the top left box of the table and F's appear in all the other boxes. It is to be noted that the table is uniquely determined by the definition of $p \land q$.

We now proceed to the three-valued logic and again agree to take the conjunction $p \land q$ to be true when and only when both p and q are true. Denoting the three possible truth values of a proposition by T, $?$, and F, we begin to construct a truth table as in Figure 9.3. By our agreement concerning the meaning of $p \land q$, the top left box in the table must contain a T, and no other box in the table is allowed to contain a T. Since there are eight remaining boxes and each may be filled in either of two possible ways, namely, with either an F or a $?$, there are altogether $2^8 = 256$ possible ways of filling the eight boxes. It follows that there are 256 different ways of developing a truth table for conjunction in a three-valued logic!

Figures 9.4a and 9.4b illustrate two of the possible 256 truth tables for conjunction in a three-valued logic. The truth table of Figure 9.4a is the one favored by Lukasiewicz, Post, and J. B. Rosser, and is formed by agreeing that $?$ is falser than T, F is falser than $?$, and $p \land q$ is exactly as true as the falser of its two factors. The truth table in Figure 9.4b is favored by D. A. Bočvar, and is obtained by considering $?$ to mean *undecidable* and by agreeing that $p \land q$ is undecidable when and only when either p or q is undecidable.

Let us now consider a truth table for negation. Our only restriction will be that when p is true, p' must fail to be true, and when p is false, p' must fail to be

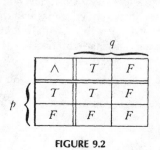

FIGURE 9.2

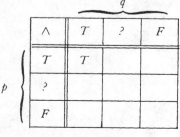

FIGURE 9.3

∧	T	?	F
T	T	?	F
?	?	?	F
F	F	F	F

FIGURE 9.4a

∧	T	?	F
T	T	?	F
?	?	?	?
F	F	?	F

FIGURE 9.4b

false. Whereas this restriction completely determines the truth table for negation in a two-valued logic, it permits twelve possible tables in a three-valued logic. Figures 9.5a and 9.5b illustrate two of these possible tables. The truth table in Figure 9.5a is favored by Post and is based on the assumption that, since in a two-valued logic $(p')'$ is the same as p, in a three-valued logic we ought to have $[(p')']'$ the same as p. The truth table in Figure 9.5b is favored by Bočvar, Lukasiewicz, and Rosser, and is based on the assumption that $(p')'$ is the same as p.

Truth tables for the other logical connectives can now be formed by defining these other connectives in terms of conjunction and negation. Since conjunction and negation are independent operations, whereas, as we have observed, the other operations may be made to depend on these two, it follows that there are altogether $(256)(12) = 3072$ possible three-valued logics. A many-valued logic apparently possesses an enormous wealth of possible structure.

In some treatments of many-valued logics, each proposition p has associated with it a real number, $\tau(p)$, lying in the range from 0 to 1, inclusive, and called its truth value. The truth value of a proposition under such circumstances may be interpreted to mean the probability that the proposition is true, and two propositions having the same truth values can be considered to be logically equivalent. The truth value 1 means truth; the truth value 0 means falsity. Negation may be defined in terms of truth values by $\tau(p') = 1 - \tau(p)$. Thus, in a three-valued logic, the truth values may be taken as 0, 1/2, and 1. If p has the truth value 1/2, then p' also has the truth value 1/2, and p is logically equivalent

p	p'
T	?
?	F
F	T

FIGURE 9.5a

p	p'
T	F
?	?
F	T

FIGURE 9.5b

to its own negation. Such a proposition may be called *doubtful*, and its negation is then also doubtful. This approach shows the intimate connection between many-valued logics and the theory of probability.

Perhaps the above fragmentary treatment is sufficient to give the reader some idea of the nature of non-Aristotelian logics. From it all there emerges a remarkable principle of discovery and advancement—namely, the constructive doubting of a traditional belief. When Einstein was asked how he came to invent the theory of relativity he replied, "By challenging an axiom." Lobachevsky and Bolyai challenged Euclid's axiom of parallels; Hamilton and Cayley challenged the axiom that multiplication is commutative; Lukasiewicz and Post challenged Aristotle's axiom of the excluded middle. Similarly, in the field of science, Copernicus challenged the axiom that the earth is the center of the solar system; Galileo challenged the axiom that the heavier body falls the faster; Einstein challenged the axiom that of two distinct instants one must precede the other. This constructive challenging of axioms has become one of the commoner ways of making advances in mathematics, and it undoubtedly lies at the heart of Georg Cantor's famous aphorism: "The essence of mathematics lies in its freedom."

9.4   Crises in the Foundations of Mathematics

A study of the history of mathematics from Greek antiquity to the present reveals that the foundations of mathematics have undergone three profoundly disturbing crises.

The first crisis in the foundations of mathematics arose in the fifth century B.C., and indeed, such a crisis could not have occurred much earlier, for, as we have seen, mathematics as a deductive study originated not earlier than the sixth century B.C., perhaps with Thales, Pythagoras, and their pupils. The crisis was precipitated by the unexpected discovery that not all geometrical magnitudes of the same kind are commensurable with one another; it was shown, for example, that the diagonal and side of a square contain no common unit of measure. Since the Pythagorean development of magnitudes was built on the firm intuitive belief that all like magnitudes are commensurable, the discovery that like magnitudes may be incommensurable proved to be highly devastating. For instance, the entire Pythagorean theory of proportion with all of its consequences had to be scrapped as unsound. The resolution of this first crisis in the foundations of mathematics was neither easily nor quickly realized. It was finally achieved in about 370 B.C. by the brilliant Eudoxus, whose revised theory of magnitude and proportion is one of the great mathematical masterpieces of all time. Eudoxus's remarkable treatment of incommensurables is found in the fifth book of Euclid's *Elements*;[11] it coincides essentially with the modern exposition of irrational numbers that was given by Richard Dedekind in 1872. We considered this first crisis in the foundations of mathematics in Section 1.5, where, among other things, we commented on its possible relation to the formulation of the axiomatic method.

---

[11] See Appendix, Section A.6.

The second crisis in the foundations of mathematics followed the discovery of the calculus by Newton and Leibniz in the late seventeenth century.[12] We have seen how the successors of these men, intoxicated by the power and applicability of the new tool, failed to consider sufficiently the solidity of the base on which the subject was founded, so that instead of having demonstrations justify results, results were used to justify demonstrations. With the passage of time, contradictions and paradoxes arose in increasing numbers, and a serious crisis in the foundations of mathematics became evident. It was realized more and more that the edifice of analysis was being built on sand, and finally, in the early nineteenth century, Cauchy took the first steps toward resolving the crisis by replacing the hazy method of infinitesimals by the precise method of limits.[13] With the subsequent so-called arithmetization of analysis by Weierstrass and his followers, it was felt that the second crisis in the foundations of mathematics had been overcome, and that the whole structure of mathematics had been redeemed and placed upon an unimpeachable base. The origin and resolution of this second crisis in the foundations of mathematics constituted the subject matter of Chapter 7.

The third crisis in the foundations of mathematics materialized with shocking suddenness in 1897, and, though now not far from a century old, is still not resolved to the satisfaction of all concerned. The crisis was brought about by the discovery of paradoxes or antinomies in the fringe of Cantor's general theory of sets. Since so much of mathematics is permeated with set concepts and, for that matter, can actually be made to rest on set theory as a foundation, the discovery of paradoxes in set theory naturally cast into doubt the validity of the whole foundational structure of mathematics.

In 1897 the Italian mathematician, Burali-Forti, brought to light the first publicized paradox of set theory.[14] As originally conceived and stated by Burali-Forti, the paradox involves technical terms and ideas which, in our limited treatment of sets in Chapter 8, we lacked space to develop. The essence of the paradox can be given, however, by a nontechnical description of a very similar paradox found by Cantor two years later. In his theory of sets, Cantor had succeeded in proving that for any given transfinite number there is always a greater transfinite number, so that just as there is no greatest natural number, there also is no greatest transfinite number. Now consider the set whose members are all possible sets. Surely no set can have more members than this set of all sets. But if this is the case, how can there be a transfinite number greater than the transfinite number of this set?

Whereas the Burali-Forti and Cantor paradoxes involve results of set theory, Bertrand Russell discovered in 1902 a paradox depending on nothing more than just the concept of set itself. Before describing the Russell paradox, we note that sets either are members of themselves or are not members of themselves. Thus

---

[12] Forewarnings of this crisis can be seen in the renowned paradoxes of Zeno of about 450 B.C. See H. Eves [1].

[13] However, see Appendix, Section A.7, concerning a relatively recent rigorization of the method of infinitesimals.

[14] C. Burali-Forti.

the set of all abstract ideas is itself an abstract idea, but the set of all men is not a man. Again, the set of all sets is itself a set, but the set of all stars is not a star. Let us represent the set of all sets that are members of themselves by M, and the set of all sets that are not members of themselves by N. We now ask ourselves whether set N is or is not a member of itself. If N is a member of itself, then N is a member of M and not of N, and N is not a member of itself. On the other hand, if N is not a member of itself, then N is a member of N and not of M, and N is a member of itself. The paradox lies in the fact that in either case we are led to a contradiction.

A more compact and less wordy presentation of the Russell paradox may be given as follows. Let X denote *any* set. Then, by the definition of N,

$$(X \in N) \leftrightarrow (X \notin X).$$

Now take X to be N, and we have the contradiction

$$(N \in N) \leftrightarrow (N \notin N).$$

This paradox was communicated by Russell to Frege just after the latter had completed the last volume of his great two-volume treatise on the foundations of arithmetic. Frege acknowledged the communication at the end of his treatise by the following pathetic and remarkably restrained sentences. "A scientist can hardly meet with anything more undesirable than to have the foundation give way just as the work is finished. In this position I was put by a letter from Mr. Bertrand Russell as the work was nearly through the press." Thus terminated the labor of a dozen or more years.

The Russell paradox has been popularized in many forms. One of the best known of these forms was given by Russell himself in 1919 and concerns the plight of the barber of a certain village who has enunciated the principle that he shaves all those persons and only those persons of the village who do not shave themselves. The paradoxical nature of this situation is realized when we try to answer the question, "Does the barber shave himself?" If he does shave himself, then he shouldn't according to his principle; if he doesn't shave himself, then he should according to his principle.

Since the discovery of the above contradictions within Cantor's theory of sets, additional paradoxes have been produced in abundance. These modern paradoxes of set theory are related to several ancient paradoxes of logic. For example, Eubulides, of the fourth century B.C., is credited with making the remark, "This statement I am now making is false." If Eubulides' statement is true, then, by what it says, the statement must be false. On the other hand, if Eubulides' statement is false, then it follows that his statement must be true. Thus Eubulides' statement can neither be true nor false without entailing a contradiction. Still older than the Eubulides paradox may be the unauthenticated Epimenides paradox. Epimenides, who himself was a Cretan philosopher of the sixth century B.C., is claimed to have made the remark, "Cretans are always liars." A simple analysis of this remark easily reveals that it, too, is self-contradictory.

The existence of paradoxes in set theory, like those described above, clearly indicates that something is wrong. Since their discovery, a great deal of literature on the subject has appeared, and numerous attempts at a solution have been offered.

As far as mathematics is concerned, there seems to be an easy way out. One has merely to reconstruct set theory on an axiomatic basis sufficiently restrictive to exclude the known antinomies. The first such attempt was made by Zermelo in 1908, and subsequent refinements have been made by Fraenkel (1922, 1925), Skolem (1922, 1929), von Neumann (1925, 1928), Bernays (1937–1948), and others. But such a procedure has been criticized as merely avoiding the paradoxes; certainly it does not explain them. Moreover, this procedure carries no guarantee that other kinds of paradoxes will not crop up in the future.

Another procedure apparently both explains and avoids the known paradoxes. If examined carefully, it will be seen that each of the paradoxes considered above involves a set S and a member m of S whose definition depends on S. Such a definition is said to be *impredicative*, and impredicative definitions are, in a sense, circular. Consider, for instance, Russell's barber paradox. Let us designate the barber by m and the set of all members of the barber's village by S. Then m is defined impredicatively as "that member of S who shaves all those members and only those members of S who do not shave themselves." The circular nature of this definition is evident—the definition of the barber involves the members of the village and the barber himself is a member of the village.

Poincaré considered the cause of the antinomies to lie in impredicative definitions, and Russell expressed the same view in his vicious circle principle: *No set* S *is allowed to contain members* m *definable only in terms of* S, *or members* m *involving or presupposing* S. This principle amounts to a restriction on the concept of set. Cantor had attempted to give the concept of set a very general meaning by stating: *By a set* S *we are to understand any collection into a whole of definite and separate objects* m *of our intuition or our thought; these objects* m *are called the elements of* S. The theory of sets constructed on Cantor's general concept of set leads, as we have seen, to contradictions, but if the notion of set is restricted by the vicious circle principle, the resulting theory avoids the known antinomies. The outlawing of impredicative definitions would appear, then, to be a solution to the known paradoxes of set theory. There is, however, one serious objection to this solution; namely, there are parts of mathematics, which mathematicians are very reluctant to discard, that contain impredicative definitions.

An example of an impredicative definition in mathematics is that of the least upper bound of a given nonempty set of real numbers—the least upper bound of the given set is the smallest member of the set of all upper bounds of the given set. There are many similar instances of impredicative definitions in mathematics, though many of them can be circumvented. In 1918, Hermann Weyl undertook[15] to find out how much of analysis can be constructed genetically from the natural number system without the use of impredicative definitions. Although he succeeded in obtaining a considerable part of analysis, he was unable to derive the important theorem that every nonempty set of real numbers having an upper bound has a least upper bound.

Other attempts to solve the paradoxes of set theory look for the trouble in logic, and the discovery of the paradoxes in the theory of unrestricted sets has

---

[15] H. Weyl [1].

brought about a thorough investigation of the foundations of logic. Very intriguing is the suggestion that the way out of the difficulties of the paradoxes may be through the use of a three-valued logic. For example, in the Russell paradox given above, we saw that the statement, "*N* is a member of itself," can be neither true nor false. Here a third possibility would be helpful, and the situation would be saved if we could simply classify the statement as *?*.

There have arisen three main philosophies, or schools of thought, concerning the foundations of mathematics—the so-called logistic, intuitionist, and formalist schools. Naturally, any modern philosophy of the foundations of mathematics must, somehow or other, cope with the present crisis in the foundations of mathematics. In the next section we very briefly consider these three schools of thought and point out how each proposes to deal with the antinomies of general set theory.

9.5 Philosophies of Mathematics

A philosophy may be regarded as an explanation that attempts to make some kind of sense out of the natural disorder of a set of experiences. From this point of view it is possible to have a philosophy of almost anything—a philosophy of art, of life, of religion, of education, of society, of history, of science, of mathematics, even of philosophy itself. A philosophy amounts to a process of refining and ordering experiences and values; it seeks relations among things that are normally felt to be disparate and finds important differences between things normally confused as the same; it is the description of a theory concerning the nature of something. In particular, a philosophy of mathematics essentially amounts to an attempted reconstruction in which the chaotic mass of mathematical knowledge accumulated over the ages is given a certain sense or order. Clearly, a philosophy is a function of time, and a particular philosophy may become outdated or have to be altered in the light of additional experiences. We are here concerned with only contemporary philosophies of mathematics— philosophies that take account of the recent advances in mathematics and of the current crisis in the subject.

Of the three principal present-day philosophies of mathematics, each has attracted a sizable group of adherents and developed a large body of associated literature. These are referred to as the logistic school, of which Russell and Whitehead are the chief expositors; the intuitionist school, led by Brouwer; and the formalist school, developed principally by Hilbert. There are, of course, present-day philosophies of mathematics other than these three. There are some independent philosophies and some that constitute various mixtures of the principal three, but these other points of view have not been so widely cultivated or do not comprise a reconstruction of mathematics of similar extent.

This section describes each of these three principal philosophies of mathematics. Obviously, it is not possible in a work of the size and nature of the present book to do proper justice to this subject, but it is hoped that the treatment will be sufficient to give some idea of these current schools of thought that are so intimately connected with the foundations and fundamental concepts of mathematics.

Logicism

The logistic thesis is that mathematics is a branch of logic. Rather than being just a tool of mathematics, logic becomes the progenitor of mathematics. All mathematical concepts are to be formulated in terms of logical concepts, and all theorems of mathematics are to be developed as theorems of logic; the distinction between mathematics and logic becomes merely one of practical convenience.

The notion of logic as a science containing the principles and ideas underlying all other sciences dates back at least as far as Leibniz (1666). The actual reduction of mathematical concepts to logical concepts was engaged in by Dedekind (1888) and Frege (1884–1903), and the statement of mathematical theorems by means of a logical symbolism was undertaken by Peano (1889–1908). These men, then, are forerunners of the logistic school, which received its definitive expression in the monumental *Principia mathematica* of Whitehead and Russell (1910–1913). This great and complex work purports to be a detailed reduction of the whole of mathematics to logic. Subsequent modifications and refinements of the program have been supplied by Wittgenstein (1922), Chwistek (1924–1925), Ramsey (1926), Langford (1927), Carnap (1931), Quine (1940), and others.

The logistic thesis arises naturally from the effort to push down the foundations of mathematics to as deep a level as possible. We have seen how these foundations were established in the real number system, and then how they were pushed back from the real number system to the natural number system and thence into set theory. Since the theory of classes is an essential part of logic, the idea of reducing mathematics to logic certainly suggests itself. The logistic thesis is thus an attempted synthesization suggested by an important trend in the history of the application of the postulational method.

The *Principia mathematica* starts with "primitive ideas" and "primitive propositions," corresponding to the "undefined terms" and "postulates" of a formal abstract development. These primitive ideas and propositions are not to be subjected to interpretation but are restricted to intuitive concepts of logic; they are to be regarded as, or at least are to be accepted as, plausible descriptions and hypotheses concerning the real world. In short, a concrete rather than an abstract point of view prevails, and consequently no attempt is made to prove the consistency of the primitive propositions. The aim of *Principia mathematica* is to develop mathematical concepts and theorems from these primitive ideas and propositions, starting with a calculus of propositions, proceeding up through the theory of classes and relations to the establishment of the natural number system, and thence to all mathematics derivable from the natural number system. In this development the natural numbers emerge with the unique meanings that we ordinarily assign to them and are not nonuniquely defined as *any things* that satisfy a certain set of abstract postulates.

To avoid the contradictions of set theory, *Principia mathematica* employs a "theory of types." Somewhat oversimply described, such a theory sets up a hierarchy of levels of elements. The primary elements constitute those of type 0; classes of elements of type 0 constitute those of type 1; classes of elements of type 1 constitute those of type 2; and so on. In applying the theory of types, one follows the rule that all the elements of any class must be of the same type.

Adherence to this rule precludes impredicative definitions and thus avoids the paradoxes of set theory. As originally presented in *Principia mathematica*, hierarchies within hierarchies appeared, leading to the so-called ramified theory of types. In order to obtain the impredicative definitions needed to establish analysis, an "axiom of reducibility" had to be introduced. The nonprimitive and arbitrary character of this axiom drew forth severe criticism, and much of the subsequent refinement of the logistic program lies in attempts to devise some method of avoiding the disliked axiom of reducibility.

Whether or not the logistic thesis has been established seems to be a matter of opinion. Although some accept the program as satisfactory, others have found many objections to it. For one thing, the logistic thesis can be questioned on the ground that the systematic development of logic (as of any organized study) presupposes mathematical ideas in its formulation, such as the fundamental idea of iteration that must be used, for example, in describing the theory of types or the idea of deduction from given premises.

Intuitionism

The intuitionist thesis is that mathematics is to be built solely by finite constructive methods on the intuitively given sequence of natural numbers. According to this view, then, at the very base of mathematics lies a primitive intuition, allied, no doubt, to our temporal sense of before and after, that allows us to conceive a single object, then one more, then one more, and so on endlessly. In this way we obtain unending sequences, the best known of which is the sequence of natural numbers. From this intuitive base of the sequence of natural numbers, any other mathematical object must be built in a purely constructive manner, employing a finite number of steps or operations. In the intuitionist thesis we have the genetical development of mathematics pushed to its extreme.

The intuitionist school (as a school) originated about 1908 with the Dutch mathematician L. E. J. Brouwer, although one finds some of the intuitionist ideas uttered earlier by such men as Kronecker (in the 1880's) and Poincaré (1902–1906). The school has gradually strengthened with the passage of time, has won over some eminent present-day mathematicians, and has exerted a tremendous influence on all thinking concerning the foundations of mathematics.

Some of the consequences of the intuitionist thesis are little short of revolutionary. Thus the insistence on constructive methods leads to a conception of mathematical existence not shared by all practicing mathematicians. For the intuitionists, an entity whose existence is to be proved must be shown to be constructible in a finite number of steps; it is not sufficient to show that the assumption of the entity's nonexistence leads to a contradiction. This means that many existence proofs found in current mathematics are not acceptable to the intuitionists.

An important instance of the intuitionists' insistence on constructive procedures is in the theory of sets. For the intuitionists, a set cannot be thought of as a ready-made collection but must be considered as a law by means of which the elements of the set can be constructed in a step-by-step fashion. This concept of set rules out the possibility of such contradictory sets as "the set of all sets."

There is another remarkable consequence of the intuitionists' insistence on finite constructibility, and this is the denial of the universal acceptance of the law of the excluded middle! Consider, for example, the number x, which is defined to be $(-1)^k$, where k is the number of the first decimal place in the decimal expansion of π where the sequence of consecutive digits 123456789 begins, and, if no such k exists, $x = 0$. Now, although the number x is well-defined, we cannot at the moment, under the intuitionists' restrictions, say that the proposition "$x = 0$" is either true or false. This proposition can be said to be true only when a proof of it has been constructed in a finite number of steps, and it can be said to be false only when a proof of this situation has been constructed in a finite number of steps. Until one or the other of these proofs is constructed, the proposition is neither true nor false, and the law of the excluded middle is inapplicable. If, however, k is further restricted to be less than a billion, say, then it is perfectly correct to say that the proposition is now either true or false, for, with k less than a billion, the truth or falseness can certainly be established in a finite number of steps.

Thus, for the intuitionists, the law of the excluded middle holds for finite sets but should not be employed when dealing with infinite sets. This state of affairs is blamed by Brouwer on the sociological development of logic. The laws of logic emerged at a time in man's evolution when he had a good language for dealing with finite sets of phenomena; he then later made the mistake of applying these laws to the infinite sets of mathematics, with the result that antinomies arose.

In the *Principia mathematica*, the law of the excluded middle and the law of contradiction are equivalent. For the intuitionists, this situation no longer prevails, and it is an interesting problem to try, if possible, to set up the logical apparatus to which intuitionist ideas lead us. This was done in 1930 by A. Heyting, who succeeded in developing an intuitionist symbolic logic. Intuitionist mathematics thus produces its own type of logic, and mathematical logic, as a consequence, is a branch of mathematics. In the intuitionist program, mathematical logic is almost irrelevant; it is subordinate and merely useful in expression or communication.

There is the final important question: How much of existing mathematics can be built within the intuitionist restrictions? If all of it can be so rebuilt, without too great an increase in the labor required, then the present problem of the foundations of mathematics would appear to be solved. Now the intuitionists have succeeded in rebuilding large parts of present-day mathematics, including a theory of the continuum and a set theory, but a great deal is still wanting. So far, intuitionist mathematics has turned out to be considerably less powerful than classical mathematics, and in many ways it is much more complicated to develop. This is the fault found with the intuitionist approach—too much that is dear to most mathematicians is sacrificed. This situation may not exist forever because there remains the possibility of an intuitionist reconstruction of classical mathematics carried out in a different and more successful way. And meanwhile, in spite of present objections raised against the intuitionist thesis, it is generally conceded that its methods do not lead to contradictions.

Formalism

The formalist thesis is that mathematics is concerned with formal symbolic systems. In fact, mathematics is regarded as a collection of such abstract developments, in which the terms are mere symbols and the statements are formulas involving these symbols; the ultimate base of mathematics does not lie in logic but only in a collection of prelogical marks or symbols and in a set of operations with these marks. Since, from this point of view, mathematics is devoid of concrete content and contains only ideal symbolic elements, the establishment of the consistency of the various branches of mathematics becomes an important and necessary part of the formalist program. Without such an accompanying consistency proof, the whole study is essentially senseless. In the formalist thesis we have the axiomatic development of mathematics pushed to its extreme.

The formalist school was founded by David Hilbert after completing his postulational study of geometry. In his *Grundlagen der Geometrie* (1899), Hilbert had sharpened the postulational method from the material axiomatics of Euclid to the formal axiomatics of the present day. The formalist point of view was developed later by Hilbert to meet the crisis caused by the paradoxes of set theory and the challenge to classical mathematics caused by intuitionistic criticism. Although Hilbert talked in formalistic terms as early as 1904, not until after 1920 did he and his collaborators, Bernays, Ackermann, von Neumann, and others, seriously start work on what is now known as the formalist program.

The success or failure of Hilbert's program to save classical mathematics hinges upon the solution of the consistency problem. Freedom from contradiction is guaranteed only by consistency proofs, and the older consistency proofs based on interpretations and models usually merely shift the question of consistency from one domain of mathematics to another. In other words, a consistency proof by the method of models is only relative. Hilbert, therefore, conceived a new direct approach to the consistency problem. Much as one may prove, by the rules of a game, that certain situations cannot occur within the game, Hilbert hoped to prove, by a suitable set of rules of procedure for obtaining acceptable formulas from the basic symbols, that a contradictory formula can never occur. In the logical notation introduced earlier in this chapter, a contradictory formula is any formula of the type $F \wedge F'$, where F is some accepted formula of the system. If one can show that no such contradictory formula is possible, then one has established the consistency of the system.

The development of the above ideas of a direct test for consistency in mathematics is called, by Hilbert, the "proof theory." Hilbert and Bernays planned to give a detailed exposition (and application to all classical mathematics) of the proof theory in their great *Grundlagen der Mathematik*, which may be considered as the *Principia mathematica* of the formalist school. The *Grundlagen der Mathematik* was finally published in two volumes, Volume I in 1934 and Volume II in 1939, but, as the work was being written, unforeseen difficulties arose, and it was not possible to complete the proof theory. For certain elementary systems, proofs of consistency were carried out, which illustrated what Hilbert would like to have done for all classical mathematics, but, for the system *in toto*, the problem of consistency remained refractory.

As a matter of fact, the Hilbert program, at least in the form originally envisioned by Hilbert, appears to be doomed to failure; this truth was brought out by Kurt Gödel in 1931, actually before the publication of the *Grundlagen* had taken place. Gödel showed, by unimpeachable methods acceptable to the followers of any one of the three principal schools of the philosophy of mathematics, that it is impossible for a sufficiently rich formalized deductive system, such as Hilbert's system for all classical mathematics, to prove consistency of the system by methods belonging to the system. This remarkable result is a consequence of an even more fundamental one; Gödel proved the incompleteness of Hilbert's system—that is, he established the existence within the system of "undecidable" problems, of which consistency of the system is one. These theorems of Gödel are too difficult to consider in their technical details here. They are certainly among the most remarkable in all mathematics, and they reveal an unforeseen limitation in the methods of formal mathematics. They show "that the formal systems known to be adequate for the derivation of mathematics are unsafe in the sense that their consistency cannot be demonstrated by finitary methods formalized within the system, whereas any system known to be safe in this sense is inadequate."[16]

PROBLEMS

9.1.1 By the use of truth tables establish the laws of logic tabulated in Section 9.1.

9.1.2 By the use of truth tables determine whether or not the following propositions are tautologies:

(a) $(p \land q) \to p$.

(b) $(p \lor q) \to p$.

(c) $(p' \to p) \leftrightarrow p$.

(d) $(p \land q) \leftrightarrow (q \land p)$.

(e) $[p \to (p \land r)] \leftrightarrow [(p \to q) \land (p \to r)]$.

(f) $[p \land (q \lor r)] \to [(p \land q) \lor (p \land r)]$.

(g) $[(p \land q') \to p'] \leftrightarrow (p \to q)$.

(h) $[(p \to q) \to r] \to [p \to (q \to r)]$.

9.1.3 Let P denote a composite proposition involving n primary propositions. Show that the truth table for P contains 2^n lines.

9.1.4 Establish the logical equivalence of the pairs of propositions tabulated at the end of Section 9.1.

9.1.5 Express the following propositions in terms of disjunction and negation symbols only:

(a) $(p \lor p) \to p$.

(b) $q \to (p \lor q)$.

(c) $(p \lor q) \to (q \lor p)$.

(d) $(q \to r) \to [(p \lor q) \to (p \lor r)]$.

(e) $[p \lor (q \lor r)] \to [q \lor (p \lor r)]$.

---

[16] F. De Sua. Here we also find the following interesting remark: "Suppose we loosely define a *religion* as any discipline whose foundations rest on an element of faith, irrespective of any element of reason that may be present. Quantum mechanics for example would be a religion under this definition. But mathematics would hold the unique position of being the only branch of theology possessing a rigorous demonstration of the fact that it should be so classified." See also the Appendix, Section A.9.

9.1.6 Let us designate *or* when used in the exclusive sense by $\underline{\vee}$, so that "$p \underline{\vee} q$" means "p or q but not both." Show that we may express $p \underline{\vee} q$ by

    (a) $(p \vee q) \wedge (p \wedge q)'$.

    (b) $(p \wedge q') \vee (p' \wedge q)$.

9.1.7 By the use of truth tables show that the following pairs of propositions are logically equivalent:

    (a) $(p \wedge q)'$ and $p' \vee q'$.

    (b) $(p \vee q)'$ and $p' \wedge q'$.

    (c) $(p \rightarrow q)'$ and $p \wedge q'$.

    (d) $(p \leftrightarrow q)'$ and $p \leftrightarrow q'$.

    (e) $(p \leftrightarrow q)'$ and $p' \leftrightarrow q$.

9.1.8 Form the negation of each of the following propositions, reducing the result in each case to one in which the negation sign is applied only to the propositions p, q, r:

    (a) $p' \rightarrow q$.

    (b) $(p \rightarrow q) \wedge r$.

    (c) $(p \wedge q) \vee r$.

    (d) $p \rightarrow q'$.

    (e) $p' \leftrightarrow (p \wedge q')$.

    (f) $(p \vee q) \wedge p$.

9.1.9 Related to the proposition "$p \rightarrow q$" are the following three propositions: (1) the *converse*, $q \rightarrow p$, (2) the *inverse*, $p' \rightarrow q'$, (3) the *contrapositive*, $q' \rightarrow p'$. Show that

    (a) the converse of a true implication is not always true,

    (b) the inverse of a true implication is not always true,

    (c) the contrapositive of a true implication is always true,

    (d) the contrapositive of an implication is the converse of the inverse of the implication.

    (e) Is the inverse of the converse of an implication the same as the converse of the inverse of the implication?

9.1.10 Write the converse, inverse, and contrapositive of each of the following implications:

    (a) $q \rightarrow p$,   **(b)** $p' \rightarrow q'$,   **(c)** $q' \rightarrow p'$,   **(d)** $p \rightarrow q'$,   **(e)** $p' \rightarrow q$.

9.1.11 It is common in mathematics to state theorems in forms like (1) "a necessary condition that p be true is that q be true," (2) "a sufficient condition that p be true is that q be true," (3) "a necessary and sufficient condition that p be true is that q be true." By definition, these three forms mean, respectively, (1) "$p \rightarrow q$," (2) "$q \rightarrow p$," (3) "$(p \rightarrow q) \wedge (q \rightarrow p)$."

    (a) Show that (3) is equivalent to "$p \leftrightarrow q$."

    (b) Show that in order to establish a necessary and sufficient condition that p be true one must establish both a theorem and its converse.

9.1.12 If the last column of the truth table for a composite proposition m contains only F's, then m is called an *absurdity*; if the last column contains both T's and F's, then m is called a *contingency*.

    (a) Show that $p \wedge p'$ is an absurdity.

    (b) Show that $[(p \wedge q) \rightarrow q]'$ is an absurdity.

    (c) Show that if m is an absurdity, then m' is a tautology.

    (d) Show that $p' \wedge q' \wedge r'$ is a contingency.

    (e) Show that $p \vee q \vee r$ is a contingency.

    (f) Show that if m is a contingency, then m' is a contingency.

9.1.13 Show that negation is indispensable for expressing combinations of propositions.

9.1.14 Take p/q to mean $p' \vee q'$.

    (a) Show the logical equivalence of p/p and p'.

    (b) Show the logical equivalence of $(p/p)/(q/q)$ and $p \vee q$.

    (c) Express $\wedge$ in terms of $/$.

(d) Express → in terms of /.

(e) Express ↔ in terms of /.

9.2.1 Establish the following additional useful rules for obtaining new tautologies from given tautologies in the calculus of propositions:

(a) **R5.** *If* m ∨ m *is a tautology, then* m *is a tautology.*

(b) **R6.** *If* m *is a tautology and* p *is any proposition, then* p ∨ m *is a tautology.*

(c) **R7.** *If* m ∨ n *is a tautology, then* n ∨ m *is a tautology.*

(d) **R8.** *If* m → n *is a tautology and* p *is any proposition, then* (p ∨ m) → (p ∨ n) *is a tautology.*

(e) **R9.** *If* p → q *and* q → r *are tautologies, then* p → r *is a tautology.*

9.2.2 (a) Show, from the truth table for implication, that "if p is true and if $p → q$, then q is true."

(b) Because of (a), Postulate L1 asserts that "if p or p is true, then p is true." Give similar interpretations of Postulates L2, L3, L4 and of Theorems 1, 2, and 3.

9.2.3 Show that if, by definition, p means q, then $p ↔ q$ is a tautology.

9.2.4 The following rule may be proved: R10: *If* p ↔ q *is a tautology, then in any tautology involving proposition* p *we may replace* p *in any of its occurrences by* q *and thus obtain another tautology.* Using this rule, prove

(a) Theorem 15.

(b) Theorem 23.

9.2.5 (a) Establish the tautology: $(p → q') → (q → p')$.

(b) Establish Theorem 16.

(c) If $m ↔ n$ is a tautology, show that $m' ↔ n'$ is a tautology.

9.2.6 (a) Establish Theorem 21.

(b) Establish the tautology $p → (p ∨ q)$.

(c) Establish Theorem 22.

9.2.7 Verify, by truth tables, that the postulates of Boolean algebra become tautologies under the interpretations made at the end of Section 9.2.

9.2.8 Convert the following relations in Boolean algebra into tautologies in the calculus of propositions:

(a) $a \cap (b \cap c) = (a \cap b) \cap c$.

(b) $a \cap b = b \cap a$.

(c) $a \cap a = a$.

(d) $a \cap (a \cup b) = a$.

9.2.9 (a) What is the counterpart in Boolean algebra of the implication relation $a → b$ of the calculus of propositions?

(b) What is the counterpart in the calculus of propositions of the inclusion relation $a \subset b$ of Boolean algebra?

9.2.10 From each of the following tautologies find another tautology by the principle of duality:

(a) $[(p \wedge p') \vee q] ↔ q$.

(b) $(p \vee q) ↔ (p' \wedge q')'$.

(c) The answers to Problem 9.2.8.

9.2.11 Establish the absolute consistency of the calculus of propositions by showing that it is impossible to obtain from the primitive tautologies L1, L2, L3, L4, by the rules R1, R2, R3, two contradictory tautologies s and s'.

9.3.1 Using the definition, $p \vee q$ means $(p' \wedge q')'$, construct truth tables for disjunction in a three-valued logic using the truth tables of

(a) Figures 9.4a and 9.5a;

(b) Figures 9.4a and 9.5b;

      (c) Figures 9.4b and 9.5a;
      (d) Figures 9.4b and 9.5b.

9.3.2 Using the definition, $p \to q$ means $(p \wedge q')'$, construct truth tables for implication in a three-valued logic using the truth tables of
      (a) Figures 9.4a and 9.5a;
      (b) Figures 9.4a and 9.5b;
      (c) Figures 9.4b and 9.5a;
      (d) Figures 9.4b and 9.5b.

9.3.3 Verify that there are twelve possible truth tables for negation in a three-valued logic.

9.3.4 **(a)** Show that for the truth table in Figure 9.5a, $[(p')']'$ is the same as p.
      (b) Show that for the truth table in Figure 9.5b, $(p')'$ is the same as p.

9.3.5 How many possible m-valued logics are there analogous to the 3072 possible three-valued logics indicated in Section 9.3?

9.3.6 Suppose we wish to define the implication relation, $p \to q$, subject to the restriction that when p and $p \to q$ are both true, then q is also true. How many such implication relations are possible in a **(a)** two-valued logic? **(b)** three-valued logic?

9.4.1 Consider the following popularizations of the Russell paradox:
      (a) Every municipality of a certain country must have a mayor, and no two municipalities may have the same mayor. Some mayors do not reside in the municipalities they govern. A law is passed compelling nonresident mayors to reside by themselves in a certain special area A. There are so many nonresident mayors that A is proclaimed a municipality. Where shall the mayor of A reside?
      (b) An adjective in the English language is said to be *autological* if it applies to itself; otherwise the adjective is said to be *heterological*. Thus the adjectives *short*, *English*, and *polysyllabic* all apply to themselves and hence are autological, whereas the adjectives *long*, *French*, and *monosyllabic* do not apply to themselves and hence are heterological. Now is the adjective *heterological* autological or heterological?
      (c) Suppose a librarian compiles, for inclusion in his library, a bibliography of all those bibliographies in his library which do not list themselves.

9.4.2 Examine the following paradox. Every natural number can be expressed in simple English, without the use of numerical symbols. Thus 5 can be expressed as "five," or as "half of ten," or as "the second odd prime," or as "the positive square root of twenty-five," and so on. Now consider the expression, "the least natural number not expressible in fewer than twenty-three syllables." This expression expresses in twenty-two syllables a natural number that cannot be expressed in fewer than twenty-three syllables.

9.4.3 Consider the following logical dilemmas:
      (a) A crocodile, which has stolen a child, promises the child's father to return the child provided the father guesses whether the child will be returned or not. What should the crocodile do if the father guesses that the child will not be returned?
      (b) An explorer has been captured by cannibals who offer the explorer the opportunity to make a statement under the condition that if it is true he will be boiled and if it is false he will be roasted. What should the cannibals do if the explorer states, "I will be roasted"?

9.4.4 Consider the following questions:
      (a) What would happen if an irresistible force should collide with an immovable body?
      (b) If Zeus can do anything, can he make a stone that he cannot lift?

APPENDIX
BIBLIOGRAPHY
SOLUTION SUGGESTIONS
FOR SELECTED PROBLEMS

APPENDIX

A.1 The First Twenty-Eight Propositions of Euclid

Following are the statements, in modern terminology, of the first twenty-eight propositions of Book I of Euclid's *Elements*. The proofs of these propositions do not require the parallel postulate.

1. To construct an equilateral triangle on a given straight line segment.

2. From a given point to draw a straight line segment equal to a given straight line segment.

3. From the greater of two given straight line segments to cut off a part equal to the smaller.

4. Two triangles are congruent if two sides and the included angle of one are equal to two sides and the included angle of the other.

5. The base angles of an isosceles triangle are equal.

6. If two angles of a triangle are equal, the sides opposite these angles are equal.

7. There cannot be two different triangles on the same side of a common base and having their other pairs of corresponding sides equal.

8. Two triangles are congruent if the three sides of one are equal to the three sides of the other.

9. To bisect a given angle.

10. To bisect a given straight line segment.

11. To erect a perpendicular to a given straight line at a given point on the line.

12. To drop a perpendicular from a given point to a given straight line.

13. Any pair of adjacent angles formed by two intersecting straight lines are either two right angles or are together equal to two right angles.

14. If, at any point on a straight line, two rays, on opposite sides of it, make a pair of adjacent angles together equal to two right angles, then these two rays lie in the same straight line.

15. Any pair of vertical angles formed by two intersecting straight lines are equal.

16. An exterior angle of a triangle is greater than either remote interior angle.

17. Any two angles of a triangle are together less than two right angles.

18. In a triangle the greater side is opposite the greater angle.

19. In a triangle the greater angle is opposite the greater side.

20. The sum of any two sides of a triangle is greater than the third side.

21. If two straight line segments are drawn from the ends of a side of a triangle to a point within the triangle, these two segments together will be less than the other two sides together but will contain a greater angle.

22. To construct a triangle having sides equal to three given straight line segments, any two of which are together greater than the third.

23. At a given point on a given straight line to construct an angle equal to a given angle.

24. If two triangles have two sides of one equal to two sides of the other, but the included angle of the first is greater than the included angle of the second, then the third side of the first is greater than the third side of the second.

25. If two triangles have two sides of one equal to two sides of the other, but the third side of the first is greater than the third side of the second, then the included angle of the first is greater than the included angle of the second.

26. Two triangles are congruent if two angles and a side of one are equal to two angles and the corresponding side of the other.

27. If a transversal of two straight lines makes a pair of alternate interior angles equal, then the two lines are parallel.

28. If a transversal of two straight lines makes a pair of corresponding angles equal, or a pair of interior angles on the same side of the transversal equal to two right angles, then the two lines are parallel.

A.2 Euclidean Constructions

We noted, in Section 2.3, that the first three of Euclid's postulates state the primitive constructions from which all other constructions in the *Elements* are to be compounded. The first two of these postulates tell us what we can do with a Euclidean straightedge; we are permitted to draw as much as may be desired of the straight line determined by two given points. The third postulate tells us what we can do with Euclidean compasses; we are permitted to draw the circle of given center and passing through a given point. Note that neither instrument is to be used for transferring distances. This means that the straightedge cannot be marked, and the compasses must be regarded as having the characteristic that if one or both points be lifted from the paper, the instrument immediately collapses. For this reason, Euclidean compasses are often referred to as *collapsing compasses*; they differ from *modern compasses*, which retain their opening and hence can be used as dividers for transferring distances. It would seem that modern compasses might be more powerful than the collapsing compasses. Curiously enough, such turns out not to be true; any construction performable with the modern compasses can also be carried out (in perhaps a longer way) by means of the collapsing compasses. We prove this fact as our first theorem.

Theorem 1 *The collapsing and modern compasses are equivalent.*

The circle with center O and passing through a given point C will be denoted by $O(C)$, and the circle with center O and radius equal to a given segment AB will be denoted by $O(AB)$. To prove the theorem it suffices to show that we may, with collapsing compasses, construct any circle $O(AB)$. This may be accomplished as follows (see Figure A.1). Draw circles $A(O)$ and $O(A)$ to intersect in D and E; draw circles $D(B)$ and $E(B)$ to intersect again in F; draw circle $O(F)$. It is an easy matter to prove that $OF = AB$, whence circle $O(F)$ is circle $O(AB)$.

Note: In view of Theorem 1, we may dispense with the Euclidean, or collapsing, compasses, and in their place employ the simpler modern compasses. We are assured that the set of constructions performable with straightedge and modern compasses is the same as the set performable with straightedge and Euclidean compasses.

We now proceed to establish a chain of theorems that will furnish us with a criterion for Euclidean constructibility. It turns out that the criterion is algebraic in nature. After the criterion is established, we shall apply it to prove that the three famous construction problems of antiquity (trisection of a general angle, duplication of a cube, and squaring of a circle) are impossible with Euclidean tools. A proof of this fact was not discovered until the nineteenth century, more than two thousand years after the original problems had been proposed.[1]

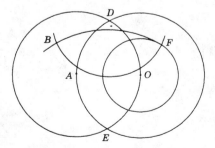

FIGURE A.1

[1] For a history of these problems and allied topics, see H. Eves [1], chap. 4.

Theorem 2   *Given line segments of lengths* a *and* b, *and a unit segment, we can construct with Euclidean tools segments of lengths* a + b, |a − b|, ab, a/b, *and* $\sqrt{a}$.

The first two constructions are trivial; the last three are apparent from Figures A.2, A.3, and A.4.

Definition 1   Let $a_1, \ldots, a_n$ be a given nonempty set of distinct real nonzero numbers, and let F_0 denote the set of all numbers each of which can be obtained from $a_1, \ldots, a_n$ by a finite number of additions, subtractions, multiplications, and permissible divisions. Since F_0 is closed under addition, subtraction, multiplication, and permissible division, F_0 is a number field (see Section 5.1); it will be called the *number field generated by* $a_1, \ldots, a_n$. The number field generated by the number 1 will be denoted by R_0.

Theorem 3 R_0 *is the field of all rational numbers. Every number field contains* R_0 *as a subfield.*

The proof is simple and is left to the reader.

Definition 2   Let w_1 be a positive number in F_0 (generated by $a_1, \ldots, a_n$) such that $\sqrt{w_1}$ is not in F_0. Then the number field F_1 generated by $a_1, \ldots, a_n, \sqrt{w_1}$ will be called a *square root extension* of field F_0. Let $F_0, F_1, F_2, \ldots, F_m$ be a sequence of number fields such that F_i is a square root extension of F_{i-1} ($i = 1, \ldots, m$). Then this sequence of fields will be called a *square root extension chain* of F_0, and the fields $F_0, F_1, \ldots, F_m$ will be called the *links* of the chain. We shall let $\sqrt{w_i}$ denote the radical that extends F_{i-1} to F_i.

Theorem 4   *Every number of* F_i ($i \geq 1$) *has the form* r + s$\sqrt{w_i}$, *where* r *and* s *are in* F_{i-1}, *and, conversely, every number of this form is in* F_i.

The generators of F_i are clearly of the required form. To prove the direct part of the theorem it then suffices to show that if A and B are of the required form, then so are $A + B$, $A - B$, AB, and $A/B (B \neq 0)$ of the required form. To this end set

$$A = a + b\sqrt{w_i}, \qquad B = c + d\sqrt{w_i},$$

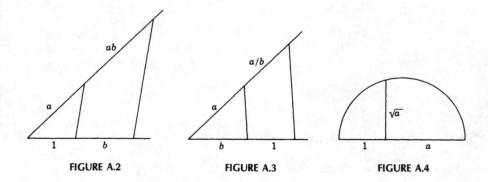

FIGURE A.2   FIGURE A.3   FIGURE A.4

where a, b, c, d are in F_{i-1}. Then

$$A + B = (a + c) + (b + d)\sqrt{w_i},$$

$$A - B = (a - c) + (b - d)\sqrt{w_i},$$

$$AB = (ac + bdw_i) + (ad + bc)\sqrt{w_i},$$

$$A/B = (a + b\sqrt{w_i})/(c + d\sqrt{w_i})$$

$$= (a + b\sqrt{w_i})(c - d\sqrt{w_i})/(c + d\sqrt{w_i})(c - d\sqrt{w_i})$$

$$= [(ac - bdw_i)/(c^2 - d^2w_i)] + [(bc - ad)/(c^2 - d^2w_i)]\sqrt{w_i},$$

and we see that $A + B$, $A - B$, AB, and $A/B(B \neq 0)$ are of the required form.

The proof to the converse part of the theorem is trivial.

Theorem 5 *Any number in* F_m *is a root of a polynomial of degree* 2^r *with coefficients in* F_{m-1}, $r = 1, 2, \ldots, m$.

Let x be a number in F_m. Then, by Theorem 4, $x = a + b\sqrt{w_m}$, where a and b are in F_{m-1}. It follows that

$$x - a = b\sqrt{w_m} \quad \text{or} \quad x^2 - 2ax + (a^2 - b^2w_m) = 0,$$

and x is a root of a quadratic polynomial with coefficients in F_{m-1}.

We now have $x^2 + px + q = 0$, where p and q are numbers in F_{m-1}. By Theorem 4, $p = c + d\sqrt{w_{m-1}}$, $q = e + f\sqrt{w_{m-1}}$, where c, d, e, f are in F_{m-2}. It follows that

or

$$x^2 + (c + d\sqrt{w_{m-1}})x + (e + f\sqrt{w_{m-1}}) = 0,$$

$$(x^2 + cx + e)^2 - (dx + f)^2w_{m-1} = 0,$$

and x is a root of a quartic polynomial with coefficients in F_{m-2}.

We now have $x^4 + rx^3 + sx^2 + tx + u = 0$, where r, s, t, u are numbers in F_{m-2}. By Theorem 4, $r = g + h\sqrt{w_{m-2}}$, $s = j + k\sqrt{w_{m-2}}$, $t = l + n\sqrt{w_{m-2}}$, $u = p + q\sqrt{w_{m-2}}$, where g, h, j, k, l, n, p, q are in F_{m-3}. Substituting these values for r, s, t, u in the quartic equation, transposing the terms containing the radical $\sqrt{w_{m-2}}$ to one side of the equation, and then squaring both sides, we find that x is a root of an octic polynomial with coefficients in F_{m-3}.

A step-by-step continuation of the above process establishes the theorem.

Definition 3 A point of the Cartesian plane is said to *belong to* F_i if its coordinates are in F_i, and a straight line or a circle is said to *belong to* F_i if the coefficients in an equation of the straight line or of the circle are proportional to numbers in F_i.

Theorem 6 (1) *If points* P *and* Q *belong to* F_i, *then the line* PQ *belongs to* F_i. (2) *If two distinct intersecting lines belong to* F_i, *then their point of intersection belongs to* F_i. (3) *If points* P *and* Q *belong to* F_i, *then the circle* P(Q) *belongs to* F_i. (4) *If an intersecting line and circle belong to* F_i, *then their points of* *intersection belong to* F_i, *or to some square root extension* F_{i+1} *of* F_i. (5) *If two distinct intersecting circles belong to* F_i, *then their points of intersection belong to* F_i *or to some square root extension* F_{i+1} *of* F_i.

1. Let $P = (a, b)$, $Q = (c, d)$, where a, b, c, d are in F_i. Then an equation of PQ is

$$(b - d)x + (c - a)y + (ad - bc) = 0,$$

the coefficients of which are in F_i.

2. Let equations of the lines be

$$ax + by + c = 0 \quad \text{and} \quad dx + ey + f = 0,$$

where a, b, c, d, e, f are in F_i and, to guarantee intersection, $ae - bd \neq 0$. Then the coordinates of the point of intersection of the lines are

$$x = \frac{bf - ce}{ae - bd}, \qquad y = \frac{dc - af}{ae - bd},$$

which are in F_i.

3. Let $P = (a, b)$, $Q = (c, d)$, where a, b, c, d are in F_i. Then an equation of circle $P(Q)$ is

$$x^2 + y^2 - 2ax - 2by - (c^2 + d^2 - 2ac - 2bd) = 0,$$

the coefficients of which are in F_i.

4. Let equations of the line and circle be

$$ax + by + c = 0 \quad \text{and} \quad x^2 + y^2 + gx + hy + k = 0,$$

where a, b, c, g, h, k are in F_i. Then the coordinates of the points of intersection of the line and circle are given by

$$x = \frac{-A \pm b\sqrt{R}}{2M}, \qquad y = \frac{-B \pm a\sqrt{R}}{2M},$$

where $M = a^2 + b^2$ and

$$A = 2ac + b^2g - abh, \qquad B = 2bc + a^2h - bag,$$

$$R = a^2h^2 + b^2g^2 - 4a^2k - 4b^2k - 4c^2 + 4acg + 4bch - 2abgh.$$

These coordinates are in F_i (if the radical belongs to F_i) or in some square root extension F_{i+1} of F_i (if the radical does not belong to F_i).

5. Let equations of the circles be

$$x^2 + y^2 + gx + hy + k = 0 \quad \text{and} \quad x^2 + y^2 + px + qy + r = 0,$$

where g, h, k, p, q, r are in F_i. Then the points of intersection of the two circles are the points of intersection of either circle with their radical axis

$$(g - p)x + (h - q)y + (k - r) = 0,$$

whose coefficients are in F_i. Part (5) is now reduced to part (4).

Theorem 7   *Let* $a_1, \ldots, a_n$ *be the distinct nonzero Cartesian coordinates of a set of at least two given points. Then any point which can be constructed from the given points by straightedge alone belongs to the field* F_0 *generated by* $a_1, \ldots, a_n$.

This is a consequence of Theorem 6.

Note: The converse of this theorem is not true. That is, there are points in F_0 which cannot be constructed from the given points with straightedge alone.

Theorem 8   *Let* $a_1, \ldots, a_n$ *be the distinct nonzero Cartesian coordinates of a set of at least two given points. Then any point which can be constructed from the given points by the Euclidean tools belongs to some square root extension chain of the field* F_0 *generated by* $a_1, \ldots, a_n$. *Conversely, given a unit segment, any point belonging to a square root extension chain of* F_0 *can be constructed with Euclidean tools.*

The direct part of the theorem is a consequence of Theorem 6. The converse part of the theorem follows from Theorems 4 and 2.

Definition 4   The set F of all numbers belonging to square root extension chains of F_0 is known as the *constructible set generated by* $a_1, \ldots, a_n$. The constructible set generated by the number 1 will be designated by R.

Definition 5   A number is said to be *algebraic* if it is a root of a polynomial equation with coefficients in R_0.[2]

Theorem 9   *All numbers in* R *are algebraic.*

This is an immediate consequence of Theorem 5 and Definition 5.

Theorem 10   *If a cubic polynomial equation with rational coefficients has no rational root, then none of its roots belongs to* R.

Let the cubic equation be

$$x^3 + ax^2 + bx + c = 0,$$

where a, b, c are in R_0 but x is not. Suppose x is in R. Then x lies in the last link R_k of some square root extension chain $R_0, R_1, \ldots, R_k$ of the rational field R_0. Let k be the *smallest* integer such that a root of the cubic lies in a square root extension link, like R_k, of R_0. Since x is not in R_0, $k > 0$. We then have (Theorem 4) $x = p + q\sqrt{w_k}$, where p, q, w_k are in R_{k-1}, but $\sqrt{w_k}$ is not in R_{k-1}. Substituting $x = p + q\sqrt{w_k}$ in the cubic equation, we find $P + Q\sqrt{w_k} = 0$, where

$$P = p^3 + 3pq^2 w_k + ap^2 + aq^2 w_k + bp + c,$$

$$Q = 3p^2 q + q^3 w_k + 2apq + bq.$$

Since P and Q are in R_{k-1}, whereas $\sqrt{w_k}$ is not in R_{k-1}, it follows that we must have $P = Q = 0$, and hence $P - Q\sqrt{w_k} = 0$. But if we replace x by $p - q\sqrt{w_k}$ in the left member of the cubic equation, we obtain $P - Q\sqrt{w_k}$. Therefore $y = p - q\sqrt{w_k}$ is also a root of the cubic equation. Since $q \neq 0$ (inasmuch as otherwise x would be in R_{k-1}), we see that $x \neq y$. Let z be the

---

[2] The equivalence of this definition with that given in Section 8.4 is the subject matter of Problem 8.4.4 (g).

third root of the cubic equation. Then, from the elementary theory of equations, $x + y + z = -a$, or

$$z = -a - x - y = -a - 2p,$$

a number in R_{k-1}. But this is a contradiction of the assumption that k is the *smallest* integer such that a root of the cubic equation lies in a square root extension link, like R_k, of R_0. The theorem now follows by *reductio ad absurdum*.

Theorem 11 *Given a line segment* AB, *it is impossible to construct with Euclidean tools a line segment* CD *such that* $(CD)^3 = 2(AB)^3$. (Figure A.5)

Place a Cartesian frame of reference on the plane with origin at A and with positive x-axis along AB. Choose the scale so that AB is one unit. We are then given the points $(0, 0)$ and $(1, 0)$. Now assume the construction is possible. Then we can construct with Euclidean tools the point E on the positive x-axis such that $AE = CD = \sqrt[3]{2}(AB) = \sqrt[3]{2}$. That is, we can construct the point $(\sqrt[3]{2}, 0)$, whence $\sqrt[3]{2}$ is in the constructible set R generated by 1. But $\sqrt[3]{2}$ is a root of the cubic equation $x^3 - 2 = 0$. Since this cubic has rational coefficients and no rational root,[3] Theorem 10 says that $\sqrt[3]{2}$ is not in R. The contradiction proves the theorem.

Note: Theorem 11 shows that it is impossible to construct with Euclidean tools the edge of a cube having exactly twice the volume of a given cube, and thus disposes of the ancient problem of *duplicating the cube*.

Theorem 12 *It is impossible to construct with Euclidean tools the trisectors of a* 60° *angle*. (Figure A.6)

Starting with the two points $O = (0, 0)$ and $A = (1, 0)$, we may construct an angle $AOB = 60°$ by drawing circles $O(A)$ and $A(O)$ to intersect in B.

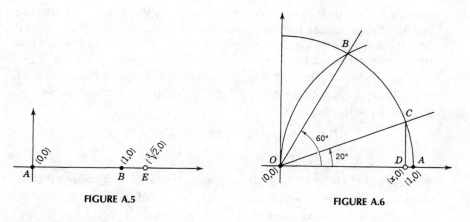

FIGURE A.5   FIGURE A.6

---

[3] If a polynomial equation

$$a_0 x^n + a_1 x^{n-1} + \cdots + a_n = 0$$

with integral coefficients $a_0, a_1, \ldots, a_n$ has a reduced rational root a/b, then a is a factor of a_n and b is a factor of a_0. Thus any rational roots of $x^3 - 2 = 0$ are among $1, -1, 2, -2$. Since, by direct testing, none of these numbers satisfies the equation, the equation has no rational root.

Now assume that it is possible with Euclidean tools to draw the trisectors of angle AOB. Let the trisector adjacent to OA cut circle $O(A)$ in C. Find, with Euclidean tools, the foot D of the perpendicular from C upon OA, and denote the coordinates of D by $(x, 0)$. Then x, which is equal to $\cos 20°$, is in the constructible set R generated by 1.

From the trigonometric identity

$$\cos \theta = 4 \cos^3 \frac{\theta}{3} - 3 \cos \frac{\theta}{3},$$

we obtain, by taking $\theta = 60°$,

$$8x^3 - 6x - 1 = 0.$$

It is easily verified that this cubic equation with rational coefficients has no rational root. Hence, by Theorem 10, x cannot lie in R, and the theorem follows by *reductio ad absurdum*.

Note: Since it is impossible with Euclidean tools to trisect an angle of 60°, it follows that there is no general procedure, using Euclidean tools, for trisecting an arbitrary given angle. Theorem 12 disposes of the ancient problem of *trisecting an angle*. Note that we have not proved that *no* angle can be trisected with Euclidean tools but only that *not all* angles can be so trisected. As a matter of fact, an angle of 90°, and an infinite number of other angles, can be trisected with Euclidean tools.

Theorem 13 *It is impossible to construct with Euclidean tools a regular polygon of nine sides.*

If such a polygon could be constructed, then the central angle of the polygon could also be constructed. But the central angle of a regular polygon of nine sides is $40° = (2/3)60°$, and a 60° angle could then be trisected.

Theorem 14 *It is impossible to construct with Euclidean tools an angle of 1°.*

If an angle of 1° could be constructed, then an angle of 20° could be constructed. But we have seen that this is impossible.

Theorem 15 *It is impossible to construct with Euclidean tools a square whose area is equal to that of a given circle.* (Figure A.7)

Take the radius of the circle as one unit and place the center of the circle at the point $(0, 1)$. Let x represent half the side of the sought square. Then

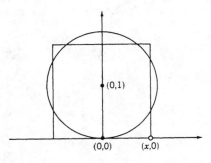

FIGURE A.7

$4x^2 = \pi$, and $x = \sqrt{\pi}/2$. Suppose the construction is possible. Then, starting with the two points $(0, 1)$ and $(0, 0)$, we can construct the point $(\sqrt{\pi}/2, 0)$, and it follows that $\sqrt{\pi}/2$, and hence π, is in R. This contradicts Theorem 9, which says that all numbers in R are algebraic, for it is known that π is nonalgebraic. The contradiction proves the theorem.

Note: Theorem 15 disposes of the ancient problem of *squaring a circle*.

A.3   Removal of Some Redundancies

In Section 7.2 we listed a postulate set for a *complete ordered field*. The first ten of these postulates characterize a *field*. We shall show that in this postulate set for a field, Postulates P1 and P9 are redundant and can therefore be deleted, thus reducing the postulate set for a field to one of eight simple postulates. The deduction of Postulate P9 from Postulates P4, P7, P10 is very simple and well known. The deduction of Postulate P1 from P2, P3, P5, P6, P7, P8 is more difficult and not so well known. For simplicity we denote the addition operation of the field by $+$ and the multiplication operation by juxtaposition.

Theorem 1 (this is P9)   *If* a, b, c *are elements of a field* S, c *is not the zero element* z *of* S, *and* ac $=$ bc, *then* a $=$ b.

For we have

$$a = au \qquad \text{(by P7)}$$
$$= a(cc^{-1}) \qquad \text{(by P10)}$$
$$= (ac)c^{-1} \qquad \text{(by P4)}$$
$$= (bc)c^{-1} \qquad \text{(substitution)}$$
$$= b(cc^{-1}) \qquad \text{(by P4)}$$
$$= bu \qquad \text{(by P10)}$$
$$= b. \qquad \text{(by P7)}$$

Theorem 2 (this is P1)   *If* a *and* b *are elements of a field* S, *then* a $+$ b $=$ b $+$ a.

Postulates P3, P6, P8 assure us that the elements of a field constitute a group with respect to addition. It follows (see Theorems 1 and 5 of Section 5.3) that both the right and the left cancellation laws for addition hold in a field. But

$$(a + b)(u + u) = (a + b)u + (a + b)u \qquad \text{(by P5)}$$
$$= (a + b) + (a + b) \qquad \text{(by P7)}$$
$$= [(a + b) + a] + b \qquad \text{(by P3)}$$
$$= [a + (b + a)] + b. \qquad \text{(by P3)}$$

Also

$$(a + b)(u + u) = (u + u)(a + b) \qquad \text{(by P2)}$$
$$= (u + u)a + (u + u)b \qquad \text{(by P5)}$$

$$= a(u + u) + b(u + u) \qquad \text{(by P2)}$$

$$= (au + au) + (bu + bu) \qquad \text{(by P5)}$$

$$= (a + a) + (b + b) \qquad \text{(by P7)}$$

$$= [(a + a) + b] + b \qquad \text{(by P3)}$$

$$= [a + (a + b)] + b. \qquad \text{(by P3)}$$

It now follows that $[a + (a + b)] + b = [a + (b + a)] + b$, whence, by an application of the right cancellation law for addition followed by an application of the left cancellation law for addition, we find $a + b = b + a$.

The examination of the truncated postulate set P2, P3, P4, P5, P6, P7, P8, P10 of Section 7.2 for independence or for further possible redundancies constitutes a study that is recommended to the interested reader.

In Section 7.3 we listed a postulate set for the natural number system, and remarked that Postulates N1, N2, N4, N7, and N8 of the set are redundant. It is interesting that the redundance of N7 and N8 was apparently observed only rather recently.[4] We now proceed to deduce Postulates N1, N2, N4, N7, and N8 from the remaining postulates N3, N5, N6, N9, and N10.

Theorem 3   *If* n *is a natural number, then* n $+ 1 = 1 +$ n.

Let M be the set of all natural numbers n such that $n + 1 = 1 + n$. Clearly 1 is in M. Suppose k is in M; that is, suppose $k + 1 = 1 + k$. Then

$$(k + 1) + 1 = (1 + k) + 1 \qquad \text{(by supposition)}$$

$$= 1 + (k + 1), \qquad \text{(by N3)}$$

and $k + 1$ is in M. It follows, by N10, that M contains all the natural numbers, and the theorem is established.

Theorem 4   (this is N1)   *If* a *and* b *are any natural numbers, then* a $+$ b $=$ b $+$ a.

Let M be the set of all natural numbers b for which $a + b = b + a$. By Theorem 3, 1 is in M. Suppose k is in M; that is, suppose $a + k = k + a$. Then

$$a + (k + 1) = (a + k) + 1 \qquad \text{(by N3)}$$

$$= (k + a) + 1 \qquad \text{(by supposition)}$$

$$= k + (a + 1) \qquad \text{(by N3)}$$

$$= k + (1 + a) \qquad \text{(by Theorem 3)}$$

$$= (k + 1) + a, \qquad \text{(by N3)}$$

and $k + 1$ is in M. It follows, by N10, that M contains all the natural numbers, and the theorem is established.

---

[4] See P. H. Diananda.

Theorem 5 *If* n *is any natural number, then* $1n = n$.

Let M be the set of all natural numbers n such that $1n = n$. Clearly 1 is in M, by N6. Suppose k is in M; that is, suppose $1k = k$. Then

$$1(k + 1) = 1k + 1 \cdot 1 \qquad \text{(by N5)}$$

$$= k + 1 \cdot 1 \qquad \text{(by supposition)}$$

$$= k + 1, \qquad \text{(by N6)}$$

and $k + 1$ is in M. It follows, by N10, that M contains all the natural numbers, and the theorem is established.

Theorem 6 *If* a, b, c *are any natural numbers, then* $(b + c)a = ba + ca$.

Let M be the set of all natural numbers a such that $(b + c)a = ba + ca$. Now

$$(b + c)1 = b + c \qquad \text{(by N6)}$$

$$= b1 + c1, \qquad \text{(by N6)}$$

and 1 is in M. Suppose k is in M; that is, suppose $(b + c)k = bk + ck$. Then

$$(b + c)(k + 1) = (b + c)k + (b + c)1 \qquad \text{(by N5)}$$

$$= (bk + ck) + (b + c)1 \qquad \text{(by supposition)}$$

$$= (bk + ck) + (b + c) \qquad \text{(by N6)}$$

$$= [(bk + ck) + b] + c \qquad \text{(by N3)}$$

$$= [bk + (ck + b)] + c \qquad \text{(by N3)}$$

$$= [bk + (b + ck)] + c \qquad \text{(by Theorem 4)}$$

$$= [(bk + b) + ck] + c \qquad \text{(by N3)}$$

$$= (bk + b) + (ck + c) \qquad \text{(by N3)}$$

$$= (bk + b1) + (ck + c1) \qquad \text{(by N6)}$$

$$= b(k + 1) + c(k + 1), \qquad \text{(by N5)}$$

and $k + 1$ is in M. It follows, by N10, that M contains all the natural numbers, and the theorem is established.

Theorem 7 (this is N2) *If* a *and* b *are any natural numbers, then* $ab = ba$.

Let M be the set of all natural numbers b such that $ab = ba$. Now

$$a1 = a \qquad \text{(by N6)}$$

$$= 1a, \qquad \text{(by Theorem 5)}$$

and 1 is in M. Suppose k is in M; that is, suppose $ak = ka$. Then

$$a(k + 1) = ak + a1 \qquad \text{(by N5)}$$

$$= ka + a1 \qquad \text{(by supposition)}$$

$$= ka + 1a \qquad \text{(since 1 is in } M\text{)}$$

$$= (k + 1)a, \qquad \text{(by Theorem 6)}$$

and $k + 1$ is in M. It follows, by N10, that M contains all the natural numbers, and the theorem is established.

Theorem 8 (this is N4)   *If* a, b, c *are any natural numbers, then* (ab)c = a(bc).

Let M be the set of all natural numbers c such that $(ab)c = a(bc)$. Now

$$(ab)1 = ab \qquad\qquad\qquad \text{(by N6)}$$

$$= a(b1), \qquad\qquad\qquad \text{(by N6)}$$

and 1 is in M. Suppose k is in M; that is, suppose $(ab)k = a(bk)$. Then

$$(ab)(k + 1) = (ab)k + (ab)1 \qquad\qquad \text{(by N5)}$$

$$= a(bk) + (ab)1 \qquad\qquad \text{(by supposition)}$$

$$= a(bk) + a(b1) \qquad\qquad \text{(since 1 is in } M)$$

$$= a[bk + b1] \qquad\qquad\qquad \text{(by N5)}$$

$$= a[b(k + 1)], \qquad\qquad\qquad \text{(by N5)}$$

and $k + 1$ is in M. It follows, by N10, that M contains all the natural numbers, and the theorem is established.

Theorem 9 (this is N7)   *If* a, b, c *are natural numbers and if* c + a = c + b, *then* a = b.

By N9 one and only one of the following holds: $a = b$, $a + x = b$, $a = b + y$, where x and y are natural numbers.

Suppose $a + x = b$. Then

$$c + b = c + (a + x) \qquad\qquad \text{(by supposition)}$$

$$= (c + a) + x. \qquad\qquad \text{(by N3)}$$

Since, by hypothesis, $c + b = c + a$, the above result contradicts N9.

Suppose $a = b + y$. Then

$$c + a = c + (b + y) \qquad\qquad \text{(by supposition)}$$

$$= (c + b) + y. \qquad\qquad \text{(by N3)}$$

Since, by hypothesis, $c + a = c + b$, the above result contradicts N9.

It now follows, by *reductio ad absurdum*, that $a = b$.

Theorem 10 (this is N8)   *If* a, b, c *are natural numbers and if* ca = cb, *then* a = b.

By N9 one and only one of the following holds: $a = b$, $a + x = b$, $a = b + y$, where x and y are natural numbers.

Suppose $a + x = b$. Then

$$cb = c(a + x) \qquad\qquad \text{(by supposition)}$$

$$= ca + cx. \qquad\qquad \text{(by N5)}$$

Since, by hypothesis, $cb = ca$, the above result contradicts N9.

Suppose $a = b + y$. Then

$$ca = c(b + y) \qquad\qquad \text{(by supposition)}$$

$$= cb + cy. \qquad\qquad \text{(by N5)}$$

Since, by hypothesis, $ca = cb$, the above result contradicts N9.

It now follows, by *reductio ad absurdum*, that $a = b$.

In view of Theorems 4, 7, 8, 9, 10 it follows that the natural number system can be made to rest on the five postulates N3, N5, N6, N9, N10. In fact, we can even weaken N3 and N5 to

N3'. *If* a *and* b *are in* N, *then* $(a + b) + 1 = a + (b + 1)$.

N5'. *If* a *and* b *are in* N, *then* $a(b + 1) = ab + a1$.

To prove this we deduce N3 from N3' and N10, and then N5 from N3, N5', and N10.

Theorem 11   N3 *is implied by* N3' *and* N10.

Let M be the set of all natural numbers c for which $(a + b) + c = a + (b + c)$. By N3', 1 is in M. Suppose k is in M; that is, suppose $(a + b) + k = a + (b + k)$. Then

$$(a + b) + (k + 1) = [(a + b) + k] + 1 \qquad \text{(by N3')}$$

$$= [a + (b + k)] + 1 \qquad \text{(by supposition)}$$

$$= a + [(b + k) + 1] \qquad \text{(by N3')}$$

$$= a + [b + (k + 1)], \qquad \text{(by N3')}$$

and $k + 1$ is in M. It follows, by N10, that M contains all the natural numbers, and the theorem is established.

Theorem 12   N5 *is implied by* N3, N5', *and* N10.

Let M be the set of all natural numbers c such that $a(b + c) = ab + ac$. Clearly, by N5', 1 is in M. Suppose k is in M; that is, suppose $a(b + k) = ab + ak$. Then

$$a[b + (k + 1)] = a[(b + k) + 1] \qquad \text{(by N3)}$$

$$= a(b + k) + a1 \qquad \text{(by N5')}$$

$$= (ab + ak) + a1 \qquad \text{(by supposition)}$$

$$= ab + (ak + a1) \qquad \text{(by N3)}$$

$$= ab + a(k + 1), \qquad \text{(by N5')}$$

and $k + 1$ is in M. It follows, by N10, that M contains all the natural numbers, and the theorem is established.

It is interesting that the reduced postulate set N3', N5', N6, N9, N10 for the natural number system is independent. The student might care to show this by considering the following five interpretations, wherein we designate the addition operation by $*$ and the multiplication operation by $\circ$.[5]

Independence of N3'.   Let N be the set $\{1, 2, 3\}$ with $1 * x = 2$, $2 * x = 3$, and $3 * x = 1$ for all x in N, $1 \circ 1 = 2 \circ 3 = 3 \circ 2 = 1$, $2 \circ 1 = 1 \circ 2 = 3 \circ 3 = 2$, and $3 \circ 1 = 1 \circ 3 = 2 \circ 2 = 3$.

---

[5]The interpretations are due to Clayton W. Dodge. See C. W. Dodge, pp. 140, 141.

Independence of N5'. Let N be the set of natural numbers with $a * b = a + b$ and $a \circ b = a^b$.

Independence of N6. Let N be the set of natural numbers with $a * b = a + b$ and $a \circ b = a(b + 1)$.

Independence of N9. Let N be the set $\{0, 1\}$ with $0 * 0 = 1 * 1 = 0, 0 * 1 = 1 * 0 = 1.$ $0 \circ 0 = 0 \circ 1 = 1 \circ 0 = 0, 1 \circ 1 = 1.$

Independence of N10. Let N be the set of all positive rational numbers with $a * b = a + b$ and $a \circ b = ab$.

A.4 Membership Tables

If A and B are sets, then an element is in $A \cup B$ if and only if it is in A, or in B, or in both A and B, and it is in $A \cap B$ if and only if it is in A and in B. If we denote *in* by I and *out of* by O, these conditions can be summarized by the two tables

A	B	$A \cup B$
I	I	I
I	O	I
O	I	I
O	O	O

A	B	$A \cap B$
I	I	I
I	O	O
O	I	O
O	O	O

(For example, the third line in the $A \cap B$ table reads: "If an element is out of A and in B, then it is out of $A \cap B$.")

To establish the relation $(A \cup B)' = A' \cap B'$ by means of membership tables, consider the four possible ways elements can be in and out of A and B. For each of these we tabulate whether or not it is then in, or out of, $(A \cup B)'$ and $A' \cap B'$.

A	B	$A \cup B$	$(A \cup B)'$
I	I	I	O
I	O	I	O
O	I	I	O
O	O	O	I

A	B	A'	B'	$A' \cap B'$
I	I	O	O	O
I	O	O	I	O
O	I	I	O	O
O	O	I	I	I

Observe that the two final columns are identical—that is, an element is in $(A \cup B)'$ if it is in $A' \cap B'$, and is out of $(A \cup B)'$ if it is out of $A' \cap B'$. This implies that the sets $(A \cup B)'$ and $A' \cap B'$ are equal.

As another example, let us establish the law

$$(A \cap B) \cup C = (A \cup C) \cap (B \cup C).$$

Considering the eight possible ways elements can be in or out of A, B, and C, we obtain the following two membership tables:

A	B	C	$A \cap B$	$(A \cap B) \cup C$
I	I	I	I	I
I	I	O	I	I
I	O	I	O	I
I	O	O	O	O
O	I	I	O	I
O	I	O	O	O
O	O	I	O	I
O	O	O	O	O

A	B	C	$A \cup C$	$B \cup C$	$(A \cup C) \cap (B \cup C)$
I	I	I	I	I	I
I	I	O	I	I	I
I	O	I	I	I	I
I	O	O	I	O	O
O	I	I	I	I	I
O	I	O	O	I	O
O	O	I	I	I	I
O	O	O	O	O	O

The identity of the two final columns establishes the law.

A.5 A Constructive Proof of the Existence of Transcendental Numbers

Cantor's proof of the existence of transcendental numbers, as given in Theorem 4 of Section 8.4, can scarcely be called a constructive proof; it does not lead to a number whose decimal expression, for example, can actually be written down. Many mathematicians, among whom are the members of the intuitionist school of philosophy of mathematics, do not accept nonconstructive existence proofs but demand that an existence proof should produce an actual example of one of the entities whose existence is being demonstrated. Such a proof of the existence of transcendental numbers was given by Joseph Liouville (1809–1882) in 1851, twenty-some years before Cantor published his nonconstructive demonstration. We give here a modern and simplified version of Liouville's approach, which, though it employs only high school mathematics, will require careful reading and close attention. We begin by recalling once more the definition of an algebraic number and of a transcendental number.

Definition 1 A number α of the complex number system is said to be an *algebraic number* if there exists a positive integer n and a set of integers $a_0, a_1, \ldots, a_n (a_0 \neq 0)$ such that

$$a_0 \alpha^n + a_1 \alpha^{n-1} + \cdots + a_{n-1} \alpha + a_n = 0.$$

If n is the smallest such positive integer, then α is said to be of *degree n*.

Definition 2   A number of the complex number system that is not algebraic is said to be a *transcendental number*.

Theorem 1 (Liouville's)   *If α is an algebraic number of degree* n > 1, *there exists a number* c, 0 < c < 1, *such that for all integers* p *and* q, *where* q > 0, *we have*

$$|\alpha - p/q| > c/q^n. \tag{1}$$

We are given that

$$a_0 \alpha^n + a_1 \alpha^{n-1} + \cdots + a_{n-1} \alpha + a_n = 0,$$

where $n > 1$, the a_i's are integers, and $a_0 \neq 0$. Let us set

$$f(x) = a_0 x^n + a_1 x^{n-1} + \cdots + a_{n-1} x + a_n.$$

Then $f(\alpha) = 0$, and we have

$$
\begin{aligned}
-f(p/q) &= f(\alpha) - f(p/q) \\
&= a_0[\alpha^n - (p/q)^n] + a_1[\alpha^{n-1} - (p/q)^{n-1}] \\
&\quad + \cdots + a_{n-2}[\alpha^2 - (p/q)^2] + a_{n-1}(\alpha - p/q) \\
&= (\alpha - p/q)\{a_0[\alpha^{n-1} + (p/q)\alpha^{n-2} + \cdots + (p/q)^{n-1}] \\
&\quad + \cdots + a_{n-2}(\alpha + p/q) + a_{n-1}\}.
\end{aligned}
\tag{2}
$$

Since $n > 1$, α must be irrational, for a rational number is algebraic of degree 1. Therefore $\alpha - p/q \neq 0$, and we must have $f(p/q) \neq 0$, inasmuch as otherwise the second factor in the right member of (2) would have to vanish, making α of degree less than n, which contradicts the hypothesis of the theorem. Consequently, we have

$$|f(p/q)| = |a_0 p^n + a_1 p^{n-1} q + \cdots + a_n q^n|/q^n \geq 1/q^n. \tag{3}$$

But from equation (2)

$$
\begin{aligned}
|f(p/q)| \leq |\alpha - p/q|\{|a_0|(|\alpha|^{n-1} + |p/q| \, |\alpha|^{n-2} + \cdots + |p/q|^{n-1}) \\
+ \cdots + |a_{n-2}|(|\alpha| + |p/q|) + |a_{n-1}|\}.
\end{aligned}
\tag{4}
$$

Since (1) is obviously true if $|\alpha - p/q| \geq 1$, we may assume $|\alpha - p/q| < 1$. Then $|p/q| < |\alpha| + 1$. Therefore, replacing both $|p/q|$ and $|\alpha|$ in the second factor of the right member of (4) by $|\alpha| + 1$, we have

$$
\begin{aligned}
|f(p/q)| < |\alpha - p/q|\{n|a_0|(|\alpha| + 1)^{n-1} + \cdots \\
+ 2|a_{n-2}|(|\alpha| + 1) + |a_{n-1}|\} = |\alpha - p/q|K,
\end{aligned}
$$

say, where K is clearly a number greater than 1. Combining this result with (3), we have

$$|\alpha - p/q| > |f(p/q)|/K \geq c/q^n,$$

where $c = 1/K$, and the theorem is proved.

Theorem 2   *The real number*

$$t = \sum_{k=1}^{\infty} a_k / r^{k!},$$

where r is an integer greater than 1 and the a_k's are integers such that $0 < a_k < r$, is a transcendental number.

Suppose t is algebraic. Then t is irrational (since its radix fraction for base r is nonperiodic) and hence of degree $n > 1$. By Theorem 1, then, there exists a real number c, $0 < c < 1$, such that

$$|t - p/q| > c/q^n$$

for every rational number p/q with $q > 0$. Let us choose for p/q the rational number

$$\sum_{k=1}^{m} a_k / r^{k!} = p/r^{m!}.$$

Then we have

$$c/r^{m!n} < |t - p/q| = \sum_{k=1}^{\infty} a_k / r^{k!} - \sum_{k=1}^{m} a_k / r^{k!}$$

$$= \sum_{k=m+1}^{\infty} a_k / r^{k!} < (1 + 1/r + 1/r^2 + \cdots)/r^{(m+1)!-1}$$

$$= [r/(r-1)]/r^{(m+1)!-1} \leq r/r^{(m+1)!-1},$$

or

$$c < r r^{m!n}/r^{(m+1)!-1} = 1/r^{m!(m+1-n)-2},$$

which, since $c > 0$, is impossible for large enough m. Hence the theorem by *reductio ad absurdum*.

Definition 3   Numbers like t of Theorem 2 are called *Liouville numbers to the base* r. If we take $r = 10$, we obtain the Liouville numbers to base 10,

$$t = \sum_{k=1}^{\infty} a_k / 10^{k!} = 0.a_1 a_2 000 a_3 00000000000000000 a_4 000 \ldots,$$

where the a_k's are arbitrary digits from 1 to 9.

═══════ **A.6   The Eudoxian Resolution of the First Crisis** ═══════
in the Foundations of Mathematics[6]

In Book V of Euclid's *Elements* appears a masterly exposition of Eudoxus's theory of proportion. It was this theory, applicable to incommensurable as well as to commensurable magnitudes, that resolved the "logical scandal" created by the Pythagorean discovery of irrational numbers. The remarkable Eudoxian definition of proportion, or equality of two ratios, is cited at the start of Book V

---

[6] Material in this section has been adapted from H. Eves [1].

and runs as follows: *Magnitudes are said to be in the same ratio, the first to the second and the third to the fourth, when, if any equimultiples whatever be taken of the first and third, and any equimultiples whatever of the second and fourth, the former equimultiples alike exceed, are alike equal to, or are alike less than the latter equimultiples taken in corresponding order.* In other words, if A, B, C, D are any four unsigned magnitudes, A and B being of the same kind (both line segments, or angles, or areas, or volumes) and C and D being of the same kind, then the ratio of A to B is equal to that of C to D when, for arbitrary positive integers m and n, $mC \gtreqless nD$ according as $mA \gtreqless nB$.

Let us indicate the differences between the Pythagorean, the Eudoxian, and the modern textbook proofs of a simple proposition involving proportions. We select Proposition VI 1: *The areas of triangles having the same altitude are to one another as their bases.* We shall permit ourselves to use Proposition I 38, which says that *triangles having equal bases and equal altitudes have equal areas*, and a consequence of I 38 to the effect that *of any two triangles having the same altitude, that one has the greater area which has the greater base.*

Let the triangles be ABC and ADE, the bases BC and DE lying on the same straight line MN, as in Figure A.8. Now the Pythagoreans, before the discovery of irrational numbers, tacitly assumed that any two line segments are commensurable. Thus BC and DE were assumed to have some common unit of measure, going, say, p times into BC and q times into DE. Mark off these points of division on BC and DE and connect them with vertex A. Then triangles ABC and ADE are divided, respectively, into p and q smaller triangles, all having, by I 38, the same area. It follows that $\triangle ABC : \triangle ADE = p : q = BC : DE$, and the proposition is established. With the later discovery that two line segments need not be commensurable, this proof, along with many others, became inadequate, and the very disturbing "logical scandal" came into existence.

Eudoxus's theory of proportion cleverly resolved the "scandal," as we now illustrate by reproving VI 1 in the manner found in Book VI of the *Elements*. On CB produced mark off, successively from B, $m - 1$ segments equal to CB, and connect the points of division, $B_2, B_3, \ldots, B_m$, with vertex A, as shown in Figure A.9. Similarly, on DE produced mark off, successively from E, $n - 1$ segments equal to DE, and connect the points of division, $E_2, E_3, \ldots, E_n$, with vertex A. Then $B_m C = m(BC)$, $\triangle AB_m C = m(\triangle ABC)$, $DE_n = n(DE)$, $\triangle ADE_n = n(\triangle ADE)$. Also, by I 38 and its corollary, $\triangle AB_m C \gtreqless \triangle ADE_n$ according as $B_m C \gtreqless DE_n$. That is, $m(\triangle ABC) \gtreqless n(\triangle ADE)$ according as

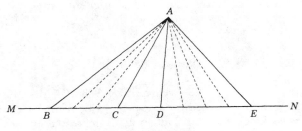

FIGURE A.8

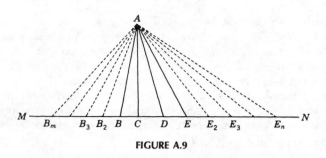

FIGURE A.9

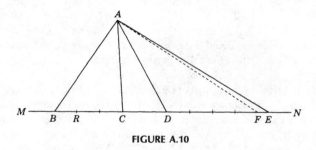

FIGURE A.10

$m(BC) \gtreqless n(DE)$, whence by the Eudoxian definition of proportion, $\triangle ABC : \triangle ADE = BC : DE$, and the proposition is established. No mention was made of commensurable and incommensurable quantities since the Eudoxian definition applies equally to both situations.

Many present-day high school textbooks advocate a proof of this theorem involving two cases, according as BC and DE are or are not commensurable. The commensurable case is handled as in the Pythagorean solution above, and simple limit notions are used to deal with the incommensurable case. Thus, suppose BC and DE are incommensurable. Divide BC into n equal parts, BR being one of the parts (see Figure A.10). On DE mark off a succession of segments equal to BR, finally arriving at a point F on DE such that $FE < BR$. By the commensurable case, already established, $\triangle ABC : \triangle ADF = BC : DF$. Now let $n \to \infty$. Then $DF \to DE$ and $\triangle ADF \to \triangle ADE$. Hence, in the limit, $\triangle ABC : \triangle ADE = BC : DE$. This approach uses the fact that any irrational number may be regarded as the limit of a sequence of rational numbers, an apporoach that was rigorously developed in modern times by Georg Cantor (1845–1918).

A.7 Nonstandard Analysis

The calculus, as developed in the seventeenth and eighteenth centuries, involved infinitely small quantities called *infinitesimals*, which, in the words of Johann Bernoulli, are so small that "if a quantity is increased or decreased by an infinitesimal, then that quantity is neither increased nor decreased." This seemingly contradictory situation offended mathematicians' feelings of rigor, and accordingly infinitesimals became banned from mathematics and their use replaced by epsilon-delta procedures and a theory of limits.

In 1960 Abraham Robinson succeeded in putting infinitesimals and their application to the calculus on a rigorous basis. This accomplishment, which greatly simplifies the calculus, has both pedagogical and theoretical values, and may prove to be one of the major mathematical achievements of the present century. It can be used to justify much of the intuitive "loose" reasoning of Euler and other early writers, and it can be employed in teaching a first course in the calculus.

In Section 7.2 it was shown that the real number system R constitutes a complete ordered field (actually, to within an isomorphism, the *only* complete ordered field). Robinson postulated the existence of a system $R^\star$ that is a proper ordered field extension of R. Whereas the elements of the field R are called *real numbers*, those of the field $R^\star$ are called *hyperreal numbers*, which contain the real numbers embedded among them.

One defines a hyperreal number x to be an *infinitesimal* if $|x| < r$ for all positive real numbers r, *finite* if $|x| < r$ for some real number r, and *infinite* if $|x| > r$ for all real numbers r. The only real infinitesimal is 0; all others are nonreal hyperreal numbers.

Two hyperreal numbers are said to be *infinitely close* if their difference is an infinitesimal. It can be shown that every finite hyperreal number x is infinitely close to a unique real number r; this unique real number r is called the *standard part* of the hyperreal number x and the relation is denoted symbolically by $r = \text{st}(x)$. If x is infinite, $\text{st}(x)$ is undefined.

To develop the calculus by the method of infinitesimals (or, as it is more commonly called, by nonstandard analysis), Robinson needs two further postulates, which we lack space to go into here. At any rate, the following definitions are formulated:

S is said to be the *slope* of a real function f at a real point a if

$$S = \text{st}\left(\frac{f(a + \Delta x) - f(a)}{\Delta x}\right)$$

for every nonzero infinitesimal Δx.

The *derivative* of a real function f is the real function f' such that

$$f'(x) = \text{slope of } f \text{ at } x \text{ if it exists,}$$

$$f'(x) \text{ is undefined otherwise.}$$

A calculus textbook based on the method of infinitesimals, designed for beginning college students, was written in 1976 by H. Jerome Keisler.[7] More advanced studies of the infinitesimal approach have been written.[8]

---

[7] H. J. Keisler [1].

[8] See A. Robinson; C. C. Chang and H. J. Keisler; H. J. Keisler [2]; and K. Stroyan.

A.8 The Axiom of Choice

It will be recalled (see Problem 5.5.11 or Section 6.2) that a set S of elements $a, b,$ $c, \ldots$ is said to be *simply ordered* with respect to a dyadic relation "$<$" (which we may read "precedes") if the following three postulates are satisfied:

O1. *If* $a \neq b$, *then either* $a < b$ *or* $b < a$.

O2. *If* $a < b$, *then* $a \neq b$.

O3. *If* $a < b$ *and* $b < c$, *then* $a < c$.

The set S is said to be *well-ordered* with respect to the relation "$<$" if the following fourth postulate also holds:

O4. *If* T *is any nonempty subset of* S, *then there exists an element* a *of* T *such that if* b *is any other element of* T, *then* $a < b$.

In short, a set S is well-ordered with respect to a dyadic relation "$<$" if it is simply ordered with respect to "$<$" and if every nonempty subset of S has a first element.

An obvious example of a well-ordered set is the set of all natural numbers, 1, 2, 3, $\ldots$, where "$<$" means "is less than" in the ordinary sense. For a more recondite example, consider the natural numbers, 1, 2, 3, $\ldots$, where "$<$" is defined as follows: (1) if a is odd and b is even, then $a < b$; (2) if a and b are both odd or both even, then $a < b$ denotes the natural order. Schematically, the order of the natural numbers in this second example is

$$1, 3, 5, 7, \ldots; 2, 4, 6, 8, \ldots,$$

and it is an easy matter to show that all four order postulates are satisfied.

Now there is a surprising theorem concerning well-ordering that was proved by E. Zermelo in 1904 and which may be stated as follows:

Zermelo's Well-ordering Theorem *If* S *is any set whatever, there exists a dyadic relation "$<$" for the set* S *with respect to which* S *is well-ordered.*[9]

The reaction of many mathematicians to Zermelo's theorem was that there must be something wrong with the proof; the theorem seemed too incredible to be accepted. Consider, for example, the set S of all real numbers x such that $0 \leq x \leq 1$. If we take "$<$" to mean "is less than" in the ordinary sense, then this set of real numbers is not well ordered. To see this, we have only to let T be the subset obtained from S by deleting the single number 0. Since there is no first real number greater than 0, this subset T has no first element, and Postulate O4 fails to hold. Now S is a nondenumerable set of cardinal number c. No one has ever been able to effect a well-ordering of such a set, much less a well-ordering of a set of higher cardinal number. Consequently, many mathematicians firmly felt that no definition of "$<$" can exist with respect to which such a set would be well-ordered. Not so, however, according to Zermelo's theorem. Even if we have a set S of cardinal number higher than c—say, the set of all single-valued

---

[9] This theorem had been conjectured by Cantor as early as 1883.

functions $f(x)$ defined on the interval $0 \leqq x \leqq 1$—Zermelo's theorem applies. There exists a definition of "<" with respect to which the set is well-ordered! This, if true, is very remarkable.

In the search for a flaw in Zermelo's proof, E. Borel discovered that Zermelo had based his demonstration on a seemingly obvious principle that mathematicians had been using for a long time but that no one had ever considered listing explicitly as an assumption. This principle may be stated as follows:

Zermelo's Postulate or the Axiom of Choice *If a set* S *is divided into a collection of mutually disjoint nonempty subsets* A, B, C, . . . , *there exists at least one set* R *which has as its elements exactly one element from each of the subsets* A, B, C, ...

Since the set R can evidently be formed if one may choose an element from each of the subsets $A, B, C, \ldots$ of S, the principle asserts that such a choice may be made; this accounts for the name, "axiom of choice."

To illustrate the meaning of the principle, let S be the set of all people residing in the United States, and let $A, B, C, \ldots$ be the subsets of these people residing in the various fifty states, where we assume that each resident of the United States lives in one and only one state. Then the axiom of choice claims that there exists at least one set R consisting of one and only one resident of each state.

Borel pointed out that not only is Zermelo's theorem based on the axiom of choice but that it is actually equivalent to the axiom of choice. For we can easily show that if we admit that every set possesses a well-ordering, then the axiom of choice must follow. Thus, let S be any set that is divided into a collection of mutually disjoint subsets $A, B, C, \ldots$. Granting Zermelo's theorem, there exists a definition of "<" with respect to which S is well-ordered. By Postulate O4, each of the subsets $A, B, C, \ldots$ of S possesses a first element. We now establish the axiom of choice by choosing for R the set of all these first elements. Accepting Zermelo's proof of his theorem from the axiom of choice, it follows that the theorem and the axiom are equivalent.

The publication of the papers by Zermelo and Borel calling attention to the above facts touched off a spirited controversy concerning the hitherto harmless axiom of choice. Sharp differences of opinion on the matter were expressed by many eminent mathematicians. Subsequent discovery of further consequences of the axiom of choice has not altered the situation, and today one finds attitudes toward the axiom of choice ranging from complete acceptance, through various degrees of skepticism, to total rejection. Modern researchers in topology apparently assume the axiom with no hesitation, for there seems little evidence that any significant portion of topology can be derived without its use. In algebra, the matter is quite different. Though some operations in algebra are quite awkward without the axiom of choice, so much can be accomplished without it that algebraists are inclined to proceed as far as possible without its use. A great deal of analysis can be established without the use of the axiom of choice, but when one reaches measure theory and those portions of modern analysis that rely extensively on topological ideas, its evasion becomes difficult if not impossible. Today, when anyone publishes a proof in which the axiom of choice is used, it is customary to call attention to the fact. There are a number of famous theorems of mathematics, particularly in analysis and in set theory, for which proofs avoiding the use of the axiom of choice are zealously sought.

Much of the objection to the axiom of choice lies in one's conception of mathematical existence. The axiom asserts the existence of a certain set R but says nothing about ways of finding R, or even that it is possible to find it; the assertion is merely that set R exists. Now some mathematicians, among whom are the members of the intuitionist school of philosophy of mathematics, deny the mathematical existence of a set if there is no known way of ascertaining the members of the set. From this point of view, a set is a fit subject for mathematical investigation only if a rule of construction is given that yields the members of the set, and any statement dealing with a set whose claim to existence is not validated by this uncompromising criterion of construction is not acceptable as a mathematical statement.

It should now be clear why it is desirable to label those theorems whose proofs employ the axiom of choice. For some mathematicians such theorems have not been established at all, while even for those who accept the proofs, the question remains as to whether the theorems can perhaps be proved without the use of the axiom of choice and thus be made acceptable to their more particular colleagues of the other school.

In order further to clarify some of the above remarks, consider the following often-quoted example, the original version of which was given by Bertrand Russell. In an infinite collection of pairs of shoes, no axiom of choice is needed to establish the existence of a set of shoes containing exactly one shoe from each pair; such a set may be effectively constructed by the simple rule of selecting all right shoes from the collection. But suppose we have an infinite collection of pairs of socks, all socks alike as to size, color, etc., and thus completely indistinguishable from one another. Does there exist a set containing exactly one sock from each pair? If we are to accept such a set, we must do so by the axiom of choice, for now there apparently is no possible rule by which the desired set may be constructed.

As a more mathematical example, consider the following. Let S be the set of all real numbers x such that $0 \leq x \leq 1$. Decompose S into subsets by putting into the same subset all members of S that differ from one another by a rational number. We then have a collection of subsets of S that are mutually disjoint and nonempty. Without the aid of the axiom of choice there is no way of demonstrating the existence of a set R containing exactly one number from each of these subsets. Such a set R exists only by special sanction.

We thus see that if we can constructively define the elements of set R in a perfectly definite way, we can avoid the axiom of choice; if we cannot so define the elements of R, we must have recourse to the axiom of choice.

Even one whose conception of mathematical existence does not seem to clash with the axiom of choice can easily develop considerable skepticism with regard to the axiom, for the axiom is remarkably far-reaching, and with its aid some very startling theorems can be deduced. As an example, let us cite an astounding result obtained in 1924 by S. Banach and A. Tarski.

The Banach-Tarski Paradox *In any Euclidean space of dimension* $n > 2$, *any two arbitrary bounded sets containing interior points are equivalent by finite decomposition.*

Without attempting to explain the meanings of all the technical terms in this statement, we shall merely illustrate by an example. Consider two solid spheres, P and S, where P is the size of a pea and S is the size of the sun. Then, according to the statement, it is possible to decompose the set of points making up P into a finite number of mutually disjoint subsets such that, by ordinary rigid motions (translations and rotations), these sets may be made to fill out the entire sphere S. In the notation of set theory this means that we have subsets $P_1, \ldots, P_n$ of P and subsets $S_1, \ldots, S_n$ of S such that

$$P = P_1 \cup P_2 \cup \cdots \cup P_n, \qquad P_i \cap P_j = \varnothing \quad \text{for} \quad i \neq j,$$

$$S = S_1 \cup S_2 \cup \cdots \cup S_n, \qquad S_i \cap S_j = \varnothing \quad \text{for} \quad i \neq j,$$

and P_i is congruent to S_i for each i. Surely such a state of affairs must appear astonishing even to the most sophisticated mathematician.

For many years a great open problem of set theory has concerned the possibility of deriving the axiom of choice from the postulates of certain existing postulate sets of set theory. Though the derivation of the axiom is easily accomplished for the case where the set S (in our statement of the axiom of choice) is divided into a *finite* collection of mutually disjoint nonempty subsets, no one as yet has been able to derive the axiom for the case where S is divided into an *infinite* collection of such subsets, and the feeling is that this latter task is impossible. Kurt Gödel proved, however, about 1940, that if the concerned postulate sets for set theory are consistent, then the addition of the axiom of choice cannot lead to any contradictions.

A.9 A Note on Gödel's Incompleteness Theorem

In 1931 the journal *Monatshefte für Mathematik und Physik* published a paper entitled, "Über formal unentscheibare Sätze der Principia Mathematica und verwandter Systeme" ("On formally undecidable propositions of Principia mathematica and related systems"). The author of the paper was a twenty-five-year-old Austrian mathematician and logician named Kurt Gödel, who was, at the time, at the University of Vienna. When the paper appeared, it received only scattered and scant attention, for it concerned itself with a highly specialized area of study that had not yet attracted many researchers and used a type of proof that was so technically novel as to be incomprehensible to most readers. Within a very few years, however, the paper became widely recognized professionally as one of the truly epoch-making contributions to the foundations of mathematics and logic. It led, in 1938, to the appointment of the author as a permanent member of the Institute for Advanced Study at Princeton and, in 1952, to the rather rare event of a mathematician receiving an honorary degree from a top-flight American university—Harvard University.

Gödel's paper revealed certain unforeseen limitations in axiomatic procedure. In particular, it accomplished the following: (1) It upset a strong belief that all the important areas of mathematics can be completely axiomatized. (2) It annihilated all hopes of establishing the inner consistency of mathematics along the lines that had been envisioned by David Hilbert. (3) It led to a reappraisal, not yet completed, of certain widely held philosophies of mathematics. (4) It introduced into foundational studies a new, powerful, and fertile technique of analysis that has suggested and initiated many new avenues of investigation.

This note comments on the above four points, but because of the advanced nature of the material and methods involved, the discussion is brief and skirts the formidable technicalities of the subject.

1. Following the development of the concept of formal axiomatics by David Hilbert and others in the early part of the twentieth century, the axiomatic method was vigorously exploited. A number of new and old branches of mathematics were supplied with sets of axioms that were thought to be adequate for the settlement of any question involving the primitive terms of those branches of mathematics; that is, the branches of mathematics were supplied with what were thought to be complete sets of axioms. For example, it was thought that the Peano axiom set for the natural number system was complete, or, if not complete, that it could surely be made so by the addition of one or more further axioms. This belief was shattered by Gödel's paper, for in that paper Gödel proved the following theorem:

Gödel's First Theorem   *For any consistent formal system* L *that contains the natural number system* (such as that of *Principia mathematica*, for example), *there are undecidable propositions in* L; *that is, there are propositions* F *in* L *such that neither* F *nor not-*F *is provable in* L.

It follows that any postulate set for the natural number system must, if consistent, be incomplete. In other words, no matter what consistent postulate set one might adopt for the natural number system, there will be statements *F* about the natural numbers such that neither *F* nor not-*F* is provable from the postulates.

There are many famous conjectures in number theory that, in spite of extended and strenuous effort, have never been established or refuted. Among these are, for example, the Goldbach conjecture (that every even integer greater than 2 can be expressed as the sum of two primes) and the conjecture known as Fermat's last theorem (that there do not exist positive integers x, y, z such that $x^n + y^n = z^n$ when n is a positive integer greater than 2). It could be that the statements about natural numbers involved in these conjectures are among those propositions that are undecidable in, say, the Peano development of the natural number system, and this could be the reason no one has been able to establish or refute the conjectures. It would surely be some gain if of a given statement about the natural numbers one could at least determine whether or not it is a provable proposition of the system. But here, too, the outlook is melancholy, for in 1936 Alonzo Church[10] proved the following theorem:

Church's Theorem   *For any consistent formal system* L *containing the natural number system, there exists no effective method of deciding which propositions of* L *are provable in* L.

2. We have seen, in our discussion of the formalist school of philosophy of mathematics in Chapter 9, that David Hilbert attempted to establish the inner

---

[10] See A. Church [7].

consistency of classical mathematics. He had hoped to accomplish this by showing that the permitted rules of logical procedure (which are, of course, part and parcel of the formal development) could never lead one to deduce two contradictory propositions F and not-F. Though Hilbert was able to illustrate with certain simple systems what he hoped to do for the more complicated system of all classical mathematics, he was unable to carry out his consistency proof for the latter system. This was shown by Gödel to be inevitable, for in his famous paper he also managed to prove the following theorem:

Gödel's Second Theorem *For any consistent formal system* L *which contains the natural number system, the consistency of* L *cannot be proved in* L.

It follows that among the undecidable problems in L, that of the consistency of L is one. This shattered Hilbert's hopes as originally envisioned, and it now seems that the internal consistency of classical mathematics cannot be attained unless one adopts principles of reasoning of such complexity that the internal consistency of these principles is as open to doubt as that of classical mathematics itself.

3. Gödel's two theorems have shown that no complete axiomatic development of certain important sectors of mathematics is attainable, and that no truly impeccable guarantee can be given that certain important sectors of mathematics are free of internal contradiction. These are severe limitations of the axiomatic method, and they point up that the processes of mathematical proof may not, indeed probably do not, coincide with formal axiomatic procedures. No prior limits can be imposed on the inventiveness of mathematicians in devising new procedures of proof. Apparently human intellectual resources cannot be fully formalized, and new principles of proof await discovery and invention. All this has obvious repercussions in any discussion of a philosophy of mathematics, and shows that some presently widely held philosophies of mathematics must be revamped or scrapped.

4. Though the proofs of Gödel's theorems (as supplied by Gödel himself, or later by others, like Barkley Rosser and S. C. Kleene) are too technical to be given here, something should be said of Gödel's remarkable device for "arithmetizing" his treatment. The device has proven to be very powerful and surprisingly applicable, and it somewhat compares, in its area of applicability, with the fertile method of Descartes in geometry.

Gödel's device lies in a method of numbering his primitive symbols, his formulas, and his proofs. In any completely formalized system L, there are, first of all, basic symbols in terms of which propositions and proofs are expressed. For instance, in Gödel's system L, appear (among others) the symbols

$$f, \sim, (,),$$

and a denumerably infinite set of variables in each of a denumerably infinite set of types. Gödel chose to assign the numbers 3, 5, 11, and 13, respectively, to the above symbols, and

$$17^n, 19^n, 23^n, 29^n, \ldots$$

to the variables of type n, where 17, 19, 23, 29, . . . is the sequence of consecutive primes after the prime 13. We shall call these numbers, assigned to the basic symbols, *Gödel numbers*.

Now a proposition in L, expressed in the symbolism of L, is nothing but a finite ordered sequence of basic symbols of L. Such sequences are called *formulas*, and a unique number is assigned to each formula as follows: Let $n(1)$, $n(2)$, ..., $n(s)$ be the Gödel numbers of the symbols of a formula F in the order in which they occur in F, and let $p_1, p_2, ..., p_s$ be the first s consecutive primes, starting with $p_1 = 2$. Then the number assigned to the formula F, which we will call the *Gödel number* of F, is the product

$$p_1^{n(1)} p_2^{n(2)} \cdots p_s^{n(s)}.$$

For example, consider the formula (or part of a formula)

$$\sim (x(fy)),$$

where x and y are variables of types 2 and 1, respectively, possessing the Gödel numbers 17^2 and 17. Then the Gödel numbers of the consecutive symbols of the formula are 5, 11, 289, 11, 3, 17, 13, 13, whence the Gödel number of the formula itself is the product

$$2^5 3^{11} 5^{289} 7^{11} 11^3 13^{17} 17^{13} 19^{13}.$$

It follows that for every proposition, or formula, there is a unique Gödel number. Also, given the Gödel number of a formula, the formula itself can be regained. All one has to do is to factor the number of the formula into its unique product of prime factors. Then the number of 2's occurring in the factorization is the Gödel number of the first symbol of the formula, the number of 3's occurring in the factorization is the Gödel number of the second symbol of the formula, the number of 5's occurring in the factorization is the Gödel number of the third symbol in the formula, and so on.

Finally, a proof is merely a finite ordered sequence of formulas, and a device can be used to assign a Gödel number to a *proof* similar to that used to assign a Gödel number to a formula. Suppose, for example, that a proof is made up of the ordered sequence of formulas $F_1, F_2, ..., F_t$. Let $f(i)$ be the Gödel number of formula F_i. Then the Gödel number of the proof will be the product

$$p_1^{f(1)} p_2^{f(2)} \cdots p_t^{f(t)},$$

where $p_1, p_2, ..., p_t$ is the sequence of the first t consecutive primes, starting with $p_1 = 2$. As before, given the Gödel number of a proof, the steps, or consecutive formulas, of the proof can be regained by factorization.

A very interesting and important feature of Gödel's device is that it permits metamathematical statements—that is, statements about the formal system L—to be "translated" into statements about numbers. For when Gödel numbers are assigned to the formulas of L, then a statement A about these formulas can be replaced by a statement B about the associated Gödel numbers of the formulas, such that statement B is true if and only if statement A is true. For example, suppose statement A says, "Formula F_1 consists of formula F_2 with some more symbols added at the end." For statement B we may then take, "The Gödel number of F_1 is a factor of the Gödel number of F_2." Now it happens that an important class of these B statements can themselves be expressed as formulas in the system L, with their own Gödel numbers, and the formal system L can, to this extent, "talk about itself." It is this ability to arithmetize the metamathematics of the formal system L that permitted Gödel to prove his two remarkable theorems, which are, of course, metamathematical statements about his system L.

Before closing this section of the appendix, we will give a simple application of the Gödel numbering device. Let us use the device to give another proof of the fact that the set of all rational numbers is denumerable. To this end, we first note that any rational number can be uniquely written in the form $(-1)^n p/q$, where n, p, q are nonnegative integers chosen to be as small as possible and $q \neq 0$. As examples, we have

$$-18/60 = (-1)^1 3/10 \quad \text{and} \quad 8/12 = (-1)^0 2/3.$$

Now set up the correspondence

$$(-1)^n p/q \leftrightarrow 2^n 3^p 5^q,$$

and call the latter number the *Gödel index* of the rational number $(-1)^n p/q$. As an example, the Gödel index of $-3/4$ is $2^1 3^3 5^4 = 33{,}750$. There is clearly a one-to-one correspondence between the rational numbers and their Gödel indices. The enumeration of the rational numbers is now obtained by listing the rational numbers in the order of magnitude of their Gödel indices.

BIBLIOGRAPHY

AGNEW, R. P. *Differential Equations*. New York: McGraw-Hill, 1942.

ALLMAN, G. J. *Greek Geometry from Thales to Euclid*. Dublin: University Press, 1889.

AMBROSE, A. "Finitism in Mathematics." *Mind* 44 (1935): 186–203, 317–340.

APOSTLE, H. G. *Aristotle's Philosophy of Mathematics*. Chicago: University of Chicago Press, 1952.

ARCHIMEDES. *Geometrical Solutions Derived from Mechanics*. Trans. by J. L. Heiberg. Chicago: Open Court, 1909.

ARISTOTLE. *The Works of Aristotle*. Trans. under the editorship of W. D. Ross. 11 vols. Oxford: Clarendon Press, 1931.

ARNOLD, B. H. *Logic and Boolean Algebra*. Englewood Cliffs, N.J.: Prentice-Hall, 1962.

ASHTON, J. "Mathematicians and the Mysterious Universe." *Thought* 6 (1931): 258–274.

BALLANTINE, J. P. "What Is a Calculus?" *American Mathematical Monthly* 29 (1922): 213–217.

BARANKIN, E. W. "Heat Flow and Non-Euclidean Geometry." *American Mathematical Monthly* 49 (1942): 4–14.

BARKER, S. F. *Philosophy of Mathematics*. Englewood Cliffs, N.J.: Prentice-Hall, 1964.

BEGLE, E. G. *Introductory Calculus with Analytic Geometry*. New York: Holt, Rinehart & Winston, 1954.

BELL, E. T. [1]. "On Proofs by Mathematical Induction." *American Mathematical Monthly* 27 (1920): 413–415.

——— [2]. "The Place of Rigor in Mathematics." *American Mathematical Monthly* 41 (1934): 599–607.

——— [3]. *The Development of Mathematics*. 2nd ed. New York: McGraw-Hill, 1945.

——— [4]. *Mathematics, Queen and Servant of Science*. New York: McGraw-Hill, 1951.

BELTRAMI, E. "Saggio di interpretazione della geometria non-Euclidea." *Giornale di Matematiche* 6 (1868): 74–105.

BENACERRAF, P., and H. PUTNAM, eds. *Philosophy of Mathematics: Selected Readings.* Englewood Cliffs, N.J.: Prentice-Hall, 1964.

BERKELEY, E. C. "Boolean Algebra and Applications to Insurance." *The Record, American Institute of Actuaries* 26, Part II (1937): 373–414.

BERNAYS, P. [1]. "Axiomatische Untersuchung des Aussagen-Kalkuls der Principia Mathematica." *Mathematische Zeitschrift* 25 (1926): 305–320.

——— [2]. "A System of Axiomatic Set Theory." *Journal of Symbolic Logic* 2 (1937): 65–77; 6 (1941): 1–17; 7 (1942): 65–89, 135–145; 8 (1943): 89–106; 13 (1948): 65–79.

——— [3]. *Axiomatic Set Theory.* Amsterdam: North-Holland, 1958.

BERNSTEIN, B. A. "Postulates for Boolean Algebras Involving the Operation of Complete Disjunction." *Annals of Mathematics* 37 (1936): 317–325.

BETH, E. W. [1]. *The Foundations of Mathematics.* Amsterdam: North-Holland, 1959.

——— [2]. *Formal Methods.* Dordrecht, Holland: Reidel, 1962.

BIRKHOFF, G. [1]. "What Is a Lattice?" *American Mathematical Monthly* 50 (1943): 484–487.

——— [2]. *Lattice Theory.* Rev. ed. New York: American Mathematical Society Colloquium, Vol. 25, 1948.

———, and S. MACLANE. *A Survey of Modern Algebra.* 3rd ed. New York: Macmillan, 1965.

———, and J. VON NEUMANN. "The Logic of Quantum Mechanics." *Annals of Mathematics* 37 (1936): 823–843.

BIRKHOFF, G. D. [1]. "A Set of Postulates for Plane Geometry, Based on Scale and Protractor." *Annals of Mathematics* 33 (1932): 329–345.

——— [2]. "Principle of Sufficient Reason." *Rice Institute Pamphlets,* No. 28 (1941): 24–50.

——— [3]. "The Mathematical Nature of Physical Theories." *American Scientist* 31 (1943): 282–310.

———, and R. BEATLEY. *Basic Geometry.* Chicago: Scott, Foresman, 1940; New York: Chelsea, 1959.

BLACK, M. [1]. "The Relevance of Mathematical Philosophy to the Teaching of Mathematics." *Mathematical Gazette* 22 (1938): 149–163.

——— [2]. *The Nature of Mathematics: A Critical Survey.* 2nd ed. New York: Humanities Press, 1950.

BLANCHÉ, R. *Axiomatics.* Trans. by G. B. Kleene. London: Routledge & Kegan Paul, 1962.

BLISS, G. A. [1]. "Mathematical Interpretation of Geometrical and Physical Phenomena." *American Mathematical Monthly* 40 (1933): 472–480.

——— [2]. "The Function Concept and the Fundamental Notions of the Calculus." *See* J. W. A. YOUNG.

BLUMBERG, H. [1]. "On the Technique of Generalization." *American Mathematical Monthly* 47 (1940): 451–462.

——— [2]. "On the Change of Form." *American Mathematical Monthly* 53 (1946): 181–192.

BLUMENTHAL, L. M. [1]. "A Paradox, a Paradox, a Most Ingenious Paradox." *American Mathematical Monthly* 47 (1940): 346–353.

——— [2]. *A Modern View of Geometry.* San Francisco: Freeman, 1961.

BOCHER, M. [1]. "Fundamental Conceptions and Methods of Mathematics." *Bulletin of the American Mathematical Society* 2 (1904): 115–135.

——— [2]. "Brouwer's Contribution to the Foundations of Mathematics." *Bulletin of the American Mathematical Society* 30 (1924): 31–40.

BOČVAR, D. A. "On a Three-Valued Calculus and Its Application to the Analysis of Contradictions." *Matematiceskij sbornik* 4 (1939): 287–308.

BOLYAI, J. "The Science of Absolute Space." Trans. by G. B. Halsted. *See* BONOLA.

BONOLA, R. *Non-Euclidean Geometry: A Critical and Historical Study of Its Developments.* Trans. by H. S. Carslaw. Containing BOLYAI and LOBACHEVSKI. New York: Dover, 1955.

BOOLE, G. *An Investigation of the Laws of Thought.* New York: Dover, 1951.

BORSUK, K., and W. SZMIELEW. *Foundations of Geometry.* Trans. by E. Marquit. Amsterdam: North-Holland, 1960.

BOURBAKI, N. "The Architecture of Mathematics." Trans. by A. Dresden. *American Mathematical Monthly* 57 (1950): 221–232.

BOYER, C. B. *The Concepts of the Calculus: A Critical and Historical Discussion of the Derivative and the Integral.* New York: Hafner, 1949.

BRADIS, V. M., V. L. MINKOVSKII, and A. K. KHARCHEVA. *Lapses in Mathematical Reasoning.* New York: Macmillan, 1959.

BROAD, C. D. "Kant's Theory of Mathematical and Philosophical Reasoning." *Proceedings of the Aristotelian Society* 42 (1941): 1–24.

BROSSARD, R. "Birkhoff's Axioms for Space Geometry." *American Mathematical Monthly* 7 (1964): 593–606.

BROUWER, L. E. J. "Intuitionism and Formalism." Trans. by A. Dresden. *Bulletin of the American Mathematical Society* 30 (1913–1914): 81–96.

BRUCK, R. H. "Recent Advances in the Foundations of Euclidean Geometry." Herbert Ellsworth Slaught Memorial Paper No. 4 (*Contributions to Geometry*), pp. 2–17. Buffalo, N.Y.: Mathematical Association of America, 1955.

BRUMFIEL, C. F., R. E. EICHOLZ, and M. E. SHANKS. *Geometry.* Reading, Mass.: Addison-Wesley, 1960.

BRUNSCHIVCE, L. *Le rôle du pythagorisme dans l'évolution des idées.* Paris: Hermann, 1937.

BUCHANAN, S. *Poetry and Mathematics.* New York: Day, 1929.

BURALI-FORTI, C. "Una questione sui numeri transfiniti." *Rendiconti del Circolo Matematico di Pálermo* 11 (1897): 154–164.

BURINGTON, R. S. "On the Nature of Applied Mathematics." *American Mathematical Monthly* 41 (1949): 221–242.

BYRNE, L. "Two Brief Formulations of Boolean Algebra." *Bulletin of the American Mathematical Society* 52 (1946): 269–272.

CAJORI, F. [1]. *A History of the Conceptions of Limits and Fluxions in Great Britain, from Newton to Woodhouse.* Chicago: Open Court, 1919.

———[2]. *A History of Mathematics.* 2nd ed. New York: Macmillan, 1919.

———[3]. "Grafting of the Theory of Limits on the Calculus of Leibniz." *American Mathematical Monthly* 30 (1923): 223–234.

———[4]. "Origins of Fourth Dimension Concepts." *American Mathematical Monthly* 33 (1926): 397–406.

CAMPBELL, A. D. "A Note on the Sources of Mathematical Reality." *American Mathematical Monthly* 31 (1927): 263–265.

CAMPBELL, N. *What Is Science?* New York: Dover, 1952.

CANTOR, G. *Contributions to the Founding of the Theory of Transfinite Numbers.* Trans. by P. E. B. Jourdain. New York: Dover, 1915.

CARMICHAEL, R. D. *The Logic of Discovery.* Chicago: Open Court, 1930.

CARNAP, R. [1]. *The Logical Syntax of Language.* New York: Harcourt, Brace & World, 1937.

———[2]. "Foundations of Logic and Mathematics." *International Encyclopedia of Unified Science*, Vol. 1, no. 3. Chicago: University of Chicago Press, 1939.

CARSLAW, H. S. *Elements of Non-Euclidean Geometry and Trigonometry.* London: Longmans, Green, 1916.

CARTWRIGHT, M. L. *The Mathematical Mind.* New York: Oxford University Press, 1955.

CAYLEY, A. "A Memoire on the Theory of Matrices." *Transactions of the London Philosophical Society* 148 (1858): 17–37.

CHACE, A. B., L. S. BULL, H. P. MANNING, and R. C. ARCHIBALD, eds., *The Rhind Mathematical Papyrus.* Buffalo, N.Y.: Mathematical Association of America, 1927–1929.

CHANG, C. C., and H. J. KEISLER. *Model Theory.* Amsterdam and London: North-Holland, 1973.

CHURCH, A. [1]. "Alternatives to Zermelo's Assumption." *Transactions of the American Mathematical Society* 29 (1927): 178–208.

———[2]. "On the Law of Excluded Middle." *Bulletin of the American Mathematical Society* 34 (1928): 75–78.

———[3]. "A set of Postulates for the Foundation of Logic." *Annals of Mathematics* 33 (1932): 346–366.

———[4]. "A Set of Postulates for the Foundation of Logic (Second Paper)." *Annals of Mathematics* 34 (1933): 839–864.

———[5]. "The Richard Paradox." *American Mathematical Monthly* 41 (1934): 356–361.

———[6]. "A Proof of Freedom from Contradiction." *National Academy of Sciences* 21 (1935): 275–281.

———[7]. "An Unsolvable Problem of Elementary Number Theory." *American Journal of Mathematics* 58 (1936): 345–363.

CLARK, J. T. "Contemporary Science and Deductive Methodology." *Proceedings of the American Catholic Philosophical Association.* Washington, D.C.: Catholic University of America, 1952.

CLIFFORD, W. K. *The Common Sense of the Exact Sciences.* New York: Dover, 1955.

COHEN, P. J. "The Independence of the Continuum Hypothesis." *Proceedings of the National Academy of Science* 50 (1963): 1143–1148; 51 (1964): 105–110.

CONFERENCE BOARD OF THE MATHEMATICAL SCIENCES. *The Role of Axiomatics and Problem Solving in Mathematics.* Boston: Ginn, 1966.

COURANT, R., and H. ROBBINS. *What Is Mathematics? An Elementary Approach to Ideas and Methods.* New York: Oxford University Press, 1941.

COURT, N. A. [1]. "Geometry and Experience." *Scientific Monthly* 60 (1945): 63–66.

———[2]. "Mathematics in the History of Civilization." *Mathematics Teacher* 41 (1948): 104–111.

COXETER, H. S. M. *The Real Projective Plane.* New York: McGraw-Hill, 1949.

CURRY, H. B. [1]. "Some Aspects of the Problem of Mathematical Rigor." *Bulletin of the American Mathematical Society* 47 (1941): 221–241.

———[2]. *Outlines of a Formalist Philosophy of Mathematics.* Amsterdam: North-Holland, 1951.

DANTZIG, T. *Number, the Language of Science: A Critical Survey Written for the Cultured Non-Mathematician.* 3rd ed., rev. and augmented. New York: Macmillan, 1946.

DARBON, A. *La philosophie des mathématiques: Etude sur la logistique de Russell.* Paris: Presses Universitaires, 1949.

DARKOW, M. D. "Interpretations of the Peano Postulates." *American Mathematical Monthly* 64 (1957): 207–271.

DAVIS, P. J., and R. HERSH. *The Mathematical Experience.* Boston: Houghton Mifflin, 1981.

DEDEKIND, R. [1]. *Stetigkeit und irrationale Zahlen*. Brunswick, Germany: Vieweg, 1872.

———[2]. *Was sind und was sollen die Zahlen?* Brunswick, Germany: Vieweg, 1888.

DE MORGAN, A. "On the Foundations of Algebra." *Cambridge Philosophical Transactions* VII (1841, 1842); VIII (1844, 1847).

DENBOW, C. H. "Postulates and Mathematics." *American Mathematical Monthly* 62 (1955): 233–236.

DESCARTES, R. *The Geometry*. Trans. by D. E. Smith and M. L. Latham. New York: Dover, 1954.

DE SUA, F. "Consistency and Completeness—A résumé." *American Mathematical Monthly* 63 (1956): 295–305.

DIANANDA, P. H. "Postulates for the Positive Integers." *American Mathematical Monthly* 69 (1962): 147–148.

DICKSON, L. E. [1]. "Why It Is Impossible to Trisect an Angle or to Construct a Regular Polygon of 7 or 9 Sides by Ruler and Compasses." *Mathematics Teacher* 14 (1921): 217–223.

———[2]. "Constructions with Ruler and Compasses: Regular Polygons." See J. W. A. YOUNG.

DODGE, C. W. *Numbers and Mathematics*. 2nd ed. Boston: Prindle, Weber & Schmidt, 1975.

DUBBEY, J. M. *Development of Mathematics*. London: Butterworth, 1970.

DUBISCH, R. [1]. *The Nature of Number: An Approach to Basic Ideas of Modern Mathematics*. New York: Ronald Press, 1952.

———[2]. "An Axiomatic Approach to the g.c.d." *American Mathematical Monthly* 69 (1962): 653.

DUHEM, P. "Dominique Soto et la scolastique parisienne." *Annales de la Faculté des Lettres de Bordeaux, Bulletin Hispanique* (1911): 454–467.

EINSTEIN, A. *The Method of Theoretical Physics*. New York: Oxford University Press, 1933.

———, and L. INFELD. *The Evolution of Physics: The Growth of Ideas from Early Concepts to Relativity and Quanta*. New York: Simon & Schuster, 1942.

EISENHART, L. P. *Coordinate Geometry*. Boston: Ginn, 1939.

ELLIS, D. "Notes on the Foundations of Lattice Theory." *Publicationes Mathematicae* 1 (1950): 205–208.

ELLIS, J. W. "Another Very Independent Axiom System." *American Mathematical Monthly* 68 (1961): 992.

EMCH, A. F. "Implication and Deducibility." *Journal of Symbolic Logic* 1 (1936): 26–35.

ENRIQUES, F. *The Historic Development of Logic: The Principles and Structure of Science in the Conception of Mathematical Thinkers*. Trans. by J. Rosenthal. New York: Holt, Rinehart & Winston, 1929.

EUCLID. *The Thirteen Books of Euclid's Elements*. 3 vols., 2nd ed., with an introduction and commentary by T. L. Heath. New York: Dover, 1956.

EVANS, H. P., and S. C. KLEENE. "A Postulational Basis for Probability." *American Mathematical Monthly* 46 (1939): 141–148.

EVES, H. [1]. *An Introduction to the History of Mathematics*. 6th ed. Philadelphia: Saunders College, 1990.

———[2]. *A Survey of Geometry*. Rev. ed. Boston: Allyn & Bacon, 1972.

———[3]. *In Mathematical Circles*. Boston: Prindle, Weber & Schmidt, 1969.

———, and V. E. HOGGATT. "Hyperbolic Trigonometry Derived from the Poincaré Model." *American Mathematical Monthly* 48 (1951): 469–474.

FEARNSIDE, W. W., and W. B. HOLTHER. *Fallacy, the Counterfeit of Argument*. Englewood Cliffs, N.J.: Prentice-Hall, 1959.

FEFERMAN, S. *The Number Systems, Foundations of Algebra and Analysis*. Reading, Mass.: Addison-Wesley, 1964.

FÉLIX, L. *The Modern Aspect of Mathematics*. Trans. by J. H. Hlavaty and F. H. Hlavaty. New York: Basic Books, 1960.

FIRESTONE, C. "A Semantical Approach to the Study of Mathematics." *Pentagon* (Spring 1943): 53–68.

FORDER, H. G. [1]. *The Foundations of Euclidean Geometry*. New York: Cambridge University Press, 1927.

———[2]. *The Calculus of Extension*. New York: Cambridge University Press, 1941.

FRAENKEL, A. A. [1]. *Einleitung in die Mengenlehre*. 3rd ed. Berlin: Verlag Julius Springer, 1928; New York: Dover, 1946.

———[2]. "Sur la notion d'existence dans les mathématiques." *L'Enseignement Mathématique* 34 (1935): 18–32.

———[3]. "Problems and Methods in Modern Mathematics." *Scripta Mathematica* 9 (1943): 5–18.

———[4]. "The Recent Controversies About the Foundation of Mathematics." *Scripta Mathematica* 13 (1947): 17–36.

———[5]. *Abstract Set Theory*. Amsterdam: North-Holland, 1953.

FREGE, G. [1]. "Class, Function, Concept, Relation." *Monist* 27 (1917): 114–127.

———[2]. *The Foundations of Arithmetic*. Oxford: Basil Blackwell & Mott, 1950.

FREUND, J. E. *A Modern Introduction to Mathematics*. Englewood Cliffs, N.J.: Prentice-Hall, 1956.

FRINK, O. "New Algebras of Logic." *American Mathematical Monthly* 45 (1938): 210–219.

GANS, D. "Axioms for Elliptic Geometry." *Canadian Journal of Mathematics* 4 (1952): 81–92.

———[2]. "Models of Projective and Euclidean Space." *American Mathematical Monthly* 65 (1958): 749–756.

GARDNER, M. *Logic Machines and Diagrams*. New York: McGraw-Hill, 1958.

GILMORE, P. C. "An Alternative to Set Theory." *American Mathematical Monthly* 67 (1960): 621–632.

GÖDEL, K. [1]. "Uber formal unentscheidbare Sätze der Principia Mathematica und verwandter Systeme I." *Monatshefte für Mathematik und Physik* 38 (1931): 173–198.

———[2]. *On Undecidable Propositions of Formal Mathematical Systems*. Princeton, N.J.: Princeton University Press, 1934.

———[3]. *The Consistency of the Continuum Hypothesis*. Annals of Mathematics Studies, No. 3. Princeton, N.J.: Princeton University Press, 1940.

———[4]. "What Is Cantor's Continuum Problem?" *American Mathematical Monthly* 54 (1947): 515–525.

GOFFMAN, C. *Real Functions*. Boston: Prindle, Weber & Schmidt, 1953.

GONSETH, F. [1]. *Les fondements des mathématiques: de la géométrie d'Euclide à la relativité générale et à l'intuitionisme*. Paris: Albert Blanchard, 1926.

———[2]. *Qu'est-ce que la logique?* Paris: Hermann, 1937.

———[3]. *Philosophie mathématique*. Paris: Hermann, 1939.

GOODMAN, N. D. "Mathematics as an Objective Science." *American Mathematical Monthly* 86 (1979): 540–551.

GOODSTEIN, R. L. *Boolean Algebra*. New York: Macmillan, 1963.

GRADSHTEIN, I. S. *Direct and Converse Theorems*. Trans. by T. Boddington. New York: Macmillan, 1963.

GRASSMANN, H. G. *Die Wissenschaft der extensiven Grösse oder die Ausdehnungslehre*. Leipzig, 1844; completed in 1862.

GREEN, L. C. "Maximum Uncertainty as a Simple Example of a Non-distributive Algebra." *American Mathematical Monthly* 55 (1948): 363–364.

GREGORY, D. F. "On the Real Nature of Symbolic Algebra." *Transactions of the Royal Society of Edinburgh* 14 (1840): 280.

GRELLING, K. [1]. *Mengenlehre*. Leipzig: Teubner, 1924.

——[2]. *Die Paradoxien der Mengenlehre*. Stuttgart: Math. Büchlein, 1925.

GUY, W. T. JR. "On Equivalence Relations." *American Mathematical Monthly* 62 (1955): 179–180.

HADAMARD, J. *An Essay on The Psychology of Invention in the Mathematical Field*. Princeton, N.J.: Princeton University Press, 1945.

HAHN, H. "Geometry and Intuition." *Scientific American* 190, No. 4 (1954): 84–91.

HALLDEN, S. "A Note Concerning the Paradoxes of Strict Implication and Lewis's System S1." *Journal of Symbolic Logic* 13 (1948): 138–139.

HALMOS, P. R. [1]. "The Foundations of Probability." *American Mathematical Monthly* 51 (1944): 493–510.

——[2]. "The Basic Concepts of Algebraic Logic." *American Mathematical Monthly* 63 (1956): 363–387.

HALSTED, G. B. [1]. *Rational Geometry*. New York: Wiley, 1904.

——[2]. *Girolamo Saccheri's Euclides Vindicatus*. Chicago: Open Court, 1920.

HAMILTON, W. R. [1]. *Lectures on Quaternions*. Dublin: 1853.

——[2]. *Elements of Quaternions*. London: 1866.

HANKEL, H. *Vorlesungen über die komplexen Zahlen und ihre Functionen*. Leipzig: 1867.

HARARY, F. "A Very Independent Axiom System." *American Mathematical Monthly* 68 (1961): 159–162.

HARDY, G. H. [1]. "Mathematical Proof." *Mind* 30 (1929): 1–25.

——[2]. *A Mathematician's Apology*. New York: Cambridge University Press, 1941.

HEATH, T. L. *The Thirteen Books of Euclid's Elements*. 3 vols., 2nd ed. New York: Dover, 1956.

HEDRICK, E. R. "The Function Concept in Elementary Teaching and in Advanced Mathematics." *American Mathematical Monthly* 45 (1938): 448–455.

HEIJENOORT, J. van (ed.). *From Frege to Gödel*. Cambridge, Mass.: Harvard University Press, 1967.

HELMER, O. "On the Theory of Axiom-Systems," *Analysis* 3 (1935): 1–11.

HEMPLE, C. G. [1]. "A Purely Topological Form of Non-Aristotelian Logic." *Journal of Symbolic Logic* 2 (1937): 97–112.

——[2]. "Geometry and Empirical Science." *American Mathematical Monthly* 52 (1945): 7–17.

——[3]. "On the Nature of Mathematical Truth." *American Mathematical Monthly* 52 (1945): 543–556.

HENKIN, L. "On Mathematical Induction." *American Mathematical Monthly* 67 (1960): 323–338.

HENLE, P. "The Independence of the Postulates of Logic." *Bulletin of the American Mathematical Society* 38 (1932): 409–414.

HEYTING, A. [1]. "Die intuitionistische Grundlegung der Mathematik." *Erkenntnis* 2 (1931): 106–115.

——[2]. *Intuitionism: An Introduction*. Amsterdam: North-Holland, 1956.

HILBERT, D. [1]. *The Foundations of Geometry*. 2nd ed. Trans. by E. J. Townsend. Chicago: Open Court, 1902.

——[2]. "On the Foundations of Logic and Arithmetic." *Monist* 15 (1905): 338–352.

——[3]. "Axiomatische Denken." *Mathematische Annalen* 78 (1918): 405–415.

——[4]. "Die Logischen Grundlagen der Mathematik." *Mathematische Annalen* 88 (1923): 151–165.

——[5]. *Grundlagen der Geometrie*. 7th ed. Leipzig: Teubner, 1930.

——[6]. *Grundlagen der Geometrie*. 8th ed. with revision and a supplement by P. Bernays. Stuttgart: Teubner, 1956.

———— [7]. *Grundlagen der Geometrie.* 9th ed. with revision and a supplement by P. Bernays. Stuttgart: Teubner, 1962.

———— [8]. *Foundations of Geometry.* 10th (1965) ed. of *Grundlagen der Geometrie,* revised and enlarged by P. Bernays, trans. by L. Unger. La Salle, Ill.: Open Court, 1971.

————, and W. ACKERMANN. *Principles of Mathematical Logic.* Trans. by L. M. Hammond, *et al.* New York: Chelsea, 1950.

————, and P. BERNAYS. *Grundlagen der Mathematik.* Berlin: Springer, 1934 (vol. I), 1939 (vol. II). Reprinted in lithoprint by Edward Brothers, Ann Arbor, 1944.

————, and S. COHN-VOSSEN. *Geometry and the Imagination.* Trans. by P. Nemenyi. New York: Chelsea, 1952.

HILDEBRANT, T. H. "A Simple Continuous Function with a Finite Derivative at No Point." *American Mathematical Monthly* 40 (1933): 547.

HOAR, R. S. "On Proofs by Mathematical Induction." *American Mathematical Monthly* 29 (1922): 162–164.

HOHN, F. E. [1]. "Some Mathematical Aspects of Switching." *American Mathematical Monthly* 62 (1955): 75–90.

———— [2]. *Applied Boolean Algebra.* New York: Macmillan, 1960.

HUNTINGTON, E. V. [1]. "Simplified Definition of a Group." *Bulletin of the American Mathematical Society* 8 (1902): 296–300.

———— [2]. "Sets of Independent Postulates for the Algebra of Logic." *Transactions of the American Mathematical Society* 5 (1904): 288–309.

———— [3]. "A Set of Postulates for Abstract Geometry, Expressed in Terms of the Simple Relation of Inclusion." *Mathematische Annalen* 73 (1913): 522–559.

———— [4]. "The Logical Skeleton of Elementary Dynamics." *American Mathematical Monthly* 24 (1917): 1–16.

———— [5]. "A New Set of Postulates for Betweenness, with Proof of Complete Independence." *Transactions of the American Mathematical Society* 26 (1924): 257–282.

———— [6]. "The Postulational Method in Mathematics." *American Mathematical Monthly* 41 (1934): 84–92.

———— [7]. "Postulates for Assertion, Conjunction, Negation, and Equality." *Proceedings of the American Academy of Arts and Sciences* 72 (1937): 1–44.

———— [8]. "The Method of Postulates." *Philosophy of Science* 4 (1937): 482–495.

———— [9]. *The Continuum and Other Types of Serial Order* 2nd ed. New York: Dover, 1955.

———— [10]. "The Fundamental Propositions of Algebra." *See* J. W. A. YOUNG.

IVINS, W. M. JR. *Art and Geometry.* Cambridge, Mass.: Harvard University Press, 1946.

JACOBSON, N. *Lectures in Abstract Algebra.* Vol. 1. *Basic Concepts.* Princeton, N.J.: Van Nostrand, 1951.

JOHNSTON, F. E. "The Postulational Treatment of Mathematics as Exemplified in the Theory of Groups." *American Scientist* 33 (1945): 39–48, 54.

JOHNSTON, L. S. "Another Form of the Russell Paradox." *American Mathematical Monthly* 47 (1940): 474.

JONES, B. W. *Elementary Concepts of Mathematics.* 2nd ed. New York: Macmillan, 1963.

KATTSOFF, L. O. [1]. "Postulational Methods, I." *Philosophy of Science* 2 (1935): 139–163.

———— [2]. "Postulational Techniques, III." *Philosophy of Science* 3 (1936): 375–417.

———— [3]. "Undefined Concepts in Postulate Sets." *Philosophical Review* 47 (1938): 293–300.

———— [4]. "Philosophy, Psychology, and Postulational Techniques." *Psychological Review* 46 (1939): 62–74.

————[5]. *A Philosophy of Mathematics.* Ames: Iowa State College Press, 1948.

————[6]. "The Independence of the Associative Law." *American Mathematical Monthly* 65 (1958): 620–622.

KEISLER, H. J. [1]. *Elementary Calculus.* Boston: Prindle, Weber & Schmidt, 1976.

————[2]. *Foundations of Infinitesimal Calculus.* Boston: Prindle, Weber & Schmidt, 1976.

KEMPNER, A. J. "Remarks on 'Unsolvable' Problems." *American Mathematical Monthly* 43 (1936): 467–473.

KERSHNER, R. B., and L. R. WILCOX. *The Anatomy of Mathematics.* New York: Ronald Press, 1950.

KEYSER, C. J. [1]. *Mathematical Philosophy: A Study of Fate and Freedom (Lectures for Educated Laymen).* New York: Dutton, 1922.

————[2]. *Mathematics and the Question of Cosmic Mind, with Other Essays.* Scripta Mathematica Library, No. 2. New York: Scripta Mathematica, 1935.

KLEENE, G. B. *Abstract Sets and Finite Ordinals.* New York: Pergamon Press, 1961.

KLEENE, S. C. [1]. "A Theory of Positive Integers in Formal Logic." *American Journal of Mathematics* 57 (1935): 153–173, 219–244.

————[2]. *Introduction to Metamathematics.* Princeton, N.J.: Van Nostrand, 1952.

————, and J. B. ROSSER. "The Inconsistency of Certain Formal Logics." *Annals of Mathematics* 36 (1935): 630–636.

KLEIN, F. "A Comparative Review of Recent Researches in Geometry." Trans. by M. W. Haskell. *Bulletin of the New York Mathematical Society* 2 (1892–1893): 215–249.

KLINE, J. R. "What Is the Jordan Curve Theorem?" *American Mathematical Monthly* 49 (1942): 281–286.

KLINE, M. [1]. *Mathematics: The Loss of Certainty.* New York: Oxford University Press, 1980.

————[2]. *Mathematics and the Search for Knowledge.* New York: Oxford University Press, 1985.

KÖRNER, S. *The Philosophy of Mathematics: An Introduction.* New York: Harper & Row, 1962.

KORZYBSKI, A. *Science and Sanity: An Introduction to Non-Aristotelian Systems and General Semantics.* 3rd ed. Lakeville, Conn.: International Non-Aristotelian Library, 1948.

KRAMER, E. E. [1]. *The Main Stream of Mathematics.* Greenwich, Conn.: Fawcett, 1964.

————[2]. *The Nature and Growth of Modern Mathematics.* New York: Hawthorn, 1978.

LAND, F. *The Language of Mathematics.* London: Murray, 1960.

LANDAU, E. G. H. *Foundations of Analysis: The Arithmetic of Whole, Rational, Irrational and Complex Numbers.* Trans. by F. Steinhardt. New York: Chelsea, 1951.

LANGER, R. E. "Alexandria—Shrine of Mathematics." *American Mathematical Monthly* 48 (1941): 109–125.

LANGFORD, C. H. "Analytic Completeness for Postulate Sets." *Proceedings of the London Mathematical Society* 25 (1926): 115–142.

LARGUIER, E. H. [1]. "Schools of Thought in Modern Mathematics." *Thought* 12 (1937): 225–240.

————[2]. "A Theory of Mathematical Reality." *Modern Schoolman* 16 (1939): 88–91.

————[3]. "Brouwerian Philosophy of Mathematics." *Scripta Mathematica* 7 (1940): 69–78.

————[4]. "Concerning Some Views on the Structure of Mathematics." *Thomist* 4 (1942): 431–445.

LASLEY, J. W. JR. "The Revolt Against Aristotle." *American Scientist* 30 (1942): 275–287.

LEFSCHETZ, S. "The Structure of Mathematics." *American Scientist* 38 (1950): 105–111.

LE LIONNAIS, F. *Les grandes courants de la pensée mathématique.* Seine: Imprimeries Bellenaud, 1948.

LENNES, N. J. "The Foundation of Arithmetic." *American Mathematical Monthly* 45 (1938): 70–75.

LEVI, H. *Elements of Algebra.* New York: Chelsea, 1954.

LIGHTSTONE, A. H. "A Remark Concerning the Definition of a Field." *Mathematics Magazine* 37 (1964): 12–13.

LOBACHEVSKI, N. "Geometrical Researches on the Theory of Parallels." Trans. by G. B. Halsted. See BONOLOA.

LUKASIEWICZ, J. *Elements of Mathematical Logic.* Trans. by O. Wojtasiewicz. New York: Macmillan, 1963.

LUNEBURG, R. K. *Mathematical Analysis of Binocular Vision.* Princeton, N.J.: Princeton University Press, published for the Dartmouth Eye Institute, 1947.

MACDUFFEE, C. C. *An Introduction to Abstract Algebra.* New York: Wiley, 1940.

MACH, E. *Space and Geometry.* Trans. by T. J. McCormack. Chicago: Open Court, 1943.

MACLANE, S. [1]. "Hilbert-Bernays on Proof-Theory." *Bulletin of the American Mathematical Society* 41 (1935): 162–165.

―――― [2]. "Metric Postulates for Plane Geometry." *American Mathematical Monthly* 66 (1959): 543–555.

MCTAGGERT, J. E. "Propositions Applicable to Themselves." *Mind* 32 (1923): 462–464.

MAZIARZ, E. A. *The Philosophy of Mathematics.* New York: Philosophical Library, 1950.

MENGER, K. [1]. "The New Logic." Trans. by H. B. Gottlieb and J. K. Senior, *Philosophy of Science* 4 (1937): 299–336.

―――― [2]. "What Is Dimension?" *American Mathematical Monthly* 50 (1943): 2–7.

―――― [3]. "Self-Dual Postulates in Projective Geometry." *American Mathematical Monthly* 55 (1948): 195.

MESERVE, B. E. [1]. *Fundamental Concepts of Algebra.* Reading, Mass.: Addison-Wesley, 1953.

―――― [2]. *Fundamental Concepts of Geometry.* Reading, Mass.: Addison-Wesley, 1955.

MEYER, B. *An Introduction to Axiomatic Systems.* Boston: Prindle, Weber & Schmidt, 1974.

MOORE, E. H. [1]. "On the Projective Axioms of Geometry." *Transactions of the American Mathematical Society* 3 (1902): 142–158.

―――― [2]. *Introduction to a Form of General Analysis.* New Haven, Conn.: Yale University Press, 1910.

MOORE, R. L. [1]. "A Note Concerning Veblen's Axioms for Geometry." *Transactions of the American Mathematical Society* 13 (1912): 74–76.

―――― [2]. "On the Foundations of Plane Analysis Situs." *Transactions of the American Mathematical Society* 17 (1916): 131–164.

MORITZ, R. E. *Memorabilia Mathematica, or the Philomath's Quotation-Book.* New York: Macmillan, 1914. Reprinted as *On Mathematics and Mathematicians.* New York: Dover, 1958.

NAGEL, E. "The Formation of Modern Conceptions of Formal Logic in the Development of Geometry." *Osiris* 7 (1939): 142–224.

―――― , and J. R. NEWMAN. *Gödel's Proof.* New York: New York University Press, 1958.

NEWSOM, C. V. [1]. "An Introduction to Mathematics: A Study of the Nature of Mathematical Knowledge." *University of New Mexico Bulletin* 295, Philosophical Series 1, no. 2 (1936).

——— [2]. "Mathematics and the Sciences." *Science* 94 (1941): 27–31.

——— [3]. "The Mathematical Method." *Pentagon* (Spring 1947): 37–46.

——— [4]. *Mathematical Discourses: The Heart of Mathematical Science.* Englewood Cliffs, N.J.: Prentice-Hall, 1964.

———, and H. EVES. *Introduction to College Mathematics.* 2nd ed. Englewood Cliffs, N.J.: Prentice-Hall, 1954.

NICOD, J. [1]. "A Reduction in the Number of the Primitive Propositions of Logic." *Proceedings of the Cambridge Philosophical Society* 19 (1916): 32–42.

——— [2]. *Foundations of Geometry and Induction.* New York: Humanities Press, 1950.

NIVEN, I. "The Transcendence of π." *American Mathematical Monthly* 46 (1939): 469–471.

OLMSTED, J. H. M. *The Real Number System.* New York: Appleton-Century-Crofts, 1962.

PEACOCK, G. [1]. "Report on Recent Progress in Analysis." Printed in the *Reports of the British Association* (1833).

——— [2]. *A Treatise on Algebra.* New York: Scripta Mathematica, 1940.

PEIRCE, B. "Linear Associative Algebra." *American Journal of Mathematics* 4 (1881): 97–229.

PERKINS, F. W. "An Elementary Example of a Continuous Non-differentiable Function." *American Mathematical Monthly* 34 (1927): 476–478.

PIERPONT, J. [1]. *Lectures on the Theory of Functions of Real Variables.* 2 vols. Boston: Ginn, 1905, 1912.

——— [2]. "Mathematical Rigor, Past and Present." *Bulletin of the American Mathematical Society* 34 (1928): 23–53.

POINCARÉ, H. *The Foundations of Science: Science and Hypothesis, Science and Method, the Value of Science.* Trans. by G. B. Halsted. Lancaster, Pa.: Science Press, 1946.

PÓLYA, G. [1]. *How to Solve It: A New Aspect of Mathematical Method.* Princeton, N.J.: Princeton University Press, 1945.

——— [2]. "Generalization, Specialization, Analogy." *American Mathematical Monthly* 55 (1948): 241–243.

——— [3]. *Induction and Analogy in Mathematics.* Princeton, N.J.: Princeton University Press, 1954.

——— [4]. *Patterns of Plausible Inference.* Princeton, N.J.: Princeton University Press, 1954.

——— [5]. *Mathematical Discovery.* Vol. 1. New York: Wiley, Inc., 1962.

PONCELET, J. V. *Traité des propriétés projectives des figures.* Paris: 1822.

QUINE, W. V. *Mathematical Logic.* New York: Norton, 1940.

RADO, T. "What Is the Area of a Surface." *American Mathematical Monthly* 50 (1943): 139–141.

RAMSEY, F. P. *The Foundations of Mathematics and Other Logical Essays.* New York: Humanities Press, 1950.

RAYNOR, G. E. "Mathematical Induction." *American Mathematical Monthly* 33 (1926): 376, 377.

REICHENBACH, H. *The Theory of Probability: An Inquiry into the Logical and Mathematical Foundations of the Calculus of Probability.* Berkeley: University of California Press, 1949.

RITT, J. F. *Theory of Functions.* Rev. ed. New York: King's Crown Press, 1947.

ROBB, A. A. *A Theory of Time and Space.* New York: Cambridge University Press, 1914.

ROBERTS, J. B. *The Real Number System in an Algebraic Setting.* San Francisco: Freeman, 1962.

ROBINSON, A. *Non-standard Analysis.* Amsterdam and London: North-Holland, 1970.

ROBINSON, G. DE B. *The Foundations of Geometry*, 2nd ed. Mathematical Expositions No. 1. Toronto: University of Toronto Press, 1946.

ROSENBAUM, R. A. "Remark on Equivalence Relations." *American Mathematical Monthly* 62 (1955): 650.

ROSENFELD, A. "An Axiomatic Triangular Geometry." Herbert Ellsworth Slaught Memorial Paper No. 4 (*Contributions to Geometry*), pp. 52–58. Buffalo, N.Y.: Mathematical Association of America, 1955.

ROSENTHAL, A. [1]. "What Are Set Functions?" *American Mathematical Monthly* 55 (1948): 14–20.

———[2]. "The History of Calculus." *American Mathematical Monthly* 58 (1951): 75–86.

ROSSER, J. B. [1]. "Extensions of Some Theorems of Gödel and Church." *Journal of Symbolic Logic* 1 (1936): 87–91.

———[2]. "Gödel Theorems for Non-constructive Logics." *Journal of Symbolic Logic* 2 (1937): 129–137.

———[3]. "An Informal Exposition of Proofs of Gödel's Theorems and Church's Theorem." *Journal of Symbolic Logic* 4 (1939): 53–60.

———[4]. "On the Many-Valued Logics." *American Journal of Physics* 9 (1941): 207–212.

———[5]. "The Burali-Forti Paradox." *Journal of Symbolic Logic* 7 (1942): 1–17.

———, and A. R. TURQUETTE. *Many-Valued Logics.* Amsterdam: North-Holland, 1951.

ROTH, W. E. "A Simplified Proof of the Distributive Law of Multiplication Given in Hilbert's 'The Foundations of Geometry.'" *American Mathematical Monthly* 23 (1916): 202–204.

RUSSELL, B. [1]. "Recent Work on the Principles of Mathematics." *International Monthly* 4 (1901): 83–101.

———[2]. "Mathematical Logic as Based on the Theory of Types." *American Journal of Mathematics* 30 (1908): 222–262.

———[3]. *Introduction to Mathematical Philosophy.* 2nd ed. New York: Macmillan, 1924.

SCHAAF, W. L. ed. *Mathematics, Our Great Heritage: Essays on the Nature and Cultural Significance of Mathematics.* Rev. ed. New York: Collier, 1963.

SCHOLZ, H. *Concise History of Logic.* Trans. by K. F. Leidecker. New York: Philosophical Library, 1961.

SEIDENBERG, A. [1]. "The Ritual Origin of Geometry." *Archive for History of Exact Sciences* 1 (1962): 488–527.

———[2]. "The Ritual Origin of Counting." *Archive for History of Exact Sciences* 2 (1962): 1–40.

SHAW, J. B. *Lectures on the Philosophy of Mathematics.* Chicago: Open Court, 1918.

SHEFFER, H. M. "A Set of Five Independent Postulates for Boolean Algebras." *Transactions of the American Mathematical Society* 14 (1913): 481–488.

SHOLANDER, M. "Postulates for Commutative Groups." *American Mathematical Monthly* 66 (1959): 93–95.

SIERPINSKI, W. *Introduction to General Topology.* Toronto: University of Toronto Press, 1934.

SINGH, J. *Great Ideas of Modern Mathematics: Their Nature and Use.* New York: Dover, 1959.

SLATER, M. "A Single Postulate for Groups." *American Mathematical Monthly* 68 (1961): 346–347.

SMITH, D. E. [1]. *History of Mathematics.* 2 vols. Boston: Ginn, 1923, 1925.

———[2]. *A Source Book in Mathematics.* New York: McGraw-Hill, 1929.

SNAPPER, E. "What Is Mathematics?" *American Mathematical Monthly* 86 (1979): 551–557.

STABLER, E. R. [1]. "Sets of Postulates for Boolean Rings." *American Mathematical Monthly* 48 (1941): 20–28.

———— [2]. "Boolean Algebra as an Introduction to Postulational Methods." *American Mathematical Monthly* 51 (1944): 106–110.

———— [3]. "Demonstrative Algebra." *Mathematics Teacher* 39 (1946): 255–260.

———— [4]. *An Introduction to Mathematical Thought.* Reading, Mass.: Addison-Wesley, 1953.

STEEN, L. A., ed. *Mathematics Today, Twelve Informal Essays.* New York: Springer-Verlag, 1978.

STEIN, S. K. *Mathematics, the Man-Made Universe.* San Francisco: Freeman, 1963.

STOLL, R. R. [1]. "Equivalence Relations in Algebraic Systems." *American Mathematical Monthly* 56 (1949): 372–377.

———— [2]. *Sets, Logic and Axiomatic Theories.* San Francisco: Freeman, 1961.

STONE, M. H. [1]. "Postulates for Boolean Algebras and Generalized Boolean Algebras." *American Journal of Mathematics* 57 (1935): 703–732.

———— [2]. "The Theory of Representations for Boolean Algebras." *Transactions of the American Mathematical Society* 40 (1936): 37–111.

STROYAN, K., with the collaboration of W. A. J. LUXEMBURG. *Introduction to the Theory of Infinitesimals.* New York: Academic Press, 1976.

SUPPES, P. *Axiomatic Set Theory.* Princeton, N.J.: Van Nostrand, 1960.

SZELE, T. "On Zorn's Lemma." *Publicationes Mathematicae* 1 (1950): 254–256; erratum, *ibid.*, p. 257.

TARSKI, A. [1]. *A Decision Method for Elementary Algebra and Geometry.* Project RAND, pub. R-109. Santa Monica, Calif.: Rand, 1948.

———— [2]. *Introduction to Logic and to the Methodology of Deductive Sciences.* Trans. by O. Helmer. 2nd ed. New York: Oxford University Press, 1954.

THIELMAN, H. P. *Theory of Functions of Real Variables.* Englewood Cliffs, N.J.: Prentice-Hall, 1953.

THRON, W. J. *Introduction to the Theory of Functions of a Complex Variable.* New York: Wiley, 1953.

THURSTON, H. A. *The Number-System.* New York: Interscience, 1956.

ULAM, S. M. "What Is Measure?" *American Mathematical Monthly* 50 (1943): 597–602.

USPENSKY, J. V. "A Curious Case of the Use of Mathematical Induction in Geometry." *American Mathematical Monthly* 34 (1927): 247–250.

VEBLEN, O. [1]. "A System of Axioms for Geometry." *Transactions of the American Mathematical Society* 5 (1904): 343–384.

———— [2]. "The Foundations of Geometry." See J. W. A. YOUNG.

————, and J. H. M. WEDDERBURN, "Non-Desarguesian and Non-Pascalian Geometries." *Transactions of the American Mathematical Society* 8 (1907): 379–383.

————, and J. H. C. WHITEHEAD, *The Foundations of Differential Geometry.* Cambridge Tracts in Mathematics and Mathematical Physics, No. 29. New York: Cambridge University Press, 1932.

VERRIEST, G. *Introduction à la géométrie non-Euclidienne par la méthode élémentaire.* Paris: Gauthier-Villars, 1951.

WAISMANN, F. *Introduction to Mathematical Thinking.* Trans. by T. J. Benac. New York: Ungar, 1951.

WAVRE, R. "Is There a Crisis in Mathematics?" *American Mathematical Monthly* 41 (1934): 488–499.

WEDBERG, A. *Plato's Philosophy of Mathematics.* Stockholm: Almqvist & Wiksell, 1955.

WEIL, A. "The Future of Mathematics." Trans. by A. Dresden. *American Mathematical Monthly* 57 (1950): 295–306.

WEYL, H. [1]. *Das Kontinuum: Kritische Untersuchungen über die Grundlagen der Analysis.* Leipzig: Gruyer, 1918, reprinted 1932.

————[2]. "Consistency in Mathematics." *Rice Institute Pamphlets*, No. 16 (1929): 245–265.

————[3]. "The Mathematical Way of Thinking." *Science* 92 (1940): 437–446.

————[4]. *Philosophy of Mathematics and Natural Science*. Rev. and augmented English ed. based on a trans. by O. Helmer. Princeton, N.J.: Princeton University Press, 1949.

WHITEHEAD, A. N., and B. RUSSELL. *Principia mathematica*. 2 vols. New York: Cambridge University Press, 1925, 1927.

WHITTAKER, E. T. [1]. "Aristotle, Newton, Einstein." *Science* 98 (1943): 249–254, 267–270.

————[2]. *From Euclid to Eddington: A Study of Conceptions of the External World*. New York: Dover, 1958.

WHITTAKER, J. V. "On the Postulates Defining a Group." *American Mathematical Monthly* 62 (1955): 636–640.

WHYBURN, G. T. "What Is a Curve?" *American Mathematical Monthly* 49 (1942): 493–497.

WILDER, R. L. [1]. "Concerning R. L. Moore's Axioms Σ_1 for Plane Analysis Situs." *Bulletin of the American Mathematical Society* 34 (1928): 752–760.

————[2]. "The Nature of Mathematical Proof." *American Mathematical Monthly* 51 (1944): 309–323.

————[3]. *Introduction to the Foundations of Mathematics*. 2nd ed. New York: Wiley, 1965.

————[4]. "The Cultural Basis of Mathematics." *Proceedings of the International Congress of Mathematicians, Cambridge, Mass., 1950*, Vol. 1, pp. 258–271. Providence, R.I.: American Mathematical Society, 1952.

————[5]. "The Origin and Growth of Mathematical Concepts." *Bulletin of the American Mathematical Society* 59 (1953): 423–448.

WINGER, R. M. "Some Applications of Groups to Geometry." *American Mathematical Monthly* 37 (1930): 1–16.

WOLFE, H. E. *Introduction to Non-Euclidean Geometry*. New York: Dryden Press, 1945.

WOODGER, J. H. *The Axiomatic Method in Biology*. New York: Cambridge University Press, 1937.

WYLIE, C. R. JR. [1]. "Hilbert's Axioms of Plane Order." *American Mathematical Monthly* 51 (1944): 371–376.

————[2]. *Foundations of Geometry*. New York: McGraw-Hill, 1964.

YOUNG, J. W. *Lectures on Fundamental Concepts of Algebra and Geometry*. New York: Macmillan, 1936.

YOUNG, J. W. A., ed. *Monographs on Topics of Modern Mathematics Relevant to the Elementary Field*. New York: Dover, 1955.

YOUNGS, J. W. T. "Curves and Surfaces." *American Mathematical Monthly* 51 (1944): 1–11.

═══ **SOLUTION SUGGESTIONS FOR SELECTED PROBLEMS** ═══

1.1.3   Let the quadrilateral be $ABCD$, with $AB = a$, $BC = b$, $CD = c$, $DA = d$. Show that $2K = ad \sin A + bc \sin C \leq ad + bc$; similarly, $2K \leq ab + cd$; therefore $4K \leq ad + bc + ab + cd = (a + c)(b + d)$. This formula, like the ancient Babylonian formula for the volume of a frustum of a cone or square pyramid, illustrates the prevalent idea in early empirical geometry of averaging.

1.1.7 Show that angle $AOF = 10°\ 00'\ 53''$. An angle of $30°$ cannot be trisected exactly with straightedge and compasses.

1.1.9 Use the rule of false position by taking BC to be any convenient length, say 1 unit, and on this basis calculate, by the law of sines, BA and BD, and then, by the law of cosines, AD; etc. This is a highly practicable procedure for solving this problem and is actually the method a surveyor would probably use.

1.1.11 By *fraction*, Ahmes means a fraction having unit numerator. Only the denominators of these unit fractions were written.

1.1.13 Take $\pi = 3$.

1.1.15 (a) Set $x = 2y$. (b) Eliminate x and y, obtaining a cubic equation in z.

1.2.2 It is fallacious to assume that a high degree of correlation indicates a causal relationship. The high correlation may indicate some common cause, or it may be purely accidental.

1.2.4 Instead of isogonal conjugate lines of a plane angle, consider *isogonal conjugate planes* of a dihedral angle.

1.2.5 Use diagrams.

1.2.10 Examples can be constructed for every case except the second.

1.3.2 Use the fact that $(\sqrt{a} - \sqrt{b})^2 \geqq 0$.

1.3.4 Construct a second triangle which is a right triangle and has the same legs.

1.3.6 (b) ab is the fourth proportional to $1, a, b$. (c) a/b is the fourth proportional to b, $1, a$. (d) $\sqrt{a}$ is a mean proportional between 1 and a.

1.3.7 A convex polyhedral angle must contain at least three faces, and the sum of its face angles must be less than $360°$.

1.4.5 Let the first player place his first cigar so that its center is directly over the center of the table, and let him thenceforth place his cigars in the positions centrally symmetrical to those occupied by the second player's cigars.

1.5.2 Follow the proof, given in the text, of the irrationality of $\sqrt{2}$.

1.5.3 See, e.g., H. Eves [1], Section 3–5.

1.5.5 Show, from a figure, that $ab = 2rs$ and $a + b = r + s$, and then solve simultaneously. This problem, as well as the previous one, illustrates the arithmetical nature of Heron's geometry, which is more akin to oriental than to typical Greek geometry.

1.5.6 (a) Let a ray of light emanating from a point A hit the mirror at point M and reflect toward a point B. If B' is the image of B in the mirror, then BB' is perpendicularly bisected by the plane of the mirror, and we must have AMB' a straight line. (b) Apply (a).

2.1.4 (c) Suppose $a > b$. Then the algorithm may be summarized as follows:

$$a = q_1 b + r_1, \qquad 0 < r_1 < b,$$

$$b = q_2 r_1 + r_2, \qquad 0 < r_2 < r_1,$$

$$r_1 = q_3 r_2 + r_3, \qquad 0 < r_3 < r_2,$$

$$\cdots, \qquad\qquad \cdots,$$

$$r_{n-2} = q_n r_{n-1} + r_n, \qquad 0 < r_n < r_{n-1},$$

$$r_{n-1} = q_{n+1} r_n.$$

Now, from the last step, r_n divides r_{n-1}. From the next to the last step, r_n divides r_{n-2}, since it divides both terms on the right. Similarly, r_n divides r_{n-3}. Successively, r_n divides each r_i, and finally a and b.

On the other hand, from the first step, any common divisor c of a and b

divides r_1. From the second step, c then divides r_2. Successively, c divides each r_i. Thus c divides r_n.

(d) From the next to the last step in the algorithm we can express r_n in terms of r_{n-1} and r_{n-2}. From the preceding step we can then express r_n in terms of r_{n-2} and r_{n-3}. Continuing this way, we finally obtain r_n in terms of a and b.

2.1.5 **(a)** If p does not divide u, then integers P and Q exist such that $Pp + Qu = 1$, or $Ppv + Quv = v$.

   (b) Suppose there are two prime factorizations of the integer n. If p is one of the prime factors in the first factorization, it must, by (a), divide one of the factors in the second factorization, that is, coincide with one of the factors.

2.1.6 **(c)** For each b_i in (b) may have $a_i + 1$ values.

   (f) Since b divides ac, we have $b_i \leqq a_i + c_i$. Also, since a and b are relatively prime, we have $a_i = 0$ or $b_i = 0$. In either case, $b_i \leqq c_i$.

   (h) Suppose, employing the definition of a rational number, that $\sqrt{2} = a/b$, where a and b are positive integers. Then, since $a^2 = 2b^2$, we have $(2a_1, 2a_2, \ldots) = (1 + 2b_1, 2b_2, \ldots)$, whence $2a_1 = 1 + 2b_1$, which is impossible.

2.1.7 Use the indirect method, and assume there are only a finite number of prime numbers, $a, b, \ldots, k$. Set $P = (a)(b) \ldots (k)$, and consider the number $P + 1$.

2.1.8 **(a)** Show that $2^{mn} - 1$ contains the factor $2^m - 1$.

   (c) If $a_1, a_2, \ldots, a_n$ represent all the divisors of N, then $N/a_1, N/a_2, \ldots, N/a_n$ also represent all the divisors of N.

2.3.3 **(b)** Let A be the given point and BC the given line segment. Construct, by Proposition 1, an equilateral triangle ABD. Draw circle $B(C)$, and let DB, produced if necessary, cut this circle in G. Now draw circle $D(G)$ to cut DA, produced if necessary, in L. Then AL is the sought segment.

2.4.3 Verification of the first four postulates presents little difficulty. To verify the fifth postulate it suffices to show that two ordinarily intersecting straight lines, each determined by a pair of restricted points, intersect in a restricted point. This may be accomplished by showing that the equation of a straight line determined by two points having rational coordinates has rational coefficients, and that two such lines, if they intersect, must intersect in a point having rational coordinates. For the last part of the problem, consider the unit circle with center at the origin, and the line through the origin having slope one.

2.4.5 **(a)** Let the line enter the triangle through vertex A. Take any point U on the line and lying inside the triangle. Let V be any point on the segment AC, and draw line UV. By Pasch's postulate, UV will (1) cut AB, or (2) cut BC, or (3) pass through B. If UV cuts AB, denote the point of intersection by W and draw WC; now apply Pasch's postulate, in turn, to triangles VWC and BWC. If UV cuts BC, denote the point of intersection by R; now apply Pasch's postulate to triangle VRC. If UV passes through B, apply Pasch's postulate to triangle VBC.

3.1.2 To deduce Euclid's fifth postulate, let AB and CD be cut by the transversal ST, and suppose $\angle BST + \angle DTS < 180°$. Through S draw QSR, making $\angle RST + \angle DTS = 180°$. Now apply, in turn, I 28, Playfair's postulate, I 17.

3.1.4 **(b)** Draw PQ perpendicular to m, and on m mark off $QQ_1 = PQ$. Then $\angle QPQ_1 = \angle QQ_1P = 180°/2^2$. Now on m mark off, in the same direction as QQ_1, $Q_1Q_2 = PQ_1$. Then $\angle Q_1PQ_2 = \angle Q_1Q_2P = 180°/2^3$. Repeat the construction obtaining $\angle Q_{n-2}PQ_{n-1} = \angle Q_{n-2}Q_{n-1}P = 180°/2^n$. By the postulate of Archimedes, there exists an integer k such that $k\alpha > 180°$. Take n so large that $2^n > k$. Then $180°/2^n < \alpha$. Etc.

   (c) $\angle QRP + \angle RPQ = \angle APR + \angle RPQ = 90°$.

3.1.5   To deduce Playfair's postulate, let P be a given point and m a given line not passing through P. Draw PQ perpendicular to m, and then draw PA perpendicular to PQ. By I 28, PA is parallel to m. Now use 3.1.4 (b), 3.1.4 (c), and Pasch's postulate to show that PA is the only line through P parallel to m.

3.1.6   Try to construct a proof along the following lines. Let ABC be any right triangle and draw the perpendicular CD to the hypotenuse. Then triangles ABC and ACD have two angles of one triangle equal to two angles of the other triangle, whence the third angles must also be equal. That is, $\angle B = \angle ACD$. Similarly, $\angle A = \angle BCD$. It now follows that $\angle A + \angle B + \angle C = 180°$. Now, if ABC is any triangle, divide it into two right triangles by an altitude.

3.1.7   Try the same experiment and reasoning on a spherical triangle, using a great circle arc in place of the straightedge.

3.2.1   **(a)** We are given $(BD/DC)(CE/EA)(AF/FB) = 1$, and we wish to prove that AD, BE, CF are concurrent. Suppose the contrary. Let BE and CF intersect in O, and let OA intersect BC in D'. Then, by Ceva's theorem, $(BD'/D'C)(CE/EA)(AF/FB) = 1$. Etc.

3.2.3   **(c)** Drop a perpendicular from the vertex of the triangle on the line joining the midpoints of the two sides of the triangle.

3.3.1   See Problem 3.2.3 (b).

3.3.2   Consult any standard text on elementary solid geometry.

3.3.6   **(b)** Let ABC be a triangle having the sum of its angles equal to two right angles. If ABC is not already right isosceles, draw the altitude BD. If neither of the resulting triangles is right isosceles, mark off on the longer leg of one of them a segment equal to the shorter leg. By (a), the resulting isosceles right triangle has the sum of its angles equal to two right angles. By putting together two such congruent isosceles right triangles, a quadrilateral can be formed having all its sides equal and all its angles right angles. By putting together four such congruent quadrilaterals, a larger quadrilateral of the same kind can be formed. By repeating the last construction enough times, one can obtain a quadrilateral of the same kind having its sides greater in length than any given segment. A diagonal of this last quadrilateral will give an isosceles right triangle of the type desired.

   (c) Let ABC be any right triangle, right-angled at C. By (b), there exists an isosceles right triangle DEF, right-angled at E, having the sum of its angles equal to two right angles and its legs greater than either of the legs of triangle ABC. Now produce CA and CB to A' and B', respectively, so that $CA' = CB' = ED$. Then triangles DEF and $A'CB'$ are congruent. Draw $A'B$, and apply (a) to triangle $A'CB'$. The extension to *any* triangle ABC is now easily accomplished by dividing ABC into two right triangles by one of its altitudes.

3.4.1   **(b)** Divide triangle ABC into two right triangles, and apply (a).

3.5.4   **(a)** Use the relations of Problem 3.5.3.

   (c) This is an immediate consequence of (a) and (b).

   (e) $K = -(1/QP)(1/QT) = -1/(QF)^2 = -1/k^2$.

3.5.5   This follows readily from the definition of the tractrix.

3.5.8   **(a)** One might interpret *abba* as "committee" and *dabba* as "committee member," and assume that there are just two committees and that no committee member serves on more than one committee.

   (b) One might interpret a *dabba* as any one of the three letters a, b, c, and an *abba* as any one of the three pairs, ab, bc, ca, of these letters.

4.1.4   **(j)** Neither; the amounts are equal. **(k)** The final pile will be over 17 million miles high!

4.1.7   **(a)**   (2) Let the three points be A, B, C, and let T be a motion transforming A into A', B into B', C into C'. Designate the inverse of T by T^{-1}. Let R be any effective motion that leaves A and B fixed. Now consider the resultant of the three motions T^{-1}, R, T, made in this order.

(b)   (1) Let the sphere have center A and pass through B, and let P be a point on the sphere. Let T be a motion transforming A into A', B into B', P into P'. Designate the inverse of T by T^{-1}. Let S be a motion that leaves A fixed but transforms B into P. Now consider the resultant of the three motions T^{-1}, S, T, made in this order.

(c)   Pieri says AC is *perpendicular* to AB if there exists a motion that leaves A and B fixed but transforms C into another point of the straight line CA.

4.2.2   **(d)**   See D. Hilbert [8], p. 14.

(e)   See D. Hilbert [8], p. 14.

4.2.4   **(a)**   Let h be a horn angle for which $\theta = 0$, and let h' be any horn angle for which $\theta > 0$. Then, for every positive integer n, we have $nh < h'$.

(d)   If $a = a'$ and $k = k'$, set $M = M'$; if $a > a'$, set $M > M'$; if $a = a'$ but $k > k'$, set $M > M'$. Define $M' = nM$ if and only if $a' = na$, $k' = nk$.

(f)   See part (e).

4.2.5   Huntington made the following definitions. A sphere that does not include any other sphere is called a *point*. If A and B are two points, the *segment* $[AB]$ is the class of all points X such that every sphere that includes A and B also includes X. The *extension* of $[AB]$ *beyond* A is the class of all points X such that $[BX]$ contains A; similarly, the extension of $[AB]$ beyond B is the class of all points X such that $[AX]$ contains B. The *ray AB* is the class of all points belonging to $[AB]$ or to the extension of $[AB]$ beyond B. The *line AB* is the class of all points belonging to $[AB]$ or to one of its extensions.

4.3.2   **(d)**   See the verification, in Section 4.3, of Postulate I-1.

(f)   Draw from O the perpendicular OP to the line. Let P' be the inverse of P, and Q' the inverse of any other point Q on the line. Now apply (e).

(g)   Let C be the center of the circle. Draw OC, cutting the circle in Q and R, and let P be any other point on the circle. If P', Q', R' are the inverses of P, Q, R, show, by the aid of (e), that $\angle R'P'Q' = \angle QPR = 90°$.

(h)   See the verification, in Section 4.3, of Postulate I-1.

(i)   Take ρ equal to the length of the tangent from the given point to the given circle.

4.3.3   Consider two curves C_1, C_2 intersecting in point R, and let their inverses C_1', C_2' intersect in point R'. Then O, R, R' are collinear, where O is the center of inversion. Through O, and close to the ray ORR', draw a second ray cutting C_1, C_2, C_1', C_2' in P, Q, P', Q', respectively. Using Problem 4.3.2 (e), show that $\angle PRQ = \angle P'R'Q'$. Now let the second ray rotate about O and approach ray ORR' as a limiting position.

4.3.4   **(a)**   Let P, Q, S, T be four points on a circle C and let P', Q', S', T' be their inverses on the inverse curve C'. If O is the center of inversion, show that $PT/P'T' = OT/OP'$, $P'S'/PS = OP'/OS$, $QT/Q'T' = OT/OQ'$, $Q'S'/QS = OQ'/OS$, and therefore that $(PQ, TS) = (P'Q', T'S')$.

4.3.5   Let the circles along the sides AC and BC of triangle ABC intersect again in R. Take R as the center of inversion.

4.3.6   **(b)**   Use Problems 4.3.5 and 4.3.3.

4.4.2   **(b)**   We have, for any four real numbers a_1, a_2, b_1, b_2,
$$(a_1 + b_1)^2 + (a_2 + b_2)^2 = a_1^2 + a_2^2 + 2(a_1 b_1 + a_2 b_2) + b_1^2 + b_2^2 \leqq a_1^2 + a_2^2 + 2\sqrt{a_1^2 + a_2^2}\sqrt{b_1^2 + b_2^2} + b_1^2 + b_2^2,$$

or

$$\sqrt{(a_1 + b_1)^2 + (a_2 + b_2)^2} \leqq \sqrt{a_1^2 + a_2^2} + \sqrt{b_1^2 + b_2^2}.$$

Now set $a_1 = x_1 - x_2, a_2 = y_1 - y_2, b_1 = x_2 - x_3, b_2 = y_2 - y_3$.

4.4.3 Let (x_1, y_1) and (x_2, y_2) be the points A and B, (x_3, y_3) the point A' on the line m, and $a(x - x_3) + b(y - y_3) = 0$ an equation of line m. The determination of the points B' and B'' is the algebraic problem of finding the common solutions of the equations

$$(x_2 - x_1)^2 + (y_2 - y_1)^2 = (x - x_3)^2 + (y - y_3)^2,$$

$$a(x - x_3) + b(y - y_3) = 0.$$

Show that these equations have two and only two solutions. Finally, show that (x_3, y_3) is the midpoint [see Problem 4.4.1 (e)] between points B' and b''.

4.4.7 Let A, B, C be the vertices of one triangle, and denote the lengths of the three opposite sides by a, b, c. Show that the law of cosines,

$$a^2 = b^2 + c^2 - 2bc \cos A,$$

holds in the algebraic model. If we use primes for the second triangle, it now follows that $a'^2 = a^2$. Then, by the law of cosines

$$b^2 = a^2 + c^2 - 2ac \cos B,$$

applied to each triangle, it follows that $\cos B = \cos B'$, or $B = B'$.

4.4.8 Let A and B be the points $(x_1, y_1), (x_2, y_2)$. Then the coordinates of A_1 are given by 4 of Section 4.4 by taking $t = a/b$, where a and b are the lengths of the segments CD and AB, respectively, Now, in 4 of Section 4.4, replace t by $2t, 3t$, and so on, thus obtaining the coordinates of points A_2, A_3, and so on.

4.4.10 Use Problem 4.3.5.

4.5.1 **(a)** Let points 1 and 6 coincide, so that line 16 becomes the tangent to the conic at point 1. **(b)** Use (a). **(c)** Let 1, 2, 3, 4 be the four points and 45 the tangent at $4 \equiv 5$, and let 12 cut 45 in P. Through 1 draw any line 16 cutting 34 in R, and then draw the Pascal line PR to cut 23 in Q. Then $5Q$ cuts 16 in a point 6 on the conic. **(d)** Take $1 \equiv 6$ and $3 \equiv 4$, and then take $2 \equiv 3$ and $5 \equiv 6$. **(e)** Take $1 \equiv 2, 3 \equiv 4, 5 \equiv 6$. **(f)** Use (e).

4.5.2 **(a)** Apply the law of sines to triangles VAP and VPB, remembering that $\sin VPA = \sin BPV$. **(b)** Use (a). **(c)** Use (b).

4.5.3 **(b)** Dualize the solution of 4.5.1 (b). **(c)** Dualize the solution of 4.5.1 (c). **(f)** Dualize the solution of 4.5.1 (f). **(g)** The theorem is its own dual.

4.5.4 **(a)** Choose for π' a plane parallel to the plane determined by point O and line m. **(b)** Project line OU to infinity. **(c)** Project line LMN to infinity, and use the elementary fact that the joins of corresponding vertices of two similar and similarly situated triangles are concurrent. **(d)** Choose a plane π' parallel to the minor axis of the ellipse and such that the angle θ between π' and the plane of the given ellipse is $\arccos b/a$, where a and b are the semimajor and semiminor axes of the ellipse. Now project the ellipse orthogonally onto π'.

4.5.5 Let P be the given point. Construct the polar p of P with respect to the ellipse, and let p cut the ellipse in points L and M. Then PL and PM are the required tangents.

5.1.2 **(a)** No. **(b)** Yes.

5.1.3 **(e)** These operations are commutative with respect to each other.

5.1.6 **(a)** Replacing a by z' in the first equality, and by z in the second equality, we find $z' \oplus z = z'$, $z \oplus z' = z$. Therefore $z' = z' \oplus z = z \oplus z' = z$. **(b)** Add each member of the equality $a \oplus b = a \oplus c$ to $\bar{a}$, and then use the associative law for $\oplus$. **(c)** Show that $x = \bar{a} \oplus b$ is a solution; then show, by (b), that if there are two solutions, x and y, we must have $x = y$. **(d)** By the distributive law, $(a \otimes a) \oplus (a \otimes z) = a \otimes (a \oplus z) = a \otimes a = (a \otimes a) \oplus z$. Therefore, by (b), $a \otimes z = z$. **(e)** Consider the triple sum $(a \otimes b) \oplus (a \otimes \bar{b}) \oplus (\bar{a} \otimes \bar{b})$. Combine the first two summands, and use the distributive law; then combine the last two summands, and use the distributive law. **(f)** Since $a \otimes b = z = a \otimes z$, we have, if $a \neq z$, by the cancellation law for the operation $\otimes$, $b = z$. **(g)** Show that $x = a^{-1} \otimes b$ is a solution; then show, by the cancellation law for the operation $\otimes$, that if there are two solutions, x and y, we must have $x = y$. **(h)** Show that $(a \otimes b) \oplus (a \otimes \bar{b}) = z$.

5.1.7 **(e)** Since $a \ominus b$ is positive. Since $b \ominus c$, $b \oplus \bar{c}$ is positive. Therefore, by Postulate P12, $(a \oplus \bar{b}) \oplus (b \oplus \bar{c})$ is positive. But $(a \oplus \bar{b}) \oplus (b \oplus \bar{c}) = a \oplus \bar{c}$. Hence $a \oplus \bar{c}$ is positive, and $a \ominus c$. **(g)** Since $a \ominus b$, $a \oplus \bar{b}$ is positive. Since c is given positive, $(a \oplus \bar{b}) \otimes c$ is positive, by Postulate P12. But $(a \oplus \bar{b}) \otimes c = (a \otimes c) \oplus (b \otimes c)$, by (h) of Problem 5.1.6. Therefore $(a \otimes c) \ominus (b \otimes c)$. **(h)** This follows from (e) and (h) of Problems 5.1.6. **(j)** Use (i).

5.2.1 **(d)** The zero element is $(0, 0)$; the unity element is $(1, 0)$; if $x = (a, b)$, then $\bar{x} = (-a, -b)$ and $x^{-1} = [a/(a^2 + b^2), -b/(a^2 + b^2)]$.

5.2.9 Define.

$$a(x) = \begin{cases} 0 & \text{for } 0 \leq x \leq \tfrac{1}{2}, \\ x - \tfrac{1}{2} & \text{for } \tfrac{1}{2} < x \leq 1, \end{cases}$$

$$b(x) = \begin{cases} 0 & \text{for } 0 \leq x \leq \tfrac{3}{4}, \\ x - \tfrac{3}{4} & \text{for } \tfrac{3}{4} < x \leq 1, \end{cases}$$

$$c(x) = \begin{cases} -x + \tfrac{1}{2} & \text{for } 0 \leq x \leq \tfrac{1}{2}, \\ 0 & \text{for } \tfrac{1}{2} < x \leq 1. \end{cases}$$

Then, although $a(x)c(x) = b(x)c(x)$ and $c(x) \neq 0$, we do not have $a(x) = b(x)$. Thus Postulate P9 fails to hold.

5.2.12 Show that $\begin{bmatrix} 0 & 1 \\ 0 & 0 \end{bmatrix} = \begin{bmatrix} a & b \\ c & d \end{bmatrix}^2$ implies: (1) $b(a + d) = 1$, (2) $c(a + d) = 0$, (3) $a^2 + bc = 0$, (4) $cb + d^2 = 0$. From (1) it follows that $a + d \neq 0$. Therefore, from (2), $c = 0$. Hence, from (3) and (4), $a = d = 0$. This contradicts the conclusion that $a + d \neq 0$.

5.2.18 **(a)** $*$ is neither commutative nor associative; $|$ is both commutative and associative; the distributive law holds. **(b)** None of the laws holds. **(c)** Only the two commutative laws hold. **(d)** $|$ is associative and the distributive law holds.

5.2.20 **(c)** This may be shown in several ways, but each is tricky. Look up the proof as given in some textbook on vector analysis.

5.3.2 **(a)** Yes. **(b)** No. **(c)** No. **(d)** No; G1 does not hold. **(e)** Yes.

5.3.5 No.

5.3.8 **(b)** Construct the operation table for the symmetric group of degree 3.

5.3.10 By G2' we are guaranteed the existence of an element i such that, for a given element b, $b * i = b$. Now let a be any element of G. By G2' there exists an element

c such that $a = c * b$. Then

$$a * i = (c * b) * i = c * (b * i) = c * b = a,$$

and G2 is established. Finally, by G2', there exists for each element a of G an element a^{-1} of G such that $a * a^{-1} = i$, and G3 is established.

5.3.11 No. Let G be the set of real linear functions of the form $a = a_1 x + a_2, a_1 \neq 0$, and let $a * b$ mean $a(db/dx)$. Then x is a right identity element and x/a_1 is a left inverse of a.

5.3.12 See M. Slater.

5.3.13 **(c)** In $x^{-1} * (x * y) = y$, set $y = e$. **(e)** In $x^{-1} * (x * y) = y$, set $y = x^{-1}$. **(f)** In $(x^{-1})^{-1} * (x^{-1} * y) = y$, set $y = x$. **(h)** Star both sides of $x * y = z$ by x^{-1} on the left and use the left inverse property to simplify. Then star both sides of the result by z^{-1} on the right and simplify, finally on the left by y^{-1}, again simplifying.

5.4.3 $(S^{-1} T^{-1})(TS) = S^{-1}(T^{-1} T)S = S^{-1} S = I$.

5.4.4 **(b)** Rotation of $180°$ about the origin, reflection in a line, inversion in a fixed circle.

5.4.5 **(a)** $R'S' = (TRT^{-1})(TST^{-1}) = TR(T^{-1} T)ST^{-1} = T(RS)T^{-1}$.

   (b) $(TST^{-1})^{-1} = TS^{-1}T^{-1}$, by Problem 5.4.3.

   (c) $TST^{-1} = (TS)T^{-1} = (ST)T^{-1} = S(TT^{-1}) = S$.

5.4.9 **(b)** It is the tangent of the angle from the first line to the second line.

5.4.10 **(a)** Show, for example, that $RH = D'$ and $HR = D$.

   (b) I, R', H, V, D, D' are self-inverse; R and R'' are inverse.

5.5.5 See R. A. Rosenbaum.

5.5.12 See F. Harary.

6.3.3 See Example (j) of Section 5.1.

6.3.8 Interpret the elements of S as a set of all rectangular Cartesian frames of reference that are parallel to one another but with no axis of one frame coincident with an axis of another frame, and let $b F a$ mean that the origin of frame b is in the first quadrant of frame a. Or interpret the elements of S as the set of all ordered pairs of real numbers (m, n), and let $(m, n) F(u, v)$ mean $m > u$ and $n > v$.

6.3.9 **(a)** Interpret K as the set of all points of a circumference of a circle and let $R(abc)$ mean "points a, b, c lie in clockwise order."

6.3.10 Interpret the bees as six people, A, B, C, D, E, F, and the four hives as the four committees (A, B, C), (A, D, E), (B, F, E), and (C, F, D). Or, interpret the bees and the hives as six trees and four rows of trees, respectively, forming the vertices and sides of a complete quadrilateral.

6.4.1 Interpret S as the set of all points on a horizontal straight line, and let F mean "is to the right of."

6.4.2 **(a)** To show independence of P2, interpret the bees and the hives as four trees and four rows of trees forming the vertices and sides of a square. To show independence of P3, interpret the bees as four trees located at the vertices and the foot of an altitude of an equilateral triangle, and the hives as the four rows of trees along the sides and the altitude of the triangle. To show independence of P4, interpret the bees and the hives as three trees and three rows of trees forming the vertices and sides of a triangle. **(b)** Taking into account the theorems of Problem 6.3.11, let us designate the four hives by I, II, III, IV and the three bees in hive I by A, B, C. Let hive II have bee A, and only bee A, in common with hive I. Then we may designate the bees in hive II by A, D, E. Now let hive III be that hive which has bee B in common with hive I. Then we may designate the bees of hive III by either B, D, F or B, E, F. In the first case, hive IV must contain bees C, E, F, and, in the second case, hive

IV must contain bees C, D, F. There are, then, the following two ways of designating the hives and the bees:

I	A, B, C		I	A, B, C
II	A, D, E		II	A, D, E
III	B, D, F	or	III	B, E, F
IV	C, E, F		IV	C, D, F

But the second designation can be changed into the first by interchanging the two labels D and E. It follows, then, that any interpretation of the postulate set can be labeled as in the first designation, and the desired isomorphism is established.

6.4.3 See W. T. Guy, Jr.

6.4.4 Show that there is essentially only one way of labeling the four elements of K.

6.4.9 **(e)** For a solution, and an interesting discussion by O. Veblen, see Problem 2894, *American Mathematical Monthly* 29 (1922): 357–358.

6.4.10 For a solution see Problem E 1152, *American Mathematical Monthly* 62 (1955): 582–584.

6.4.11 For a solution see Problem E 1098, *American Mathematical Monthly* 61 (1954): 474.

6.5.4 See Section 9.3.

6.5.7 **(b)** In one interpretation of T let S be the set of all positive integers, let "$a R b$" mean "a is a proper divisor of b," and let "$a D b$" also mean "a is a proper divisor of b." In the second interpretation let S be the set of all positive integers, let "$a R b$" mean "a is a proper divisor of b," and let "$a D b$" mean "b is a proper divisor of a."

6.5.8 See F. Harary.

7.1.5 $-\dfrac{5}{3}$, not $-\dfrac{2}{3}$.

7.1.10 **(a)** The integral is improper, since the integrand is discontinuous at $x = 0$. **(b)** Examine for end-point maxima and minima.

7.2.4 Consider the collection N of all numbers of the form $-x$, where x is in M.

7.2.8 l.u.b. = 2, g.l.b. = -2; l.u.b. = 3/2, g.l.b. = -1.

7.2.9 No.

7.2.11 If $r > 0$, then, by the Archimedean law, there exists an integer m' such that $m'(1) > r$. Let M be the set of all integers m' such that $m' > r$. By the postulate of continuity, this set contains a smallest member m. If $r = 0$, then $m = 1$. If $r < 0$, show that $m = -n + 1$, where n is the smallest integer such that $n > -r$.

7.2.12 **(a)** Show that $c > (c + d)/2 > d$. **(b)** By hypothesis $c - d > 0$. Then, by the Archimedean law, there is a positive integer n such that $n(c - d) > 1$, or $1/n < c - d$. Let m be the smallest integer such that $m > nd$ (see Problem 7.2.11). Then $m - 1 \leqq nd$, or $(m - 1)/n \leqq d$. That is, $d < m/n = (m - 1)/n + 1/n < d + (c - d) = c$.

7.3.3 **(e)** Let $f(n) = 3^{2n+2} - 8n - 9$. Show that $f(1) = 64$ and that $f(k + 1) - f(k) = 8(9^{k+1} - 1) = 64(9^k + 9^{k-1} + \cdots + 1)$. **(f)** If $f(n) = 9^n - 8n - 1$, show that $f(k + 1) = 9f(k) + 64k$. **(h)** $a^{k+1} - b^{k+1} = a(a^k - b^k) + b^k(a - b)$.

7.3.7 Show that $u_1 = u_2 = 1$ and that $u_{k+2} = u_k + u_{k+1}$. Now employ the third principle of mathematical induction, taking $h = 2$. The sequence of numbers $u_1, u_2, \ldots$ is known as the *Fibonacci sequence*.

7.3.9 The passage from $P(k)$ to $P(k + 1)$ assumes $k \geqq 2$.

7.3.10 If $\max(a, b) = k + 1$, it does not necessarily follow that α and β are both natural numbers.

7.3.12 Let M be a set of natural numbers containing 1 and containing $k + 1$ whenever it contains k. Suppose M does not contain all the natural numbers, and let T be the set of all natural numbers not in M. By the principle of the least natural number, there is a smallest natural number t in T. Since 1 is in M, $t > 1$ and $t - 1$ exists. Since t is the smallest natural number not in M, $t - 1$ is in M. Hence, by hypothesis, t is also in M. But, since t is in T, this is impossible.

7.3.14 (a) See J. E. Freund.

7.4.2 (a) We are given $(a + c, b + d) = (a + e, b + f)$, whence $a + c + b + f = b + d + a + e$ and $c + f = d + e$, which gives $(c, d) = (e, f)$. **(b)** Suppose (x, y) is a positive integer and let $x = y + z$. Then $(y + z, y)(a, b) = (y + z, y)(c, d)$. Perform the multiplication, use the definition of equality, and finally employ the cancellation laws for natural numbers to obtain $b + c = a + d$. The proof when (x, y) is a negative integer is similar. **(c)** Show that $(c + b, d + a)$ is a solution. Now let (x', y') be any solution. Then $(a, b) + (c + b, d + a) = (a, b) + (x', y')$, whence, by (a), $(x', y') = (c + b, d + a)$.

7.4.3 (a) Let $y = (a, b)$. Then $-y = (b, a)$ and $-(-y) = (a, b) = y$. **(b)** Let $x = (a, b)$, $y = (c, d)$. Then $-x = (b, a)$, $-y = (d, c)$, and $(-x)(-y) = (b, a)(d, c) = (bd + ac, bc + ad) = (a, b)(c, d) = xy$. **(c)** Let $x = (a, b)$, $y = (c, d)$. Then $-x = (b, a)$ and $(-x)y = (b, a)(c, d) = (bc + ad, ac + bd) = -[(a, b)(c, d)] = -(xy)$, etc. **(d)** Let $x = (a, b)$, $y = (c, d)$. Then $-(x + y) = -[(a, b) + (c, d)] = -(a + c, b + d) = (b + d, a + c) = (b, a) + (d, c) = (-x) + (-y) = -x - y$.

7.4.5 (b) In proving that addition is well defined, make use of the fact that $\langle ma, mb \rangle = \langle a, b \rangle$ if $m \neq 0$.

7.5.1 (c) Choose A' to be the set of all rational numbers of the form $-x$, where x is any element of B except the possible smallest element. Let B' be all the remaining rational numbers.

7.5.6 (b) $[x, y] = [(ac + bd)/(c^2 + d^2), (bc - ad)/(c^2 + d^2)]$.

7.5.7 (b) Use (a) and mathematical induction.

8.2.2 (a) We have

$$(a \cup b) \cap (b \cup c) = [a \cap (b \cup c)] \cup [b \cap (b \cup c)]$$

$$= (a \cap b) \cup (a \cap c) \cup b.$$

Therefore

$$(a \cup b) \cap (b \cup c) \cap (c \cup a)$$

$$= [(a \cap b) \cup (a \cap c) \cup b] \cap (c \cup a)$$

$$= [(a \cap b) \cap (c \cup a)] \cup [(a \cap c) \cap (c \cup a)] \cup [b \cap (c \cup a)]$$

$$= [(a \cap b \cap c) \cup (a \cap b \cap a)] \cup [(a \cap c \cap c) \cup (a \cap c \cap a)]$$

$$\cup [(b \cap c) \cup (b \cap a)]$$

$$= [(a \cap b \cap c) \cup (a \cap b)] \cup [(a \cap c)] \cup [(b \cap c) \cup (b \cap a)]$$

$$= (a \cap b) \cup (a \cap c) \cup (b \cap c).$$

8.2.3 Suppose $x = z$. Then

$$(x \cap t') \cup (x' \cap t) = (z \cap t') \cup (u \cap t) = z \cup t = t.$$

Now suppose $t = (x \cap t') \cup (x' \cap t)$. Then

$$x \cap t = x \cap [(x \cap t') \cup (x' \cap t)] = [x \cap (x \cap t')] \cup [x \cap (x' \cap t)]$$

$$= [(x \cap x) \cap t'] \cup [(x \cap x') \cap t] = (x \cap t') \cup (z \cap t)$$

$$= (x \cap t') \cup z = x \cap t'.$$

Therefore

$$x = x \cap u = x \cap (t \cup t') = (x \cap t) \cup (x \cap t') = (x \cap t) \cup (x \cap t)$$

$$= x \cap t = x \cap (t \cap t) = (x \cap t) \cap t = (x \cap t') \cap t = x \cap (t' \cap t)$$

$$= x \cap z = z.$$

8.2.9 **(d)** $ab(a + b) = (ab)a + (ab)b = (aa)b + a(bb) = ab + ab = z.$

8.2.10 See R. L. Goodstein.

8.3.1 Let M_1 be the midpoint of AB, M_2 the midpoint of $M_1 B$, M_3 the midpoint of $M_2 B$, etc. Denote by E the set of all points on $[AB]$ with the exception of points $A, B, M_1, M_2, M_3, \dots$. Then we have

$$[AB] = E, A, B, M_1, M_2, M_3, \dots,$$

$$(AB] = E, B, M_1, M_2, M_3, \dots,$$

$$[AB) = E, A, M_1, M_2, M_3, \dots,$$

$$(AB) = E, M_1, M_2, M_3, \dots.$$

It is now apparent how we may put the points of any one of the four segments in one-to-one correspondence with the points of any other one of the four segments.

8.3.6 Let d denote the cardinal number of the set of all natural numbers. Then the one-to-one correspondence

$$
\begin{array}{ccccccc}
1 & 2 & 3 & 4 & \cdots & n & \cdots \\
| & | & | & | & & | & \\
2 & 3 & 4 & 5 & \cdots & n+1 & \cdots
\end{array}
$$

shows that $d = d + 1$. This implies that $d + 1 = (d + 1) + 1 = d + 2$, although $1 \neq 2$.

Again, the one-to-one correspondences

$$
\begin{array}{ccccccc}
1 & 2 & 3 & 4 & \cdots & n & \cdots \\
| & | & | & | & & | & \\
2 & 4 & 6 & 8 & \cdots & 2n & \cdots
\end{array}
$$

and

$$
\begin{array}{ccccccc}
1 & 2 & 3 & 4 & \cdots & n & \cdots \\
| & | & | & | & & | & \\
1 & 3 & 5 & 7 & \cdots & 2n-1 & \cdots
\end{array}
$$

show that $d + d = d$. Hence, by Theorem 3 of Section 8.3, $(1 + 1)d = 1d$, or $2d = 1d$, although $2 \neq 1$.

Essentially these results have been stated in certain elementary books in the undesirable form: Infinity is unchanged by adding on or multiplying by a finite number.

8.4.4 **(b)** Consider the polynomial $x^2 - 2$. **(c)** Algebraic, since it is a zero of the polynomial $x^2 + 1$. **(g)** Multiply the polynomial through by the lowest common denominator of the a_k's.

8.4.5 **(b)** Use the idea employed in Theorem 1 of Section 8.4.

8.4.7 **(a)** Use an indirect argument, along with Problem 8.4.5 (a).

8.4.8 **(a)** See Problem E 832, *American Mathematical Monthly* 56 (1949): 407.

(c) No, for there are c points on a straight line or a circle, and there are only d rational numbers and d algebraic numbers.

8.4.10 It is a matter of existence; there exists a greatest triangle inscribed in a circle, but there does not exist a greatest natural number. To complete argument I, we must also prove that a maximum triangle inscribed in a circle *exists*. The problem illustrates the importance in mathematics of existence theorems.

8.4.12 **(a)** In each interval choose a point with rational coordinate. These points are all distinct, and therefore in one-to-one correspondence with the intervals, and they constitute an infinite subset of the denumerable set of all rational numbers.

8.4.13 With the finite sequence $\{n_1, n_2, \ldots, n_r\}$ associate the natural number

$$n = 2^{n_1} 3^{n_2} \cdots p_r^{n_r},$$

where p_r is the rth prime number. Since factorization of a natural number into powers of primes is unique, there is a one-to-one correspondence between the set of all finite sequences of nonnegative integers and the set of all natural numbers.

8.4.14 Assume S is denumerable. Then we can arrange the members of S in a sequence $\{f_1(x), f_2(x), \ldots, f_n(x), \ldots\}$. Consider the array.

$$f_1(1) \quad f_1(2) \quad f_1(3) \quad \cdots$$

$$f_2(1) \quad f_2(2) \quad f_2(3) \quad \cdots$$

$$f_3(1) \quad f_3(2) \quad f_3(3) \quad \cdots$$

$$\cdots$$

Now form the function $f(x)$ such that $f(n) = f_n(n) + 1$. Then $f(x)$ belongs to S. But $f(x)$ cannot be in the given denumeration, for it differs from $f_1(x)$ in the value taken by $x = 1$, from $f_2(x)$ in the value taken by $x = 2$, and so on.

8.4.15 Let K be the set of all natural numbers, and let $a \, R \, b$ mean $a < b$. Let K' be the set of all natural numbers, but define $a \, R \, b$ as follows: if a is odd and b is even, we write $a \, R \, b$; if a and b are both odd or both even, then $a \, R \, b$ means $a < b$. Each of these systems satisfies Postulates P1, P2, P3, P'4, but the two systems are not isomorphic.

8.5.2 **(a)** S plus the y-axis. **(b)** S plus the circle $x^2 + y^2 = 9$. **(c)** The origin. **(d)** S plus the segment $[-1, 1]$ of the y-axis. **(e)** The x-axis.

8.5.3 For Example (3), a *circle* is a square with center at C and having its diagonals equal to $2r$ in length and lying parallel to the coordinate axes.

8.5.5 **(c)** It is only the verification of the triangle inequality that presents any difficulty. Denote $\rho(y, z)$, $\rho(z, x)$, $\rho(x, y)$ by a, b, c, respectively. Then we have

$$b/(1 + b) = 1/(1/b + 1) \leq 1/[1/(c + a) + 1] = (c + a)/(1 + c + a)$$

$$= c/(1 + c + a) + a/(1 + c + a) \leq c/(1 + c) + a/(1 + a).$$

(Notice that for this metric all *distances* are less than 1.)

9.1.2 **(e)** There will be 8 lines in the truth table.

9.1.5 Use the logical equivalences tabulated at the end of Section 9.1.

9.1.8 Use the results of Problem 9.1.7.

9.1.10 See Problem 9.1.9.

9.1.11 **(a)** Show that $(p \leftrightarrow q) \leftrightarrow [(p \rightarrow q) \wedge (q \rightarrow p)]$ is a tautology.

9.1.13 All propositions constructed by applying $\wedge$, $\vee$, $\rightarrow$, $\leftrightarrow$, to the proposition p are true propositions provided p is true; p', on the other hand, has the truth value opposite to that of p. Therefore p' cannot be expressed without the employment of negation.

9.2.1 **(a)** Substituting m for p in Postulate L1, we obtain $(m \vee m) \rightarrow m$. But we are given $m \vee m$. Hence, by Rule R3, we have m. **(b)** Substituting m for q in Postulate L2 we obtain $m \rightarrow (p \vee m)$. But we are given m. Hence, by Rule R3, we have $p \vee m$. **(c)** Substitute m for p and n for q in Postulate L3, etc. **(d)** Substitute m for q and n for r in Postulate L4, etc. **(e)** Start with Theorem 1 and use Rule R3 twice.

9.2.3 By Theorem 3, $p \rightarrow p$. Therefore, by Rule R2, $p \rightarrow q$ and $q \rightarrow p$, and by Rule R4, $(p \rightarrow q) \vee (q \rightarrow p)$, whence, by Definition 3, $p \leftrightarrow q$.

9.2.4 **(a)** Substitute p' for p in Theorem 6, and then use Rule R10 based on Theorem 11. Or, start with L1 and use Rule R10 based on Theorem 12. **(b)** By Theorem 4 and Rule R10 based on Theorem 11, $p' \vee (p')'$. By Rule R10, again based on Theorem 11, $\{[p' \vee (p')']'\}'$. Therefore, by Definition 2, $(p \wedge p')'$.

9.2.5 **(a)** Substitute p' for p, and q' for q, in Postulate L3, and then use Definition 1. **(b)** Replace q by q' in the tautology of part (a), etc. **(c)** By Theorem 16, $(m \rightarrow n) \rightarrow (n' \rightarrow m')$ and $(n \rightarrow m) \rightarrow (m' \rightarrow n')$. But $m \rightarrow n$ and $n \rightarrow m$. Therefore $n' \rightarrow m'$ and $m' \rightarrow n'$, or $m' \leftrightarrow n'$.

9.2.6 **(a)** Substitute p' for p in Postulate L2 and use Definition 1. **(b)** Substitute p for q and q for p in Postulates L2 and L3, and then use Rule R9 [see Problem 9.2.1 (e)]. **(c)** Substitute p' for p in the tautology of (b), and use Definition 1.

9.2.9 **(a)** $a' \cup b$, **(b)** $(a \vee b) \leftrightarrow b$.

9.2.11 Suppose two contradictory tautologies s and s' can be obtained. Replacing p in Theorem 22 by s we obtain, by R1, the tautology $s' \rightarrow (s \rightarrow q)$. Now applying R3 we find that $s \rightarrow q$ is a tautology. Applying R3 again, we find that q is a tautology. But q represents *any* proposition, and therefore need not be a tautology. It thus follows that the calculus of propositions cannot be inconsistent, that is, it is consistent. The proof illustrates the *direct method* of establishing absolute consistency. Note that the method has been applied to a system containing an infinite number of primitive elements.

9.3.5 $m^{m-2}(m - 1)^{m^2 + 1}$.

9.3.6 **(a)** 4. **(b)** 2916.

INDEX

Abbas and dabbas, 76, 106
Abelian group, 124
Abelian semigroup, 124
Absolute independence, 172
Absorption laws of Boolean algebra, 218
Abstract spaces, 230
  Hausdorff space, 230–231
  Hilbert space, 232
  metric space, 232
  taxi-cab space, 232
  topological space, 234
Absurdity, 272
Ackermann, W., collaboration with
  Hilbert on formalist program, 270
Adams, H., on the nature of
  mathematics (quoted), 168
Addition of points, 140
Ahmes papyrus, 4
Aleph-null, 225n
Algebraic geometry, 98
Algebraic numbers
  definition of, 226, 281, 290
  denumerability of, 227
Algebraic structure, 113–118
Alice in Wonderland (Lewis Carroll), 21
Almagest (Ptolemy), 44, 53
Analogy, reasoning by, 6
Analysis situs, 131
Analytica posteriora (Aristotle), 16, 29,
  30, 34
Analytic geometry, 92–98
  interpretation of Hilbert's postulates,
  94

invention of, 92
  method of geometry, 97
Angle of parallelism, 63
Apiary, 170
Apollonius
  Greek mathematician of antiquity, 43
  history of analytic geometry, 93
Applicable surfaces, 75
Applied mathematics
  definition of, 149–150
  examples of, 152–153
Arabs, custodians of mathematics
  during Dark Ages, 43–44
Archimedean law of real numbers,
  181–182
Archimedes
  application of axiomatic method,
  42–43
  mentioned by Proclus, 26
  On Floating Bodies, 42
  On Plane Equilibriums, 42, 50
  On the Sphere and Cylinder, 42
  postulate of, 42, 86–87
Aristotle
  Analytica posteriora, 16, 29, 30, 34
  on axiomatic method (quoted), 29–32
  on axioms and postulates, 32
  irrationality of $\sqrt{2}$, 15–16
  laws of logic, 258
  Metaphysics, 28
  petitio principii in theory of parallels,
  36
  teacher of Eudemus, 11

Arithmetic in Nine Sections, 19
Arithmetic mean, 22
Arithmetization of analysis, 178
Āryabhata, *Ganita*, 19
Associative laws, 114
　of Boolean algebra, 218
Ausdehnungslehre (Grassmann), 121
Autological adjective, 274
Axiom
　challenging an, 262
　of choice, 296–299
　independence of, axiom of choice, 299
　versus postulate, 31–34
　of reducibility, 268
　statement of, axiom of choice, 297
　systems (*See* Postulate systems)
Axiomatic, or postulational, method,
　　147–150
　Aristotle on, 29–32
　Euclid's conception of, 36–37
　origin of, 15–16
　Proclus on, 31–32
Axiomatics
　formal, 13, 150
　material, 13–14, 150

Banach-Tarski paradox, 298
Beatley, R., postulational treatment of
　high school geometry, 88
Bees and hives, 170
Begriffsschrift (Frege), 244
Bell, E. T., on the nature of
　mathematics (quoted), 68
Beltrami, E.
　consistency of non-Euclidean
　　geometry, 65
　resurrection of Saccheri's work, 58
　surfaces of constant negative
　　curvature, 65
Berkeley, Bishop G., criticism of
　foundations of the calculus, 175
Bernays, P.
　collaboration with Hilbert on
　　formalist program, 270
　on postulates for calculus of
　　propositions, 256
　reconstruction of set theory, 265
Bernays and Hilbert, *Grundlagen der
　Mathematik*, 244, 270
Bernoulli, Johann
　function concept, 234
　infinitesimals (quoted), 175, 294

Bible
　quotations from, 21
　value of π, 2n
Birkhoff, G.
　application of many-valued logics to
　　quantum theory, 259
　on modern algebra (quoted), 122
Birkhoff, G. D., postulational treatment
　of high school geometry, 88
Bočvar, D. A., three-valued logic, 260,
　　261
Bolyai, J.
　absolute science of space, 62
　challenging an axiom, 262
　discovery of non-Euclidean geometry,
　　61–62
Bolyai, W., father of Johann Bolyai, 61
Boole, G.
　Boolean algebra, 221
　*Investigation into the Laws of Thought,
　　An*, 244
　Mathematical Analysis of Logic, The,
　　244
　symbolic logic, 244
Boolean algebra, 103, 216–221
　absorption laws, 218
　associative laws, 218
　definition of, 216–217
　De Morgan's laws, 219
　idempotent laws, 217
　postulates, 216–217
　principle of duality, 217
Boolean ring, 238
Boole-Schröder algebra, 244
Borel, É., axiom of choice, 297
Boundlessness of straight lines, 39
Brianchon, C. J.
　French geometer, 99
　theorem, 101
Brouwer, L. E. J., founder of
　intuitionist school, 268
Burali-Forti, C., paradox, 263

Calculus of propositions, 250–257
　a Boolean algebra, 257
　definition of, 250
　postulates for, 251, 256
　principle of duality, 257
　rules for obtaining theorems, 251–252
　theorems, 252–255
Campanus, J., translation of Euclid's
　Elements, 44

Cancellation laws, 117
Cantor, G., aphorism, 68, 262
 denumerability of algebraic numbers,
  227
 denumerability of rational numbers,
  226
 diagonal process, 227
 existence of transcendental numbers,
  228, 290
 genetical development of real number
  system, 183, 197
 on nature of mathematics (quoted),
  168
 nondenumerability of real numbers,
  277
 originator of set theory, 221
 paradox, 263
 transfinite numbers, 224–228
Cap, 214
Cardinal number of continuum, 228
 definition of, 223
Carlyle, T., translation of Legendre's
  Éléments de géométrie, 59n
Carnap, R., contributor to logistic
  program, 267
Carroll, Lewis, quotation from *Alice in
  Wonderland*, 21
Cartesian product, 224
Categoricalness of postulates, 98, 160–161
Categorical proposition, 21
Catenary, 75
Catoptrica (Heron), 24
Cauchy, A. L.
 development of a theory of limits, 178
 resolution of the second crisis in the
  foundations of mathematics, 263
Cayley, A.
 challenging an axiom, 262
 lines, 110
 matric algebra, 122–123
 on projective geometry (quoted), 131
Center of inversion, 108
Ceva's theorem, 71, 111
Cevian, 72
Chasles, M.
 French geometer, 99
 history of geometry, 98–99
Church, A.
 on systems of logic (quoted), 258–259
 theorem, 300
Chwistek, L., contributor to logistic
  program, 267

Cigar game, 24
Circle of inversion, 108
Clairaut, A. C., function notation, 235
Cohen, P. J., independence of
  continuum hypothesis, 229
Collapsing compasses, 277
Commandino, F., translation of Euclid's
  Elements, 45
Commentary on Euclid, Book I (Proclus),
  10, 26, 31, 54
Commutative laws, 114
Commutative ring, 122, 128
Complement of a set, 214
Complete independence, 162
Completeness of postulates, 160–162
Complete ordered field, 180
Complex numbers
 classification of, 202
 genetical development of, 199–201
 Hamilton's treatment of, 119
Conclusion, 6
Condition, necessary, sufficient,
  necessary and sufficient, 272
Conformal transformation, 108
Conjunction
 definition and notation, 245, 245n
 truth table, 247, 260–261
Consistency
 absolute, 155
 direct test for, 157
 of postulate sets, 155–157
 relative, 157
Constructible set of numbers, 281
Contingency, 272
Continuity, 235
 postulate, 39, 85, 86
Continuum, 231
 hypothesis, 228
 independence of hypothesis, 229
Contrapositive proposition, 272
Contrapositive statement, 22
Converse proposition, 272
Converse statement, 22
Copernicus, N., challenging an axiom,
  262
Crises in the foundations of
  mathematics, 262–266
Cross ratio
 definition of, 89
 a descriptive property, 100
 group, 141
 invariant under projection, 110

Cup, 214
Curvature of surface, 74
    Gaussian, 74
    principal, 74
    total, 74
Cut, Dedekind, 197
Cyclic order, 170
Cycloidal arch, 19

D'Alembert, J. le R., foundations of
    analysis, 177
Dantzig, T., on the nature of
    mathematics (quoted), 168
Dark Ages, 43
De arte combinatoria (Leibniz), 243
Decision theory, 164, 271
Dedekind, R.
    contributor to logistic program, 267
    cut, 197
    definition of infinite set, 225
    Eudoxian theory of magnitude, 262,
        293
    genetical development of real number
        system, 183, 197
    postulate of, 48–49, 86
    resolution of first crisis, 292–294
    Stetigkeit und irrationale Zahlen, 197
    theorem of continuity, 182
    Was sind und was sollen die Zahlen?
        197
Deduction, 6–9
    explanations of Greek preference for, 10
Definition
    explicit, 79
    implicit, 79–80
    impredicative, 265
    by induction, 188
Dehn, M.
    non-Archimedean geometry, 69
    postulate of Archimedes and
        Legendre's two theorems, 60, 69
De Moivre's theorem, 211
De Morgan, A.
    Formal Logic, 233
    foundations of algebra, 116
    laws of, 219
Denumerability
    of algebraic numbers, 227
    of rational numbers, 226, 303
Denumerable set, 225
Dependent variable, 235
Derivative in nonstandard analysis, 295

Derived tautology, 251
Desargues, G.
    projective geometry, 98–99
    seventeenth-century French
        geometer, 45, 98
    two-triangle theorem, 110
Descartes, R., invention of analytic
    geometry, 45, 93
Descriptive properties, 99
De Sua, F., on formal systems (quoted),
    271n
Die Theorie der Parallellinien (Lambert),
    58
Diophantus
    Greek mathematician of antiquity, 43
    non-Hellenic quality of his
        mathematics, 9
Dirichlet, L., function concept, 235
Discrete topology, 234
Disjunction
    definition and notation, 245, 245n
    exclusive, 272
    truth table, 247
Distance between two subsets, 233
Distance function, 232
Distributive law, 114
Division ring, 122, 128
Dodge, C. W., independence of some
    postulates for the natural numbers,
        288n
Domain of definition, 235
Doubtful propositions, 262
Duplication of the cube, 282

Einstein, A.
    challenging an axiom, 262
    theory of relativity, 69, 132, 230
Elements (Euclid), 13, 26–41, 276,
    292–294
    contents, 28
    definitions, axioms, postulates, 32–37
    first 28 propositions, 275–276
    formal nature of, 26–28
    Heiberg text, 33
    logical shortcomings of, 37–41
    Theon's recension, 29
    translated by Campanus, 44
    translated by Commandino, 45
    translated by Robert Simson, 45
Éléments de géométrie (Legendre), 59
    translated by Thomas Carlyle, 59n
    translated by John Farrar, 59n

Elements of a subject, 27–28
Empirical reasoning, 2–4
Empty set, 213
Epimendides paradox, 264
Equivalence
  absolutely independent postulate set
    for relation, 172
  classes, 135
  logical, 245, 248
  notation, 245n
  of postulate sets, 154–155
  relation, 135
  truth table, 247
Eratosthenes
  Greek mathematician of antiquity, 43
  mentioned by Proclus, 26
Erlanger Programm, 130, 230
Eubulides paradox, 264
Euclid
  attempts to prove fifth postulate,
    52–53
  conception of the axiomatic method,
    36–37
  Elements [*See* Elements (Euclid)]
  fifth postulate, 51–54
  mentioned by Proclus, 51–52
  parallel postulate, 36
  royal road to geometry, 97
  substitutions for fifth postulate,
    52–53
Euclidean algorithm, 45–46
Euclidean compasses, 47
Euclidean constructions, 276–284
Euclidean geometry, relative consistency
  of, 97
Euclidean tools, 35
Euclides ab omni naevo vindicatus
  (Saccheri), 56
Eudemian Summary (Proclus), 10–12,
  15, 22–23, 28, 35
Eudemus, early history of geometry, 11
Eudoxus
  definition of proportion, 293
  Euclid's Book V and Book VI, 28
  postulate of Archimedes, 43
  resolution of first crisis in the
    foundations of mathematics, 16,
    262, 292–294
Euler, L.
  diagrams, 6–8
  formalism in mathematics, 176–177,
    295

function concept and notation,
  234–235
  Swiss mathematician, 6
Exclusive disjunction, 272
Explicit definition, 79

Farrar, J., translation of Legendre's
  Éléments de géométrie, 59n
Fermat, P.
  creator of modern number theory, 45
  invention of analytic geometry, 93
  last theorem, 300
Field, 128
  number, 118, 278
  operations, 118
  ordered (*See* Ordered field)
  postulates, 117
  redundancies of postulates, 284–285
  theorems, 137
Finite group, 124
Finite projective geometry, 103–104
Fitch, G. D., on the nature of
  mathematics (quoted), 167
Fluxion, 175
Forder, H. G., postulational treatment
  of Euclidean geometry, 87–88, 163
Formal axiomatics, 13, 147–166
  pattern of, 148–149
Formal Logic (De Morgan), 244
Fourier, J. B. J., function concept, 235
Fraenkel, A., reconstruction of set
  theory, 265
Fréchet, M., abstract spaces, 232
Frege, G.
  Begriffsschrift, 244
  contributor to logistic program, 267
  definition of cardinal number of a set,
    223
  Grundgesetze der Arithmetik, 244
  postulates for calculus of
    propositions, 256
  Russell paradox, 264
  symbolic logic, 244
Function concept, continuity, 235
  dependent variable, 235
  domain of definition, 235
  history of, 234–235
  independent variable, 235
  range of values, 235
Fundamental theorem
  of algebra, 201, 228
  of arithmetic, 46

Galileo Galilei
  area under arch of cycloid, 19
  challenging an axiom, 262
  founder of science of dynamics, 45
  infinite sets, 225
Ganita (Āryabhata), 19
Gauss, C. F.
  legend about sum of angles of large
      physical triangle, 77
  non-Euclidean geometry, 61
  rigor in analysis, 178
Gaussian curvature, 74
Gaussian integers, 115
General Electric Company, 17
General topology, 234
Geodesics, 65
Geometric algebra, 98
Geometrical paradoxes, 47–48
*Geometriche Untersuchungen zur Theorie
    der Parallellinien* (Lobachevsky),
    62
Geometric mean, 22
Geometries, 130*ff*
Geometry
  Klein's definition of, 130
  Lorentz, 143
  non-Archimedean, 69
  non-Desarguesian, 111
  non-Euclidean (*See* Non-Euclidean
      geometry)
  non-Riemannian, 69
  plane equiform, 130
  plane Euclidean metric, 130
  plane similarity, 130
  projective (*See* Projective geometry)
  Riemannian, 69
  [*See also Elements* (Euclid); *topic
      under* Euclidean]
Gergonne, J. D.
  French geometer, 99
  principle of duality, 101
  "proof" of parallel postulate, 70
Gherardo of Cremona, translator of
    Arabic works, 44
Gibbs, J. W., vector analysis, 139–140
Gödel, K.
  axiom of choice, 299
  consistency of continuum hypothesis,
    229
  first theorem, 300
  incompleteness theorem, 299–303
  index, 303

limitation of the formalist program,
    271, 301
  number, 302
  second theorem, 301
Goldbach conjecture, 300
Grandi, L. G., on the sum of an infinite
    series, 177
Grassmann, H. G.
  Ausdehnungslehre, 121
  liberation of algebra, 116–117
Greek mathematics
  nature of, 9–10
  origin of, 9
Gregory, D. F., foundations of algebra,
    116
Group(s), 124–127
  Abelian, or commutative, 124
  consistency of postulates, 126
  cross ratio, 141
  finite, 124
  fundamental theorems, 127
  identity element of, 124
  infinite, 124
  inverse element of, 124
  non-Abelian, 124
  operation table, 125–126
  postulates, 124, 141
  significance of, 128–132
  symmetric, 141
  symmetries of a square, 144
  transformation, 130
Grundgesetz der Arithmetik (Frege), 244
Grundlagen der Geometrie (Hilbert),
    82–87, 270
Grundlagen der Mathematik (Hilbert and
    Bernays), 244, 270

Halmos, P. R., symbol for end of a
    proof, 149
Halstead, G. B., postulational treatment
    of high school geometry, 88
Hamilton, W. R.
  challenging an axiom, 262
  liberation of algebra, 117, 118–121
  quaternions, 120–121
  treatment of complex numbers, 119
Hankel, H.
  foundations of algebra, 116
  on nature of mathematics (quoted),
    176
Hardy, G. H., *reductio ad absurdum*, as
    gambit, 55

Harmonic mean, 22
Harriot, T., algebraic notation, 45
Hauff, J. K. F., "proof" of parallel postulate, 70–71
Hausdorff space, 230–231
Heath, T. L.
  Thales' measurement, 22
  translation of Heiberg text of Euclid's *Elements*, 33
Hegel, G. W. F., three-valued logic, 259n
Height of a polynomial, 227
Helmer, O., on comparison of equivalent postulate sets, 154–155
Hermite, C., transcendentality of *e*, 228
Herodotus, origin of geometry, 1
Heron
  approximate construction of a regular heptagon, 25
  Catoptrica, 24
  geometry, 16
  Greek mathematician of antiquity, 43
  method of approximating square roots, 24
  Metrica, 24, 25
  non-Hellenic quality of his mathematics, 16
Heterological adjective, 274
Heyting, A., intuitionist symbolic logic, 256, 269
Hilbert, D.
  founder of formalist school, 270
  Grundlagen der Geometrie, 82–87, 270
  number, 228
  postulates for plane Euclidean geometry, 83–87
  remark on Euclid's *Elements*, 9
  space, 232
Hilbert and Bernays, *Grundlagen der Mathematik*, 244, 270
Hindu-Arabic numeral system, 43
Hindus, custodians of mathematics during Dark Ages, 43
Hippocrates of Chios, early demonstrative geometry, 12, 28
Hobson, E. W., on the nature of mathematics (quoted), 167
Hopkinson, J., on the nature of mathematics (quoted), 168
Horn angle, 107–108
Huntington, E. V.
  postulates for Boolean algebra, 216

treatment of Euclidean geometry, 87, 108
Huygens, C., theory of probability, 45
Hyperbolic functions, 59, 75
Hyperparallels, 63
Hyperreal number(s), 295
  infinitely close, 295
  standard part of, 295
Hypotheses, of acute, obtuse, and right angles, 57

Idempotent laws of Boolean algebra, 217
Implication
  definition and notation, 245, 245n
  material, 246n
  strict, 246n
  truth table, 247
Implicit definition, 79–80
Impredicative definition, 265
Improper integrals, 178
Incompleteness theorem of Gödel, 299–303
Independence of postulates and of postulate systems
  absolute, 172
  complete, 162, 172
  ordinary, 158–159
  very, 127
Independence
  of primitive terms, 172
  of some postulates for natural numbers, 288–289, 288n
Independent variable, 235
Index, Gödel, 303
Induction, 5–6
Inequality
  Schwarz's, 109
  triangle, 109
Infinite group, 124
Infinitesimal, 295
Infinite-valued logic, 259
Infinitude of straight lines, 39–40
Integers, genetical development of, 192–195
Integral domain, 122, 128
Interpretation of postulate system, 149
Intersection of sets, 214
Intuitionism, 268–269
Inverse proposition, 272
Inverse statement, 22
Inversion, 108
  center of, 108

Inversion (*continued*)
  circle of, 108
  power of, 108
*Investigation into the Laws of Thought,
  An* (Boole), 244
Irrationality
  of π, 59
  of $\sqrt{2}$, 15–16
Irrational numbers, 199
Isogonal conjugate lines, 20
Isomorphism, 161
Isosceles birectangle, 56

Jordan algebras, 123
Journal of Symbolic Logic, 244

Kantian theory of space, 68
Keisler, H. J., textbook on nonstandard
  analysis, 295
Kepler, J., laws of planetary motion, 45
Keyser, C. J.
  on nature of mathematics (quoted),
    167
  on propositional functions, 148
Kirkman points, 110
Kleene, S. C., Gödel's theorems, 301
Klein, F.
  application of groups to geometry,
    128*ff*
  definition of a geometry, 130
  Erlanger Programm, 130, 240
  model of Lobachevskian geometry, 76
Kronecker, L.
  forerunner of the intuitionist
    program, 268
  on whole numbers (quoted), 201
K'ui-ch'ang Suan-shu, 19
Kurple, 106

Lagrange, J. L.
  attempted rigorization of analysis,
    177, 179
  Théorie des fonctions analytiques, 177
Lambert, J. H.
  Die Theorie der Parallellinien, 58
  hyperbolic functions, 59
  irrationality of π, 59
  non-Euclidean geometry, 58–59
Landau, E., *Foundations of Analysis*,
  192
Langford, C. H., contributor to the
  logistic program, 267

Law(s)
  absorption, 218
  of Archimedes, 182
  associative, 114, 218
  cancellation, 117
  commutative, 114
  of contradiction, 54–55, 71, 166, 248,
    254, 255
  of contradiction and intuitionist
    school, 269
  of contraposition, 248, 254
  De Morgan's, 219
  distributive, 114
  of double negation, 248, 254, 261
  of excluded middle, 54–55, 71, 166,
    248, 253
  of excluded middle and intuitionist
    school, 269
  of exponents, 202
  idempotent, 217
  of logic, 248
  parallelogram, 257–258
  of planetary motion, 45
  Poretsky's, 237
  of sines in Lobachevskian geometry,
    62
  of sines in spherical geometry, 73
  of syllogism, 248, 255
  of triple negation, 261
Least upper bound, 180
Legendre, A. M.
  Éléments de géométrie, 59
  first theorem, 60, 72
  non-Euclidean geometry, 59–60
  second theorem, 60, 73
Leibniz, G. W.
  contributor to logistic program, 267
  creation of calculus, 45, 174–175,
    263
  De arte combinatoria, 243–244
  function concept, 234
  mathematical logic, 243
Leon, early demonstrative geometry,
  28
Lewis, C. I., strict implication, 246*n*
Lie algebras, 123
Limit point, 231, 233
Lindemann, C. L. F., transcendentality
  of π, 228
Line coordinates, 112
Linear continuum, 180
Linear order, 146

Liouville, J.
  existence of transcendental numbers, 290
  numbers, 292
Lobachevskian geometry, 62
  consistency of, 88–92, 97
Lobachevsky, N. I., challenging an axiom, 262
  Geometriche Untersuchungen zur Theorie der Parallellinien, 62
  non-Euclidean geometry, 62–64
  Pangéométrie, 62
Logic(s)
  laws of, 248
  many-valued, or *m*-valued, 259–262
  mathematical, 243
  non-Aristotelian, 259
  non-uniqueness of, 258–259
  other, 257–262
  part of pattern of logical discourse, 165–166, 243, 257–258
  of relations, 244
  symbolic, 243
  three-valued, 259–262
  (*See also* Calculus of propositions; Reasoning)
Logica demonstrativa (Saccheri), 56
Logical equivalence, 248
Logicism, 267–268
Loop, 141
  left inverse property, 142
  right inverse property, 142
Lorentz geometry, 143
Lukasiewicz, J.
  challenging an axiom, 262
  independence of postulates of two-valued logic, 259
  m-valued logic, 259
  postulates for calculus of propositions, 256
  three-valued logic, 259, 260, 261

MacColl, H., three-valued logic, 259*n*
MacLane, S., on modern algebra (quoted), 122
Many-valued, or *m*-valued, logics, 259–263
  applied to quantum theory, 259
Mapping, 82
Material axiomatics, 13–14, 150
  pattern of, 14

Material implication, 246*n*
Mathematical Analysis of Logic, The (Boole), 244
Mathematical induction, 184
Mathematical logic, 243
Mathematics
  applied, 149
  categories of, 173–175
  origin of, 1–2
  pure, 149
Matric algebra, 122–124
Means
  arithmetic, 22
  geometric, 22
  harmonic, 22
  subcontrary, 22
Membership tables, 215, 289–290
Menaechmus, history of analytic geometry, 93
Menelaus, Greek mathematician of antiquity, 43
Metamathematics, 154, 165
Metaphysics (Aristotle), 28
Method of similitude, 17
Metric properties, 99
Metric space
  circle, 241
  distance function, 232
  metric of, 232
  postulates for, 232
Metrica (Heron), 24, 25
Michalski, K., early history of three-valued logics, 259*n*
Mnemonic
  for the symbol ∨, 246*n*
  for the symbols ∩ and ∪, 214
Model of a postulate system, 149
  concrete, 155
  ideal, 155
Modern compasses, 47, 277
Moore, E. H.
  complete independence, 162
  Hilbert's postulate system for Euclidean geometry, 107
Moore, R. L.
  postulate set for set-theoretic topology, 159
  Veblen's revised postulate system for Euclidean geometry, 87
Moscow papyrus, 4, 18
Mystic hexagram theorem, 99–100, 101, 109–110

Napier, J., invention of logarithms, 45
Napoleon, confiscation of Italian
   manuscripts, 29
Nasir-ed-din, "proof" of Euclid's
   parallel postulate, 53
Natural number system, 183–191
   postulates for, 184, 191
   redundancies of postulates for,
   285–288
Nature of mathematics, E. T. Bell
   (quoted), 68
   Cantor's aphorism, 68, 262
   B. Peirce (quoted), 150
   B. Russell (quoted), 150
   H. Poincaré (quoted), 150
   quotations, 167–168
Necessary condition, 272
Necessary and sufficient condition, 272
Negation, definition and notation, 245,
   245n
   truth table, 247, 261
Neighborhoods, 230
Neuman, J. von
   application of many-valued logics to
     quantum theory, 259
   collaborator with Hilbert on formalist
     program, 270
   reconstruction of set theory, 265
Newton, I.
   creation of calculus, 45, 174–175, 263
   Quadrature of Curves, 175
Nicod, J., postulates for calculus of
   propositions, 256
Non-Archimedean geometry, 69
Non-Archimedean systems, 107–108
Non-Aristotelian logics, 259
Non-Desarguesian geometry, 111
Non-Euclidean geometry, 51–70
   consistency of, 65–70
   discovery of, 60–65
   Kantian theory of space, 68
   significance of, 67–68
Non-Riemannian geometries, 69
Nonstandard analysis, 294–295
   derivative of real function, 295
   Keisler's textbook, 295
   Robinson, A., 295
   slope of real function, 295
Nuclear propositions, 80
Null set, 213
Number(s)
   algebraic, 226, 281, 290

   constructible, 281
   e, 228
   Gödel, 302
   Hilbert's, 228
   Liouville, 290
   perfect, 46
   π, 2, 17, 59, 228
   transcendental, 227, 291
   transfinite, 224–227
Number field, 118, 278
   square root extension chain, 278
   square root extension of, 278

On Floating Bodies (Archimedes), 42
On Plane Equilibriums (Archimedes), 42,
   50
On the Sphere and Cylinder
   (Archimedes), 42
One-to-one correspondence, 222
Open sets, 234
Open sphere, 233
Operation(s)
   closure property, 136
   commutative, 136
   definition of, 113
   well defined, 145
Ordered field, 117–118, 180
   complete, 180
   positive and negative elements of, 117
   postulates for, 117–118, 180–181
   postulates for complete, 180–181
   theorems, 137
Oresme, N., history of analytic
   geometry, 93
Orthocentric tetrahedron, 20
Oughtred, W., algebraic notation, 45

Padoa, A., independence of primitive
   terms, 172
Pangéométrie (Lobachevsky), 62
Pappus, Greek geometer of antiquity,
   43
Paradox(es), 47–48, 203–207, 209
   Banach-Tarski, 298
   barber, 264, 265
   Burali-Forti's, 263
   Cantor's, 263
   of Epimenides, 264
   of Eubulides, 264
   geometrical, 47–48
   in logic, 274
   Russell's, 263–264, 274

in set theory, 263–265
of Zeno, 263*n*
Parallax of a star, 77
Parallelogram law of forces, 257–258
Partial order, 146
Particular affirmative, 21
Particular negative, 21
Pascal, B.
  French seventeenth-century
    mathematician, 45
  line, 110
  mystic hexagram theorem, 99–100,
    101, 109–110
  Traité du triangle arithmétique, 185*n*
Pasch, M.
  postulate, 40, 50, 84, 85
  treatment of Euclidean geometry, 78–80
Pattern of formal axiomatics, 148–149
Pattern of material axiomatics, 14
Peacock, G., foundations of algebra,
    115–116
Peano, G.
  contributor to development of logistic
    program, 267
  Formulaire de mathématiques, 244
  genetical development of real number
    system, 183
  postulates for natural number system,
    191
  treatment of Euclidean geometry,
    80–81
Peirce, B.
  father of C. S. Peirce, 244
  on nature of mathematics (quoted), 150
Peirce, C. S.
  on nature of mathematics (quoted), 168
  symbolic logic, 244
Perfect numbers, 46
Peyrard, F., tenth-century copy of
    Euclid's *Elements*, 29
Philolaus, geometrical harmony, 22
Philosophies of mathematics, 266–271
π
  Babylonian value, 2
  Bible value, 2*n*
  Egyptian value, 2
  Hindu value, 17
  irrationality of, 59
  transcendentality of, 228
Pieri, M.
  postulates for Euclidean geometry,
    81–82, 106

treatment of Euclidean geometry,
    81–82
Plato
  influence on mathematics, 12
  mentioned by Proclus, 26
  Pythagorean triples, 23
Plato's academy, 12, 27, 28
Playfair's postulate, 52, 85, 85–86
Plücker, J.
  German geometer, 99
  coordinates of a line, 112
  equation of a point, 112
  lines, 110
  proof of principle of duality, 112
Poincaré, H.
  forerunner of the intuitionist
    program, 268
  imaginary universe, 77–78
  impredicative definitions, 265
  model, 67, 89–92, 109
  on nature of mathematics (quoted), 150
Points, addition of, 140
Poles and polars, 101–102
Polycrates, tyrant of Samos, 11
Poncelet, V.
  principle of duality, 101
  projective geometry, 99
Poretsky's law, 237
Post, E. L.
  challenging an axiom, 262
  m-valued logic, 259
  three-valued logic, 260, 261
Postulate(s)
  of Archimedes, 42, 50, 85, 86
  of completeness, 85
  of congruence, 84–85
  of connection, 83
  of continuity, 39, 48–49, 85, 180–181
  of Dedekind, 48–49, 86
  definition of, 148
  of finite induction, 184, 191
  independence of, 158, 172
  of order, 83–84
  origin of, 164–165
  of parallels (Euclid's fifth), 36, 51–54,
    85
  of Pasch, 50, 84, 85
  of Playfair, 85–86
  properties of, 154–162
  of Zermelo, 297
Postulate system(s)
  abbas and dabbas, 76, 106

Postulate system(s) (*continued*)
  bees and hives, 170
  Boolean algebra, 216–217
  commutative (or Abelian) ring, 122
  complete ordered field, 180
  cyclic order, 170
  division ring, 122
  equivalence relation, 135, 172
  Euclid's, 44
  field, 117
  field, redundancies in, 284–285
  finite geometry, 171
  group, 124, 141
  Hausdorff space, 230–231
  Hilbert's, 83–85
  integral domain, 122
  Lobachevskian geometry, 88
  m-class, 170–171
  metric space, 232
  natural number system, 184, 191
    independence of, 288–289
    redundancies, 285–288
  ordered field, 117
  partial order, 146
  Pieri's, 81–82, 106
  plane projective geometry, 103, 111
  properties of, 154–162
    absolute independence, 172
    categoricalness, 160–162
    completeness, 160
    consistency, 155–157
    equivalence, 154–155
    independence, 158–159
    relative weakness, 163
    very independent, 172
  propositional calculus, 250–257
  real number system, 179–183
  ring, 122
  ring with unity, 122
  semigroup, 124
  sfield, 122
  simple, or linear, order, 146, 153, 296
  topological space, 234
  well-ordered set, 296
Power of inversion, 108
Pre-Helenic mathematics, nature of, 1–5
Premises, 6
Prime numbers, infinitude of, 46
Primitive tautologies, 251
Primitive terms, 148
Principal curvatures, 74
Principia mathematica (Whitehead and

Russell), 166, 222, 244, 250, 255, 259, 267, 269
Principle, of addition, 251
  of duality, 101*ff*
    Boolean algebra, 217
    calculus of propositions, 257
    partially ordered sets, 146
    Plücker's proof, 112
    poles and polars, 101–102
    postulational approach, 103
    spherical geometry, 112
    trigonometry, 111–112
  of mathematical induction, 184
    second, 190
    third, 190
  of permanence of equivalent forms, 116
  of permutation, 251
  of smallest natural number, 190
  of summation, 251
  of tautology, 251
  vicious circle, 265
Proclus, *Commentary on Euclid, Book I*, 10, 26, 31, 54
  on definitions, axioms, and postulates (quoted), 31–32
  on elements (quoted), 27
  about Euclid (quoted), 26
  on Euclid's fifth postulate (quoted), 52
  Eudemian Summary, 10–12, 15, 22–23, 28
  Greek mathematician of antiquity, 43
Projective geometry, 98–104
  definition of, 101
  finite, 104
  poles and polars, 101–102
  postulates for, 103, 111
  principle of duality, 101*ff*
  projecting a line to infinity, 111
Proof theory, 270
Proper subset, 213
Proposition
  contrapositive, 272
  converse, 272
  definition of, 245
  doubtful, 262
  inverse, 272
  undecidable, 260
Propositional function, 147–148
Pseudosphere, 66

Ptolemy, C.
  Almagest, 44, 53
  Greek mathematician of antiquity, 43
Ptolemy, King, royal road in geometry, 97*n*
Pure mathematics, 149
  example of, 150–152
Pythagoras, Greek mathematician of antiquity, 11–12, 262
Pythagorean school, 11
Pythagorean theorem, 22–23
  metrical aspect of, 99
Pythagorean triples, 23

Quadrature of Curves (Newton), 175
Quaternions
  definition of, 120
  multiplication table, 121
  units, 120
Quine, W. V.
  contributor to the logistic program, 267
  logic, 164
  postulates for calculus of propositions, 256

Ramsey, F. P., contributor to the logistic program, 267
Range of values, 235
Rational cut, 197
Rational numbers, density of, 225
  denumerability of, 226, 303
  genetical development of, 195–196
Real numbers, genetical development of, 197–199
  nondenumerability of, 227
Real number system, 173–202
  Archimedean law, 182
  Dedekind's theorem on continuity, 182
  postulational treatment of, 179–183
  significance of, 173–179
Reasoning
  by analogy, 6
  deductive, 6–7
  empirical, 2–4
  inductive, 5
  valid and invalid, 6
Reductio ad absurdum, 54–56, 253
  considered as a gambit, 55
  most general form, 255
Regular heptagon, Heron's approximate construction, 25

Reichenbach, H.
  application of many-valued logic to quantum theory, 259
  infinite-valued logic, 259
Relations, 132–136
  antisymmetric, 146
  asymmetric, 133
  determinate, or connected, 145
  distinctly transitive, 146
  equivalence, 135
  intransitive, 134
  irreflexive, 133
  nonreflexive, 133
  nonsymmetric, 133
  nontransitive, 134
  reflexive, 133
  symmetric, 133
  transitive, 134
Rhind, A. H., Scottish antiquarian, 4
Rhind, or Ahmes, papyrus, 4, 17, 18
Riemann, B.
  abstract spaces, 30
  distinction between boundlessness and infinitude of straight lines, 39
  improper integrals, 178
  non-Euclidean geometry, 63, 69
  Über die Hypothesen welche der Geometrie zu Grunde Liegen, 39
Riemannian geometries, 69
Ring, 122, 128
  Boolean, 238
  with unity, 122, 128
Robinson, A., nonstandard analysis, 295
Robinson, G. de B., postulates for Euclidean geometry, 89
Roman mathematics, 43
Rosser, J. B.
  Gödel's theorems, 301
  postulates for calculus of propositions, 256
  three-valued logic, 260, 261
Rule
  of adjunction, 252
  of definitional substitution, 252
  of detachment, 252
  of false position, 4
  of implication, 252
  of substitution, 251
Russell, B.
  axiom of choice, 298
  definition of cardinal number of a set, 223

Russell, B. (*continued*)
 logistic program, 267–268
 on nature of mathematics (quoted),
  150, 167
 paradox, 263–264, 274
 Principia mathematica [*See Principia
  mathematica* (Whitehead and
  Russell)]
 propositional function, 147
 theory of types, 164, 267
 vicious circle principle, 265

Saccheri, G., *Euclides ab omnia naevo
  vindicatus*, 56
 Euclid's parallel postulate, 56
 Logica demonstrativa, 56
 some theorems, 57
Salmon points, 110
Schöne, R., discovery of Heron's
  Metrica, 24
Schröder, E., *Vorlesungen über die
  Algebra der Logic*, 244
Schwarz's inequality, 109
Semigroup, 124
 Abelian, 124
Set(s)
 cardinal number, 223
 Cartesian product, 224
 closure of, 231
 compact, 231
 complement, 214
 concept of, 212–213
 connected, 231
 denumerable, 225
 diameter of, 233
 elements, 213
 empty, 213
 equality, 213
 equivalence of, 222
 finite, 223
 infinite, 223, 231
 intersection, 214
 limit point, 231, 233
 null, 213
 one-to-one correspondence between,
  222
 open, 234
 partially ordered, 146
 separable, 231
 simply ordered, 296
 union, 214

 universe, 214
 well-ordered, 296
Set theory paradoxes, 263–265
Sfield, 122, 128
Sheffer's stroke, 249, 256
Similitude, method of, 17
Simple order, 146
Simson, R., English edition of Euclid's
  Elements, 45
Skolem, T., reconstruction of set
  theory, 265
Slope of a real function in nonstandard
  analysis, 295
Spherical degree, 72
Spherical excess, 72
Spherical geometry
 polar triangles, 112
 principle of duality, 112
Square root(s)
 approximation of, 24
 extension, 278
 extension chain, 278
 links of extension chain, 278
Squaring the circle, 283–284
Staudt, K. C. G. von, German
  geometer, 99
Steiner, J.
 points, 110
 Swiss geometer, 99
Steiner-Lehmus theorem, 22
Stetigeit und irrationale Zahlen
  (Dedekind), 197
Stevin, S., sixteenth century Dutch
  mathematician and physicist, 42
Stewart, D., on nature of mathematics
  (quoted), 168
Straight lines, infinitude and
  boundlessness of, 39
Strict implication, 246*n*
Stroke, Sheffer's, 249, 256
Subcontrary mean, 22
Subset, 213
 proper, 213
Successor, 191
Sufficient condition, 272
Śulvasūtras, 17
Surfaces
 applicable, 75
 of constant curvature, 75
Susa tablets, 19
Symbolic logic, 243
Symmetric group, 141

Tarski, A., *m*-valued logic, 259
Tautologies
  definition of, 248
  derived, 251
  importance in logic, 250
  primitive, 251
Taxi-cab space, 232
Tetrahedron, orthocentric, 20
Thales
  father of deductive mathematics, 11,
    262
  measurement of distance to a ship,
    11, 22
Theaetetus, early demonstrative
  geometry, 28
Theon of Alexandria, Greek
  commentator of antiquity, 29, 43
Theon's recension, 29
Theorem(s)
  Brianchon's, 101
  Ceva's, 71, 111
  Church's, 300
  of continuity, 182
  definition of, 149
  De Moivre's, 211
  Desargues's two-triangle, 110
  field, 137
  Gödel's first, 300
  Gödel's second, 301
  group, 127
  incompleteness (Gödel), 299–303
  Legendre's first, 60, 72
  Legendre's second, 60, 73
  ordered field, 137
  Pascal's mystic hexagram, 99–100,
    101, 109–110
  of Pythagoras, 22–23, 99
  Steiner-Lehmus, 22
  well-ordering, 296
  Zermelo's, 296
Théorie des fonctions analytiques
  (Lagrange), 177
Theory
  of numbers, 185
  of relativity, 69, 230
  of types, 164, 267
  of types, ramified, 268
Theudius of Magnesia, early
  demonstrative geometry, 28
Thibaut, B. F., "proof" of parallel
  postulate, 70
Three-valued logics, 260–262

Topological space, 234
  postulates for, 234
Topology, 131–132
  discrete, 234
  general, 234
  trivial, 234
Torricelli, E., area under arch of cycloid
  curve, 19
Townsend, E. J., translation of
  Hilbert's *Grundlagen der Geometrie*,
    82
Tractoid, 66, 76
Tractrix, 66, 75–76
Traité du triangle arithmétique (Pascal),
    185*n*
Transcendental numbers
  definition of, 227, 291
  existence of, 228, 290–292
Transfinite numbers, 224–229
Transform, 142
Transformation(s), 82
  group, 130
  identical, 129
  inverse, 129
  involutoric, 142
  nonsingular, 129
  product of, 129
Translations, 125
Triangle inequality, 109
Trigonometry, principle of duality, 112
Trirectangle, 58
Trisection of an angle, 282–283
Trivial topology, 234
Truth tables
  conjunction, 247
  definition of, 247
  disjunction, 247
  equivalence, 247
  implication, 247
  negation, 247
  in three-valued logic, 260–261
Type-I sequential independence, 171
Type-II sequential independence, 172

*Über die Hypothesen welche der
  Geometrie zu Grunde liegen*
  (Riemann), 39
Undecidable propositions, 260
Union of sets, 214
Universal affirmative, 21
Universal negative, 21
Universe, 214

University of Alexandria, 26
Upper bound, 180
   least, 180

Valid and invalid reasoning, 6
Variable, 235
Veblen, O., postulational treatment of
   Euclidean geometry, 87
Vector analysis, 139–140
Vector product, 140
Venn, J.
   diagrams, 214–215
   originator of Venn diagrams, 214n
Very independent postulates, 172
Vicious circle principle, 265
Visual space, non-Euclidean, 69
Vorlesungen über die Algebra der Logic
   (Schröder), 244

Wallis, J., "proof" of parallel postulate,
   54
Was sind und was sollen die Zahlen?
   (Dedekind), 197
Weierstrass, K.
   arithmetization of analysis, 178, 263
   continuous nondifferentiable function,
   178
Weyl, H., on impredicative definitions,
   265

Whitehead, A. N.
   logistic program, 267–268
   on nature of mathematics (quoted),
   167
   theory of types, 164
Whitehead and Russell, Principia
   mathematica (See Principia
   mathematica)
Wilder, R. L., dependence of R. L.
   Moore's sixth postulate, 159
William of Occam, three-valued logics,
   259n
Wittgenstein, L., contributor to logistic
   program, 267

Young, J. W., on nature of mathematics
   (quoted), 167

Zeno
   ancient Greek philosopher, 263n
   foreshadowing second crisis in
   foundations of mathematics, 263n
   paradoxes of, 263n
Zermelo, E.
   postulate of, 297
   reconstruction of set theory, 265
   well-ordering theorem, 296
Zwicky, F., application of many-valued
   logic to quantum theory, 259

A CATALOG OF SELECTED

DOVER BOOKS
IN SCIENCE AND MATHEMATICS

Mathematics

FUNCTIONAL ANALYSIS (Second Corrected Edition), George Bachman and Lawrence Narici. Excellent treatment of subject geared toward students with background in linear algebra, advanced calculus, physics and engineering. Text covers introduction to inner-product spaces, normed, metric spaces, and topological spaces; complete orthonormal sets, the Hahn-Banach Theorem and its consequences, and many other related subjects. 1966 ed. 544pp. 6⅛ x 9¼. 0-486-40251-7

ASYMPTOTIC EXPANSIONS OF INTEGRALS, Norman Bleistein & Richard A. Handelsman. Best introduction to important field with applications in a variety of scientific disciplines. New preface. Problems. Diagrams. Tables. Bibliography. Index. 448pp. 5⅜ x 8½. 0-486-65082-0

VECTOR AND TENSOR ANALYSIS WITH APPLICATIONS, A. I. Borisenko and I. E. Tarapov. Concise introduction. Worked-out problems, solutions, exercises. 257pp. 5⅝ x 8¼. 0-486-63833-2

AN INTRODUCTION TO ORDINARY DIFFERENTIAL EQUATIONS, Earl A. Coddington. A thorough and systematic first course in elementary differential equations for undergraduates in mathematics and science, with many exercises and problems (with answers). Index. 304pp. 5⅜ x 8½. 0-486-65942-9

FOURIER SERIES AND ORTHOGONAL FUNCTIONS, Harry F. Davis. An incisive text combining theory and practical example to introduce Fourier series, orthogonal functions and applications of the Fourier method to boundary-value problems. 570 exercises. Answers and notes. 416pp. 5⅜ x 8½. 0-486-65973-9

COMPUTABILITY AND UNSOLVABILITY, Martin Davis. Classic graduate-level introduction to theory of computability, usually referred to as theory of recurrent functions. New preface and appendix. 288pp. 5⅜ x 8½. 0-486-61471-9

ASYMPTOTIC METHODS IN ANALYSIS, N. G. de Bruijn. An inexpensive, comprehensive guide to asymptotic methods–the pioneering work that teaches by explaining worked examples in detail. Index. 224pp. 5⅜ x 8½ 0-486-64221-6

APPLIED COMPLEX VARIABLES, John W. Dettman. Step-by-step coverage of fundamentals of analytic function theory–plus lucid exposition of five important applications: Potential Theory; Ordinary Differential Equations; Fourier Transforms; Laplace Transforms; Asymptotic Expansions. 66 figures. Exercises at chapter ends. 512pp. 5⅜ x 8½. 0-486-64670-X

INTRODUCTION TO LINEAR ALGEBRA AND DIFFERENTIAL EQUATIONS, John W. Dettman. Excellent text covers complex numbers, determinants, orthonormal bases, Laplace transforms, much more. Exercises with solutions. Undergraduate level. 416pp. 5⅜ x 8½. 0-486-65191-6

RIEMANN'S ZETA FUNCTION, H. M. Edwards. Superb, high-level study of landmark 1859 publication entitled "On the Number of Primes Less Than a Given Magnitude" traces developments in mathematical theory that it inspired. xiv+315pp. 5⅜ x 8½. 0-486-41740-9

CALCULUS OF VARIATIONS WITH APPLICATIONS, George M. Ewing. Applications-oriented introduction to variational theory develops insight and promotes understanding of specialized books, research papers. Suitable for advanced undergraduate/graduate students as primary, supplementary text. 352pp. 5⅜ x 8½.
0-486-64856-7

COMPLEX VARIABLES, Francis J. Flanigan. Unusual approach, delaying complex algebra till harmonic functions have been analyzed from real variable viewpoint. Includes problems with answers. 364pp. 5⅜ x 8½.   0-486-61388-7

AN INTRODUCTION TO THE CALCULUS OF VARIATIONS, Charles Fox. Graduate-level text covers variations of an integral, isoperimetrical problems, least action, special relativity, approximations, more. References. 279pp. 5⅜ x 8½.
0-486-65499-0

COUNTEREXAMPLES IN ANALYSIS, Bernard R. Gelbaum and John M. H. Olmsted. These counterexamples deal mostly with the part of analysis known as "real variables." The first half covers the real number system, and the second half encompasses higher dimensions. 1962 edition. xxiv+198pp. 5⅜ x 8½. 0-486-42875-3

CATASTROPHE THEORY FOR SCIENTISTS AND ENGINEERS, Robert Gilmore. Advanced-level treatment describes mathematics of theory grounded in the work of Poincaré, R. Thom, other mathematicians. Also important applications to problems in mathematics, physics, chemistry and engineering. 1981 edition. References. 28 tables. 397 black-and-white illustrations. xvii + 666pp. 6⅛ x 9¼.
0-486-67539-4

INTRODUCTION TO DIFFERENCE EQUATIONS, Samuel Goldberg. Exceptionally clear exposition of important discipline with applications to sociology, psychology, economics. Many illustrative examples; over 250 problems. 260pp. 5⅜ x 8½.
0-486-65084-7

NUMERICAL METHODS FOR SCIENTISTS AND ENGINEERS, Richard Hamming. Classic text stresses frequency approach in coverage of algorithms, polynomial approximation, Fourier approximation, exponential approximation, other topics. Revised and enlarged 2nd edition. 721pp. 5⅜ x 8½.   0-486-65241-6

INTRODUCTION TO NUMERICAL ANALYSIS (2nd Edition), F. B. Hildebrand. Classic, fundamental treatment covers computation, approximation, interpolation, numerical differentiation and integration, other topics. 150 new problems. 669pp. 5⅜ x 8½.   0-486-65363-3

THREE PEARLS OF NUMBER THEORY, A. Y. Khinchin. Three compelling puzzles require proof of a basic law governing the world of numbers. Challenges concern van der Waerden's theorem, the Landau-Schnirelmann hypothesis and Mann's theorem, and a solution to Waring's problem. Solutions included. 64pp. 5¾ x 8¼.
0-486-40026-3

THE PHILOSOPHY OF MATHEMATICS: AN INTRODUCTORY ESSAY, Stephan Körner. Surveys the views of Plato, Aristotle, Leibniz & Kant concerning propositions and theories of applied and pure mathematics. Introduction. Two appendices. Index. 198pp. 5⅜ x 8½.   0-486-25048-2

INTRODUCTORY REAL ANALYSIS, A.N. Kolmogorov, S. V. Fomin. Translated by Richard A. Silverman. Self-contained, evenly paced introduction to real and functional analysis. Some 350 problems. 403pp. 5⅜ x 8½.   0-486-61226-0

APPLIED ANALYSIS, Cornelius Lanczos. Classic work on analysis and design of finite processes for approximating solution of analytical problems. Algebraic equations, matrices, harmonic analysis, quadrature methods, much more. 559pp. 5⅜ x 8½.   0-486-65656-X

AN INTRODUCTION TO ALGEBRAIC STRUCTURES, Joseph Landin. Superb self-contained text covers "abstract algebra": sets and numbers, theory of groups, theory of rings, much more. Numerous well-chosen examples, exercises. 247pp. 5⅜ x 8½.   0-486-65940-2

QUALITATIVE THEORY OF DIFFERENTIAL EQUATIONS, V. V. Nemytskii and V.V. Stepanov. Classic graduate-level text by two prominent Soviet mathematicians covers classical differential equations as well as topological dynamics and ergodic theory. Bibliographies. 523pp. 5⅜ x 8½.   0-486-65954-2

THEORY OF MATRICES, Sam Perlis. Outstanding text covering rank, nonsingularity and inverses in connection with the development of canonical matrices under the relation of equivalence, and without the intervention of determinants. Includes exercises. 237pp. 5⅜ x 8½.   0-486-66810-X

INTRODUCTION TO ANALYSIS, Maxwell Rosenlicht. Unusually clear, accessible coverage of set theory, real number system, metric spaces, continuous functions, Riemann integration, multiple integrals, more. Wide range of problems. Undergraduate level. Bibliography. 254pp. 5⅜ x 8½.   0-486-65038-3

MODERN NONLINEAR EQUATIONS, Thomas L. Saaty. Emphasizes practical solution of problems; covers seven types of equations. ". . . a welcome contribution to the existing literature...."–*Math Reviews*. 490pp. 5⅜ x 8½.   0-486-64232-1

MATRICES AND LINEAR ALGEBRA, Hans Schneider and George Phillip Barker. Basic textbook covers theory of matrices and its applications to systems of linear equations and related topics such as determinants, eigenvalues and differential equations. Numerous exercises. 432pp. 5⅜ x 8½.   0-486-66014-1

LINEAR ALGEBRA, Georgi E. Shilov. Determinants, linear spaces, matrix algebras, similar topics. For advanced undergraduates, graduates. Silverman translation. 387pp. 5⅜ x 8½.   0-486-63518-X

ELEMENTS OF REAL ANALYSIS, David A. Sprecher. Classic text covers fundamental concepts, real number system, point sets, functions of a real variable, Fourier series, much more. Over 500 exercises. 352pp. 5⅜ x 8½.   0-486-65385-4

SET THEORY AND LOGIC, Robert R. Stoll. Lucid introduction to unified theory of mathematical concepts. Set theory and logic seen as tools for conceptual understanding of real number system. 496pp. 5⅜ x 8¼.   0-486-63829-4

Physics

OPTICAL RESONANCE AND TWO-LEVEL ATOMS, L. Allen and J. H. Eberly. Clear, comprehensive introduction to basic principles behind all quantum optical resonance phenomena. 53 illustrations. Preface. Index. 256pp. 5⅜ x 8½. 0-486-65533-4

QUANTUM THEORY, David Bohm. This advanced undergraduate-level text presents the quantum theory in terms of qualitative and imaginative concepts, followed by specific applications worked out in mathematical detail. Preface. Index. 655pp. 5⅜ x 8½. 0-486-65969-0

ATOMIC PHYSICS (8th EDITION), Max Born. Nobel laureate's lucid treatment of kinetic theory of gases, elementary particles, nuclear atom, wave-corpuscles, atomic structure and spectral lines, much more. Over 40 appendices, bibliography. 495pp. 5⅜ x 8½. 0-486-65984-4

A SOPHISTICATE'S PRIMER OF RELATIVITY, P. W. Bridgman. Geared toward readers already acquainted with special relativity, this book transcends the view of theory as a working tool to answer natural questions: What is a frame of reference? What is a "law of nature"? What is the role of the "observer"? Extensive treatment, written in terms accessible to those without a scientific background. 1983 ed. xlviii+172pp. 5⅜ x 8½. 0-486-42549-5

AN INTRODUCTION TO HAMILTONIAN OPTICS, H. A. Buchdahl. Detailed account of the Hamiltonian treatment of aberration theory in geometrical optics. Many classes of optical systems defined in terms of the symmetries they possess. Problems with detailed solutions. 1970 edition. xv + 360pp. 5⅜ x 8½. 0-486-67597-1

PRIMER OF QUANTUM MECHANICS, Marvin Chester. Introductory text examines the classical quantum bead on a track: its state and representations; operator eigenvalues; harmonic oscillator and bound bead in a symmetric force field; and bead in a spherical shell. Other topics include spin, matrices, and the structure of quantum mechanics; the simplest atom; indistinguishable particles; and stationary-state perturbation theory. 1992 ed. xiv+314pp. 6⅛ x 9¼. 0-486-42878-8

LECTURES ON QUANTUM MECHANICS, Paul A. M. Dirac. Four concise, brilliant lectures on mathematical methods in quantum mechanics from Nobel Prize-winning quantum pioneer build on idea of visualizing quantum theory through the use of classical mechanics. 96pp. 5⅜ x 8½. 0-486-41713-1

THIRTY YEARS THAT SHOOK PHYSICS: THE STORY OF QUANTUM THEORY, George Gamow. Lucid, accessible introduction to influential theory of energy and matter. Careful explanations of Dirac's anti-particles, Bohr's model of the atom, much more. 12 plates. Numerous drawings. 240pp. 5⅜ x 8½. 0-486-24895-X

ELECTRONIC STRUCTURE AND THE PROPERTIES OF SOLIDS: THE PHYSICS OF THE CHEMICAL BOND, Walter A. Harrison. Innovative text offers basic understanding of the electronic structure of covalent and ionic solids, simple metals, transition metals and their compounds. Problems. 1980 edition. 582pp. 6⅛ x 9¼. 0-486-66021-4

HYDRODYNAMIC AND HYDROMAGNETIC STABILITY, S. Chandrasekhar. Lucid examination of the Rayleigh-Benard problem; clear coverage of the theory of instabilities causing convection. 704pp. 5⅝ x 8¼. 0-486-64071-X

INVESTIGATIONS ON THE THEORY OF THE BROWNIAN MOVEMENT, Albert Einstein. Five papers (1905–8) investigating dynamics of Brownian motion and evolving elementary theory. Notes by R. Fürth. 122pp. 5⅜ x 8½. 0-486-60304-0

THE PHYSICS OF WAVES, William C. Elmore and Mark A. Heald. Unique overview of classical wave theory. Acoustics, optics, electromagnetic radiation, more. Ideal as classroom text or for self-study. Problems. 477pp. 5⅜ x 8½. 0-486-64926-1

GRAVITY, George Gamow. Distinguished physicist and teacher takes reader-friendly look at three scientists whose work unlocked many of the mysteries behind the laws of physics: Galileo, Newton, and Einstein. Most of the book focuses on Newton's ideas, with a concluding chapter on post-Einsteinian speculations concerning the relationship between gravity and other physical phenomena. 160pp. 5⅜ x 8½. 0-486-42563-0

PHYSICAL PRINCIPLES OF THE QUANTUM THEORY, Werner Heisenberg. Nobel Laureate discusses quantum theory, uncertainty, wave mechanics, work of Dirac, Schroedinger, Compton, Wilson, Einstein, etc. 184pp. 5⅜ x 8½. 0-486-60113-7

ATOMIC SPECTRA AND ATOMIC STRUCTURE, Gerhard Herzberg. One of best introductions; especially for specialist in other fields. Treatment is physical rather than mathematical. 80 illustrations. 257pp. 5⅜ x 8½. 0-486-60115-3

AN INTRODUCTION TO STATISTICAL THERMODYNAMICS, Terrell L. Hill. Excellent basic text offers wide-ranging coverage of quantum statistical mechanics, systems of interacting molecules, quantum statistics, more. 523pp. 5⅜ x 8½. 0-486-65242-4

THEORETICAL PHYSICS, Georg Joos, with Ira M. Freeman. Classic overview covers essential math, mechanics, electromagnetic theory, thermodynamics, quantum mechanics, nuclear physics, other topics. First paperback edition. xxiii + 885pp. 5⅜ x 8½. 0-486-65227-0

PROBLEMS AND SOLUTIONS IN QUANTUM CHEMISTRY AND PHYSICS, Charles S. Johnson, Jr. and Lee G. Pedersen. Unusually varied problems, detailed solutions in coverage of quantum mechanics, wave mechanics, angular momentum, molecular spectroscopy, more. 280 problems plus 139 supplementary exercises. 430pp. 6½ x 9¼. 0-486-65236-X

THEORETICAL SOLID STATE PHYSICS, Vol. 1: Perfect Lattices in Equilibrium; Vol. II: Non-Equilibrium and Disorder, William Jones and Norman H. March. Monumental reference work covers fundamental theory of equilibrium properties of perfect crystalline solids, non-equilibrium properties, defects and disordered systems. Appendices. Problems. Preface. Diagrams. Index. Bibliography. Total of 1,301pp. 5⅜ x 8½. Two volumes. Vol. I: 0-486-65015-4 Vol. II: 0-486-65016-2

WHAT IS RELATIVITY? L. D. Landau and G. B. Rumer. Written by a Nobel Prize physicist and his distinguished colleague, this compelling book explains the special theory of relativity to readers with no scientific background, using such familiar objects as trains, rulers, and clocks. 1960 ed. vi+72pp. 5⅜ x 8½. 0-486-42806-0

CATALOG OF DOVER BOOKS

A TREATISE ON ELECTRICITY AND MAGNETISM, James Clerk Maxwell. Important foundation work of modern physics. Brings to final form Maxwell's theory of electromagnetism and rigorously derives his general equations of field theory. 1,084pp. 5⅜ x 8½. Two-vol. set.   Vol. I: 0-486-60636-8   Vol. II: 0-486-60637-6

QUANTUM MECHANICS: PRINCIPLES AND FORMALISM, Roy McWeeny. Graduate student-oriented volume develops subject as fundamental discipline, opening with review of origins of Schrödinger's equations and vector spaces. Focusing on main principles of quantum mechanics and their immediate consequences, it concludes with final generalizations covering alternative "languages" or representations. 1972 ed. 15 figures. xi+155pp. 5⅜ x 8½.   0-486-42829-X

INTRODUCTION TO QUANTUM MECHANICS With Applications to Chemistry, Linus Pauling & E. Bright Wilson, Jr. Classic undergraduate text by Nobel Prize winner applies quantum mechanics to chemical and physical problems. Numerous tables and figures enhance the text. Chapter bibliographies. Appendices. Index. 468pp. 5⅜ x 8½.   0-486-64871-0

METHODS OF THERMODYNAMICS, Howard Reiss. Outstanding text focuses on physical technique of thermodynamics, typical problem areas of understanding, and significance and use of thermodynamic potential. 1965 edition. 238pp. 5⅜ x 8½.   0-486-69445-3

THE ELECTROMAGNETIC FIELD, Albert Shadowitz. Comprehensive undergraduate text covers basics of electric and magnetic fields, builds up to electromagnetic theory. Also related topics, including relativity. Over 900 problems. 768pp. 5⅜ x 8¼.   0-486-65660-8

GREAT EXPERIMENTS IN PHYSICS: FIRSTHAND ACCOUNTS FROM GALILEO TO EINSTEIN, Morris H. Shamos (ed.). 25 crucial discoveries: Newton's laws of motion, Chadwick's study of the neutron, Hertz on electromagnetic waves, more. Original accounts clearly annotated. 370pp. 5⅜ x 8½.   0-486-25346-5

EINSTEIN'S LEGACY, Julian Schwinger. A Nobel Laureate relates fascinating story of Einstein and development of relativity theory in well-illustrated, nontechnical volume. Subjects include meaning of time, paradoxes of space travel, gravity and its effect on light, non-Euclidean geometry and curving of space-time, impact of radio astronomy and space-age discoveries, and more. 189 b/w illustrations. xiv+250pp. 8⅜ x 9¼.   0-486-41974-6

STATISTICAL PHYSICS, Gregory H. Wannier. Classic text combines thermodynamics, statistical mechanics and kinetic theory in one unified presentation of thermal physics. Problems with solutions. Bibliography. 532pp. 5⅜ x 8½.   0-486-65401-X